Natural Capital and Exploitation of the Deep Ocean

Natural Capital and Exploitation of the Deep Ocean

EDITED BY

**Maria Baker, Eva Ramirez-Llodra
and Paul Tyler**

Great Clarendon Street, Oxford, OX2 6DP,
United Kingdom

Oxford University Press is a department of the University of Oxford.
It furthers the University's objective of excellence in research, scholarship,
and education by publishing worldwide. Oxford is a registered trade mark of
Oxford University Press in the UK and in certain other countries

First Edition published in 2020
Impression: 1

Published in the United States of America by Oxford University Press
198 Madison Avenue, New York, NY 10016, United States of America

British Library Cataloguing in Publication Data

Data available

Library of Congress Control Number: 2020935287

ISBN 978–0–19–884165–4 (hbk.)
ISBN 978–0–19–884166–1 (pbk.)

Printed and bound by
CPI Group (UK) Ltd, Croydon, CR0 4YY

We dedicate this book to the volunteers of the
Deep-Ocean Stewardship Initiative.

DEEP-OCEAN STEWARDSHIP INITIATIVE

Preface

There has never been a time like the present, when there is so much media, scientific, and economic interest in the deep waters of the world ocean and the animals that live there. It is increasingly important for students and new researchers, as well as experienced scientists, to understand how their research can help to address pressing societal challenges. It is beneficial for deep-sea scientists, social scientists, lawyers, authorities, conservationists, industry, and civil society to have broad knowledge of the issues surrounding exploitation in the deep ocean, which has gradually become an increasingly important research focus. The current and future work of deep-sea scientists in all disciplines provides rigorous scientific data and knowledge to support sound management of human activities in this highly complex and variable realm. In this volume, we have brought together internationally recognised scientists, economists and legal experts to describe the processes by which humans can benefit from the natural capital of the deep sea in a sustainable framework. For this to happen, communication between all deep-sea stakeholders is essential, and this volume aims to facilitate future discussions between the many different sectors of society who may influence the global deep ocean for future generations.

Acknowledgements

We sincerely thank the contributing chapter authors for their expertise and enthusiasm in the timely production of this book. Their commitment to this process is greatly appreciated. ER would like to acknowledge NIVA and REV Ocean for their support during the editorial process. MB would like to thank Arcadia - a charitable fund of Lisbet Rausing and Peter Baldwin. We also express our thanks for the kind assistance of the Oxford University Press publishing staff, Charles Bath, Ian Sherman and Bethany Kershaw.

The following reviewers contributed significantly to the high quality of all chapters. We acknowledge their contribution and sincerely thank them:

Dr Alan Williams—Commonwealth Scientific and Industrial Research Organisation, Tasmania
Dr Andrew Gates—National Oceanography Centre, Southampton, UK
Professor Andrew Sweetman—Heriot-Watt University, UK
Professor Anna Metaxas—Dalhousie University, Canada
Professor Bhavani Narayanaswamy—Scottish Association for Marine Science, UK
Professor Danielle Skropeta—University of Wollongong, Australia
Dr Fabio De Leo—University of Victoria, Canada
Dr Henry Ruhl—Monterey Bay Aquarium Research Institute, USA
Dr Lenaick Menot—Ifremer, France
Professor Marcel Jaspars—University of Aberdeen, UK
Professor Robin Warner—University of Wollongong, Australia
Professor Robert Carney—Louisiana State University, USA
Dr Sabine Gollner—Royal Netherlands Institute for Sea Research, Netherlands
Professor Sebastian Villastante—University of Santiago de Compostela, Spain
Professor Tony Koslow—Scripps Institution of Oceanography, University of California, USA
Professor Tim Stephens—The University of Sydney, Australia
Professor Tracey Sutton—Nova Southeastern University, USA
Torsten Thiele—Institute for Advanced Sustainability Studies (IASS), Germany

Contents

List of contributors

Dr Diva J. Amon, Natural History Museum, London, UK

Dr Maria Baker, University of Southampton, UK

Dr Stace Beaulieu, Woods Hole Oceanographic Institution, USA

Dr Angelo F. Bernardino, Universidade Federal do Espírito Santo, Brazil

Dr Abbie S. A. Chapman, University of Southampton; University College London, UK

Dr Erik E. Cordes, Temple University, USA

Duncan Currie, Globelaw, New Zealand

Professor Jeffrey C. Drazen, University of Hawaii, USA

Matthew Gianni, Deep Sea Conservation Coalition, Netherlands

Kristina Gjerde, International Union for Conservation of Nature; Middlebury Institute of International Studies, USA

Dr Harriet Harden-Davies, University of Woolongong, Australia

Dr Porter Hoagland, Woods Hole Oceanographic Institution, USA

Dr Aline Jaeckel, University of New South Wales, Australia

Dr Di Jin, Woods Hole Oceanographic Institution, USA

Dr Daniel O. B. Jones, National Oceanography Centre, Southampton, UK

Professor S. Kim Juniper, University of Victoria, Canada

Professor Nadine Le Bris, France

Professor Lisa A. Levin, Scripps Institution of Oceanography, University of California, USA

Dr Amanda N. Netburn, National Oceanic and Atmospheric Administration, USA

Dr Eva Ramirez-Llodra, Norwegian Institute for Water Research and REV Ocean, Norway

Dr Ylenia Randrianarisoa, University of Toamasina, Madagascar

Professor Thomas A. Schlacher, University of the Sunshine Coast, Australia

Dr Kate Thornborough, Independent Researcher, Manly, NSW, Australia

Dr Andrew R. Thurber, Oregon State University, USA

Emeritus Professor Paul Tyler, University of Southampton, UK

Dr Lissette Victorero, Museum National d'Histoire Naturelle, Paris, France

Professor Les Watling, University of Hawaii, USA

CHAPTER 1

Introduction: Evolution of knowledge, exploration, and exploitation of the deep ocean

Maria Baker, Eva Ramirez-Llodra, and Paul Tyler

1.1 Introduction

1.1.1 Natural capital defined

Following the harmonised definition adopted by the Natural Capital Coalition (https://naturalcapitalcoalition.org/), 'natural capital' is a term for 'the stock of renewable and nonrenewable resources (e.g. plants, animals, air, water, soils, and minerals) that combine to yield a flow of benefits to people'. For the purposes of this book, we define 'natural capital of the deep sea' as: 'any use of the deep ocean and its seabed (>200 m depth), and the stock of renewable and non-renewable biotic and abiotic resources that may be of benefit to humankind. These include economic, social, environmental, cultural, spiritual or wellbeing benefits'.

Today we face challenges as to how to manage Earth's natural capital in a sustainable way for future human prosperity. Accounting for benefits from national natural capital is emerging as an important action around the globe. Long-term economic consideration of a variety of natural resources is being undertaken by numerous countries who are now looking beyond their gross domestic product. In addition to human capital and produced capital, natural capital such as water, energy, and minerals needs to be accounted for in order to manage these resources wisely and to evaluate the trade-offs needed for making development deci-

sions. Without this information, overexploitation and deterioration of natural assets will result in their depletion, and so they will be unavailable for future generations. By fully accounting for natural capital and all the associated benefits, countries can provide more accurate information to their policy-makers, resulting in improved economic, social, environmental, and cultural decisions about development priorities and investments. The deep ocean is, by far, the planet's largest biome and holds a wealth of potential natural assets. Most of it lies beyond national jurisdiction and, hence, is the responsibility of us all. This volume is an attempt to understand past exploitations of the deep ocean, to provide a present understanding of its natural capital and how this may be exploited sustainably for the benefit of humankind, whilst maintaining its ecological characteristics.

1.1.2 Deep-ocean morphology and abiotic characteristics

Much discussion has arisen from the definition of the 'deep sea', with the result that there is no single definition, but for our purposes, we are using the traditional 200 m–depth contour as the upper limit of the deep sea, based upon the approximate water depth of the global shelf break and the loss of sunlight for photosynthesis (Gage and Tyler 1991; Ramirez-Llodra et al. 2010). We will stray into

Maria Baker, Eva Ramirez-Llodra, and Paul Tyler, *Introduction* In: *Natural Capital and Exploitation of the Deep Ocean*. Edited by: Maria Baker, Eva Ramirez-Llodra, and Paul Tyler, Oxford University Press (2020). © Oxford University Press.
DOI: 10.1093/oso/9780198841654.003.0001

shallower waters but only when there is direct relationship to the environment below 200 m depth. Because of extensive bathymetric surveying and technological development, we have a reasonable understanding of the diverse morphology of the global ocean (Figure 1; Thistle 2003), although there is still much discovery and mapping that may better inform policy- and decision-making about the use of this realm.

Deep-sea facts
The deep sea is the single largest ecosystem on Earth.
Total area: 360.10^6 km^2.
Volume of seawater: 1368.10^6 km^3.
Maximum depth: 10,994 m.
Average depth: 3800 m.
Abyssal seafloor: 75 per cent.
Area beyond national jurisdiction: 61 per cent.
Twilight: ca. 200 to 1000 m.
Dark: below 1000 m.
High pressure: increases by 101.325 kPa (1 atm) every 10 m.
Cold: −2 to 4 °C, higher in the Mediterranean Sea (13 °C) and the Red Sea (21 °C).
Oxygen: saturated, except OMZs.
Known habitat types: 28.
Number of described deep-sea species: approaching 26,000.
Estimated number of marine species: 2.2 million, 91 per cent unknown.

The deep sea is an amalgam of different pelagic and benthic habitats with environmental characteristics that support a wide variety of ecosystems (Ramirez-Llodra et al. 2010; Figure 2). The largest recognisable features are the *abyssal plains* of all the oceans (Figure 2A), often separated in each ocean basin by a *mid ocean ridge* (MOR), although this ridge may be asymmetrically placed in some oceans (Figure 1). The MORs are a semicontinuous mountain chain where new seafloor is being formed. It is at MORs and *back-arc basins* where many of the known *hydrothermal vents* are formed, through the circulation of seawater in the crust. There, the water becomes superheated and charged with metals, finally emanating from chimneys formed by the

deposition of the minerals that precipitate from the hydrothermal fluid. These hydrothermal fluids are charged with reduced compounds (e.g. H$_2$S and CH$_4$) that fuel unique faunal communities (Figure 2B–C) through microbial chemoautotrophy (Van Dover 2000). The MORs are incised by *transform faults*, allowing limited water-circulation exchange between basins on either side of the ridges. All transform faults and fracture zones are a result of the stresses of tectonic processes. The final tectonic structures of the seafloor are the *trenches*, where subduction of seafloor occurs forming the very deepest parts of each ocean. All the major oceans are connected at depth, with the exception of the Arctic and Mediterranean, which are separated from their adjacent oceans by *sills* (Figure 1; Tyler 2003). This makes the Arctic and Mediterranean concentration basins, where dense water is formed as a result of cooling or raised salinity. Most deep-ocean waters are oxygenated. The water formed in the polar regions aerates most of the deep sea. However, some regions are characterised by bathyal *oxygen minimum zones* (OMZ) that infringe the continental margins, such as the OMZs found in Eastern Pacific Ocean, West Africa, the Arabian Sea, and the Bay of Bengal (Levin 2003).

The *continental margins* border land masses and may be passive (tectonically quiescent), or active where there is regular tectonic activity as seen round the rim of the Pacific. The margin consists of the *continental shelf*, normally down to 200 m depth, after which there is an increase in gradient to form the *continental slope*, descending to the *continental rise* where the seabed starts to level out as it approaches the abyssal plain (Thistle 2003). Continental margins, with their variety of subhabitats including rock, sediment, cold seeps, pockmarks, mud volcanoes, canyons, landslides, and cold-water corals, are one of the most heterogeneous and biodiverse systems in the deep ocean (Figures 2D, E, F, G) (Menot et al. 2010; Levin and Sibuet 2012). In many parts of the ocean, the margin is incised by *canyons* (Figure 2E) that act as rapid conduits of water, sediment, and other materials from the shelf into the deep sea (Harris and Whiteway 2011; Fernandez-Arcaya et al. 2017). *Cold seeps* (Figure 2D) are found on active and passive margins, where dense sediment-hosted faunal com-

Figure 1 Bathymetry of the ocean highlighting examples of its main topographic features. Image reproduced and modified from the GEBCO world map 2014, www.gebco.net.

Figure 2 Main ecosystems in the deep sea. A: Porcupine Abyssal Plain showing part of a time-lapse camera and individuals of the holothurian *Amperima rosea* (Photo: D. Billett, NOCS); B: Mid-ocean ridge at the Cayman Trough (Photo: NOAA *Okeanos* cruise EX1104); C: hydrothermal vent community showing *Riftia pachyptila* tubeworms, bythograeid crabs and zoarcid fish on the East Pacific Rise (Photo: R. Lutz, Rutgers Uni.); D: cold seep community showing *Seepiophilia* tubeworms and bathymodiolid mussels in the Gulf of Mexico (Photo: C. Fisher, Penn State Uni); E: Stalked crinoid on a canyon wall, continental margin in the NE Atlantic (Photo: P. Tyler, NERC cruise JC10). F. Cold-water corals on a canyon in the NE Atlantic (Photo: P. Tyler, NERC cruise JC36). G: grenadier fish on the Eastern Atlantic (Photo: P. Tyler, NERC cruise JC36); H: pelagic salps in bathyal waters in the Bahamas (Photo: C.M. Young).

munities are sustained by chemosynthetic processes. The physiological processes at cold seeps are similar to those found at hydrothermal vents, but the interstitial fluids are cold and charged with reduced compounds (especially methane) formed by biogenic or thermogenic processes in the sediments (Tunnicliffe et al. 2003). Further, deep-seabed seepage in the form of *mud volcanoes* and *pockmarks* are also present and may support important biological communities whose main energy source are the chemicals released from the seabed. Pockmarks may be found in high numbers in some areas of the deep sea (e.g. Barents Sea) and increased faunal abundance and species richness have been observed in many pockmark regions (e.g. Webb et al. 2009; Ritt et al. 2011). *Cold-water corals* are found in all the ocean basins and are important not just as highly diverse systems (over 3000 different species identified so far, some many hundreds of years old), but also as providers of habitat, nursery areas, and refuge for many other species. Lastly, scattered often in large numbers over the seafloor, are *seamounts*. These are formed at the MORs or oceanic hotspots and transported on the tectonic plates across the ocean to ultimate destruction in the subduction zones. Seamounts can be as low as a few metres, often called *abyssal hills*, or can rise many thousands of metres up into the water column. Topographically modified currents affect geochemical cycles, nutrient mixing processes, and detrital fallout from the euphotic zone that deliver allochthonous energy and nutrients to these systems dominated by suspension feeders and their predators (Rogers 2018).

The water column of the deep oceans is the largest single interconnected environment on Earth and perhaps the least known. Although ambient sunlight is technically measurable down to 1000 m, above 1000 m there is a considerable twilight zone in which there have been numerous organismal adaptations to deal with the very low light levels (Aguzzi and Company 2013) and high pressures (Figure 2H). The physico-chemical characteristics of the deep pelagic zone shape the distribution of organisms, resulting in marked zonation of the meso-, bathy-, aybsso-, and hado-pelagic assemblages (Angel 2003).

Below 200 m depth, the ecosystems are almost entirely heterotrophic (excluding chemosynthetically based ecosystems such as hydrothermal vents and cold seeps), relying on the flux of organic material from surface production. That flux may be in the form of small particles or as large organic masses including dead mammals, fish, macroalgal, and wood debris (Smith and Baco 2003; Bernardino et al. 2010). The large organic falls sink to the seafloor remarkably quickly, providing a high-quantity and high-quality food source to benthic communities. Some of the later stages of organic matter degradation can include a sulphophilic phase supporting chemosynthesis-based communities on the seabed. The particulate organic matter produced in the euphotic zone by phytoplankton is consumed by zooplankton and further degraded by microbes as it is transported down through the water column. It is estimated that only 10 per cent of the organic matter produced in the euphotic zone enters deep waters (below 200 m) and only 1–3 per cent reaches the abyssal seafloor (Gage 2003). The result is that most of the deep-sea pelagic and benthic biomes are food-poor or food-limited (Herring 2002; Smith et al. 2008).

1.1.3 Diversity and biomass

Determination of the biodiversity of the deep sea has been a compelling topic since animals were first discovered at depth. Since the 1960s, this debate has had periods of intensity and calm, but it still underpins much of deep-sea biology. Howard Sanders's study of the fauna collected along the Gay Head-Bermuda transect in the 1960s awoke the scientific community to the higher-than-expected diversity of, particularly, small individuals sampled (Sanders et al. 1965). In a more intense study, Grassle and Maciolek (1992) suggested the deep sea may contain as many as 10 million species. Subsequently, Gage and May (1993) dampened enthusiasm by suggesting a figure nearer 500,000, while Mora et al. (2011) predicted 2.2 million species, with 91 per cent unknown. The debate flows back and forth, and the relationships of species diversity with depth and latitude have made the analysis more complex (Rex et al. 1993; Rex and Etter 2010). The Census of

Marine Life programme (2000 to 2010) greatly contributed to the description of new species, and there are currently almost 26,000 deep-sea species catalogued in the World Register of Deep-Sea Species (WoRDSS—www.marinespecies.org/deepsea).

There are some generalisations that can be made about biodiversity and biomass. The deep-sea benthic habitats support very high biodiversity, often compared to that of the rain forests, mostly composed of meio- and macro-fauna, and with many 'rare' species. As a result of decreasing food availability with depth, the faunal composition shifts from mega- and macro-fauna to communities dominated by meiofauna and microbes (Rex and Etter 2010). Hence, there is a general trend of decreasing biomass with depth. Nevertheless, some habitats such as cold-water corals and seamounts can support high biomass through provision of habitat and specific hydrological circulation that retains food particles. High biomass is also characteristic of hydrothermal-vent and cold-seep communities, which are supported by *in situ* primary productivity, although species diversity in chemosynthetic-based ecosystems is generally low because of the particular environmental constraints. Another important large-scale generalisation is the parabolic pattern of biodiversity with depth, with maximum biodiversity at mid-depths (2000 m). This is not a universal pattern and is shaped by complex interacting factors including biological interactions, food, and habitat heterogeneity (Rex and Etter 2010).

1.1.4 The legal framework of the ocean

Although morphology and water masses address the physical structure of the deep ocean, the ocean, as regulated in terms of human exploitation, has been divided up with differing jurisdictions based on a legal framework (Chapter 3). The United Nations Convention on the Law of the Sea (UNCLOS) grants all maritime countries an Exclusive Economic Zone (EEZ), usually 200 nautical miles from the shoreline. As a result, EEZs can include significant deep water, for example 85 per cent of the water within the US jurisdiction falls within the 'deep sea' (US Bureau of Ocean and Energy Management). Legal instruments for

exploitation and conservation within EEZs are the responsibility of the coastal state. Beyond the EEZs are areas beyond national jurisdiction (ABNJ), which include the high seas and the 'Area'. The high seas comprise the water column beyond 200 nm, while the Area is the seabed and the subsoil thereof beyond the outer limits of a coastal state's continental shelf, or its extended continental shelf. The ABNJ are those areas of ocean for which no one nation has sole responsibility for management. In all, the ABNJ make up 40 per cent of the surface of our planet, comprising 61 per cent of the surface of the oceans and nearly 95 per cent of its volume. In ABNJs, there are currently limited and somewhat haphazard legal frameworks. The seabed resources in the ABNJ are the common heritage of mankind, as declared in the 1982 UNCLOS. The common heritage of mankind principle requires not only the sharing of benefits for all nations but also the conservation and preservation of natural and biological resources for both present and future generations.

1.2 Exploration, technical development, and analysis leading to economic benefits of the deep sea

The exploration and investigation of the deep global ocean is intrinsically linked to technological developments that have enabled access to remote regions of the planet (e.g. polar regions, high seas, and abyssal plains), with complex topographies (e.g. canyons, seamounts, ridges, and hydrothermal vents) and in challenging environmental conditions (e.g. high pressure, darkness). In this section, we provide a historical summary of such technical and methodological developments and some of the major scientific discoveries that have ensued (Table 1), considerably increasing our understanding of deep-sea ecosystem composition and functioning (Ramirez-Llodra et al. 2010). These developments and new knowledge have also contributed to the development of resource exploration and exploitation in deeper waters, followed in parallel by the development of legislation (Table 1).

Table 1 Summary of key historical explorations, discoveries, technical and methodological developments, and exploitation in the deep sea

Period	Scientific Explorations/Discoveries	Significant Expeditions/Technical Innovation/International Laws	Exploitation and Other Uses
Nineteenth century	First robust evidence of deep-water benthic animals Evidence of global distribution of deep-water benthos Pelagic fauna described Increased knowledge of deep-sea bathymetry Accurate temperature measurements First deep-sea fish and fauna described in *Ichthyologie de Niece*	HMS *Cyclops* 1857 HMS *Lightning* 1868 HMS *Porcupine* 1869 HMS *Challenger* 1872–1876 SS *Blake* 1877–1880 USS *Albatross* 1891 Sigsbee/Agassiz Trawl 1880 Reversing thermometer *Traveilleur* and *Talisman* 1880–1883 *Hirondelle I* and *II* and *Princess Alice I* Fish traps for mobile fauna SS *National* 1889 SS *Valdivia* 1898–1899	First deep-sea communication cables 1858 Disposal of clinker 1850s to 1940s First known overexploitation of deep-sea fishery (Greenland halibut)
Early twentieth century	Significance of deep-water fish populations realized	MV *Michael Sars* 1910	
Interwar years	Atlantic water masses defined First *in situ* bathypelagic observations	Echo sounder first use on ships 1924 SS *Meteor* 1925–1927 Wust's Water mass identification 1927 First HOV *Nishimura-shiki Mame Sensui-tei ichi-go* 1929 Beebe's *Bathysphere* 1934 HMS *Discovery* 1930s HEMS *Mabahis* 1933–1934	Deep-sea observations widely disseminated for first time (e.g. *Bathysphere*)
1940s/1950s	Hadal invertebrate fauna collected	SS *Galathea* samples deep-sea trenches 1950–1952 Deep-water echosounding	Radioactive waste disposal starts 1946 Wartime wrecks and munitions 1939–1945 First large-scale deep-sea commercial fishing >400 m 1956
1960s	Discovery of high deep-sea biodiversity Publication of the Heezen & Tharp bathymetric map	Piccard and Walsh dive to Marianas Trench in Bathyscaphe *Trieste* 1960 Development of quantification in sampling via box dredge, Reineck and USNEL box corer Construction of DSRV *Alvin* 1964 Development of DP Development of SASS (deep-ocean bathymetry) UNCLOS —four conventions came into force	Red coral fishery 1965–1990 Large-scale deep-sea commercial fishing accelerates Deep-seabed mining interest sparked by publication of Mero's *Mineral Resources of the Sea*
1970s	Identification of varying diversity along a depth gradient Discovery of hydrothermal vents and their extraordinary fauna 1977	DP installed on ships 1971 onwards Availability of commercial swath bathymetry London Convention came into force 1975 Development of molecular techniques	Deep offshore drilling for oil and gas 1970 onwards and deeper Deep disposal of pharmaceuticals 1973–1978
1980s	Discovery of cold seeps 1984 Understanding chemosynthesis in deep-sea populations Understanding vertical flux and seasonality 1983 Large organic food falls (whale, kelp, and wood) 1984 First evidence of climate-change effects in the deep sea 1984 Describing benthic storms 1984	UNCLOS 1982 adopted ROV *Jason* built 1988 GPS paired with dynamic positioning—precise positioning for deep-sea research gear Real-time imaging MARPOL came into force 1983	Radioactive waste disposal ends 1982 Sewage and dredge sludge disposal 1986–1992 Terrestrial mine tailings disposal begins in the deep sea Anticancerigenous drug Halinchondrin B isolated from sponges

Table 1 (*continued*)

Period	Scientific Explorations/Discoveries	Significant Expeditions/Technical Innovation/International Laws	Exploitation and Other Uses
1990s	Discovery and importance of prokaryotic life Global diversity estimates and latitudinal gradients	Use of Molecular techniques UN CBD entered into force 1993 UNCLOS came into force 1994 UNFCCC came into force 1994 OSPAR came into force 1998	Lost shipping and containers Further dumping and loss of nuclear waste
2000s	Recognition of Archaea and other microorganisms including viruses Census of Marine Life: Deep-sea Field Programmes General perception of threats from climate change and ocean acidification to deep-sea ecosystem function	Sophisticated swath bathymetry Rapid expansion of scientific ROV numbers and use Vast quantities of image data acquired London Protocol came into force 2006 VME concept established 2006 UNGA adopted deep-sea bottom fisheries Resolution 61/105 2006 EBSA criteria adopted 2008	Climate change Ocean acidification First hydrothermal-vent MPAs established 2003 and 2006
2010s	Understanding diversity and connectivity High-resolution maps Fauna discovered in deepest trenches Extent of deep-sea litter (all size-fractions) starting to emerge Deep-sea biology studies increasingly related to deep-sea exploitation	Massive sequencing effort especially for taxonomy Digital revolution and AI application to data HOV *Challenger* 2012 Establishment of deep-sea observatories, streaming live data ISA established regulations for mineral exploration (2010–2013) Final testing of deep-sea mining technologies prior to exploitation Development of international legally binding instrument under UNCLOS on the conservation and sustainable use of marine biological diversity of ABNJ	Deepwater Horizon oil spill 2010 Litter prevalent, especially plastics Deep-seabed mining close to reality World's first network of high-seas MPAs established 2010 Outreach—real-time viewing of deep-sea expeditions from land Around 600 natural products derived from the deep ocean, and increasing

Abbreviations: ABNJ, Areas beyond national jurisdiction ; AI, artificial intelligence; CBD, Convention on Biological Diversity; DP, dynamic positioning; EBSA, Ecologically or Biologically Significant Area; GPS, Global Positioning System; HOV, human-occupied vehicle; ISA, International Seabed Authority; MARPOL, The International Convention for the Prevention of Pollution from Ships (Marine Pollution); MPAs, Marine Protected Areas; OSPAR, The Convention for the Protection of the Marine Environment of the North-East Atlantic; ROV, Remote operated vehicle; SASS, Sonar Array Sounding System; UN, United Nations; UNCLOS, United Nations Convention on the Law of the Sea; UNFCCC, United Nations Framework Convention on Climate Change; UNGA, United Nations General Assembly.

1.2.1 Nineteenth century

Scientific and technological advances: Early deep-sea biological observations such as that of the euryalid brittlestar accidentally caught on a sounding line in 1818 by Sir John Ross (Wyville-Thomson 1873; Koslow 2007) and during bathymetric sampling in the Antarctic by Sir James Clarke Ross (Koslow 2007) were serendipitous. In 1850, the Sars, both father and son, sampled numerous benthic species at upper bathyal depths in Norwegian fjords (Murray and Hjort 1912). And in 1857, a solitary coral and other invertebrates were discovered on a deep-water telegraph cable recovered between Sardinia and Tunisia (Wyville-Thomson 1873). All these

samples provided evidence to refute Forbes's 1843 'Azoic zone' theory, which stated that the deep sea was devoid of life (Anderson and Rice 2006). Whilst incorrect, this theory was, nevertheless, an instigator to continue exploration of the deep ocean. Such biological observations intrigued Victorian scientists, which, together with the Admiralty's interest in seabed bathymetry, resulted initially in HMS *Lightning* expeditions of 1868 to the North-East Atlantic off North-West Scotland and subsequently to a series of expeditions aboard HMS *Porcupine* round UK, Irish, and other European coasts (Wyville-Thomson 1873). These expeditions increased significantly the knowledge of deep-sea bathymetry, including finding the Faroe-Shetland

Ridge, a primitive understanding of water mass structures north and west of Scotland, and, most importantly, the occurrence of living animals in the deepest water sampled (Wyville-Thomson 1873). This limited knowledge base was expanded considerably by the voyage of HMS *Challenger* (1872–1876), which confirmed the presence of benthic and pelagic animals in the deepest waters of the ocean. From this point to the end of the century, there was a series of expeditions, many for soundings and hydrographic observations, the most important biological observations being on the SS *Blake* (1877 to 1880, Caribbean and Gulf of Mexico) and the USS *Albatross* (1891, North-West Atlantic and Panama) under the guidance of Alexander Agassiz. In the Eastern Atlantic, exploration was undertaken by the French ships *Travailleur* and *Talisman* (1880 to 1883), which were extended into the Mediterranean on *Hirondelle I* and *II* and *Princess Alice I* and *II* under the guidance of the Prince of Monaco. Most of these expeditions concentrated on the benthic fauna, and it was not until 1889 that Victor Hensen specifically sampled deep-water plankton aboard the SS *National*. Pelagic fauna was also sampled during the Italian circumnavigation by the *Vetter Pisani* (1882 to 1885), led by Chierchia and Palumbo, who discovered plankton (e.g. siphonophores) down to 2300 m depth. The *Valdivia* expedition (1898–1899) led by Carl Chun conducted extensive sampling of deep-water pelagic fauna down to and beyond 2000 m. On all these expeditions, the equipment would be considered primitive and nonquantitative by modern standards. Depth was measured by sounding lines with a small piece of tallow to collect a sediment sample as the line weight touched the seabed. More efficient methods such as Ballie's or Sigsbee's sounding tubes collected a small core of sediment (Murray and Hjort 1912). Dredges were coarse and rugged and would be lined with sacking to aid collection of animals and sediment or rock. Attached to the back of the dredge would be flails made from lengths of uncoiled rope that were remarkably efficient in trapping rugose animals such as sea urchins and brittle stars. For larger epibenthic samples, Agassiz or Sigsbee trawls were used (and are still in use today) (Clark et al. 2016a, b). Perhaps the most innovative sampling was by

Prince Albert of Monaco, who used fish traps for mobile fauna.

Submarine Cables: The nineteenth-century exploration of the deep ocean is usually thought of as an interest and enquiry-led exploration, the outcome of which was a greater knowledge of science, a central concept in the Victorian ideal. However, profit was also a central tenet of the Victorian ideal, and the deep ocean was not immune. Communication across the world was still primitive and desperately slow in the early nineteenth century. The development of armoured waterproof telegraph cables in the 1840s gave the opportunity to connect continents (Chapter 10). Britain and France were connected by cable in 1850. However, the terrain of the seabed across the Atlantic was not known, and, in 1857, HMS *Cyclops* conducted a bathymetric survey during which it discovered the Mid-Atlantic Ridge and Atlantic abyssal plains. The first exploitation of the deep sea was the laying of the submarine cable between Ireland and Newfoundland in 1858, jointly by HMS *Agamemnon* and USS *Niagara*. Rapidly, deep-sea cables were laid, especially between colonising powers and their colonies. Today, modern digital communication is between economic giants in the global economy through the cable network. Cables have also become an important component of modern deep-sea observatories. For deep-sea biology, submarine cables have been a mixed blessing. The recovery of a damaged cable laid between Sardinia and Tunisia at 2200 m showed the growth of fifteen species of benthic invertebrates, including the solitary coral *Caryophyllia borealis* (Wyville-Thomson 1873). These observations were taken as the animals may cause damage to the cable, whereas Ramirez-Llodra et al. (2011) predicted a minimal impact of cables on underwater fauna. However, more recently, there has been evidence that the electromagnetic field generated by underwater cables may have an impact on animals that settle on cables or use them as a refuge (Scott et al. 2018).

Clinker: Of benefit to maritime exploration was steam propulsion, which became generally available after the 1850s. One of the by-products of steam generation is clinker from coal burning as the fires are drawn. Clinker was most easily disposed of over the side, and there are areas of the deep seabed where clinker has significantly modified the sedimentology (Kidd and Huggett 1981).

Exploration: By the end of the nineteenth century, different parts of the deep ocean had been sampled, although the overall scale was miniscule compared to the habitat size. There was confirmation that benthic animals lived at all depths, with the exception of life in the deep trenches, which would not be discovered for many years. Science and intellect had led the way, and exploitation was extremely limited.

Deep-sea fisheries: Artisanal fishing for deep-sea fish (as opposed to commercial fisheries) has been practiced for centuries. In the North Atlantic, a line fishery for Black scabbard fish (*Aphanopus carbo*) had existed since the seventeenth century, whilst in the Pacific, Oilfish (*Ruvettus pretiosus*) had been fished at bathyal depths off oceanic islands (Koslow 2007). In both these fisheries, the catch levels were limited, and the fishery remained sustainable. In the Arctic and North-West Atlantic, Greenland halibut (*Reinhardius hippoglossoides*) was fished as an artisanal fishery in the nineteenth century, but succumbed to overexploitation by commercial fishing in the late-nineteenth century that eventually led to severely depleted stocks (Moore 1999).

International legislation: Although exploitation of the deep sea was still at its infancy, the oceans *per se* were strategic and economic areas of great importance. In 1494, Pope Alexander VI divided the Atlantic Ocean between Spain and Portugal. The seventeenth and eighteenth centuries saw the development of rules that would precede the criteria to define areas within and beyond national jurisdiction. The early seventeenth century saw a philosophical and legal debate around whether the seas were open to all, as argued by Grotius's *Mare Liberum*, or under state dominion, as claimed for example by John Selden's *Mare Calusum*. The *Mare Liberum* doctrine prevailed and built the basis for today's freedom of the high seas. A century later, among other concepts, the cannon-shot rule was introduced, dictating that coastal states had sovereign rights as far seawards as a cannon ball will reach, which at the time was 3 miles, beyond which was *Mare Liberum* (Lodge et al. 2013).

1.2.2 Early twentieth century

Scientific and technological advances: Early twentieth-century deep-sea exploration became more focused, both in terms of geography and concentrating on particular aspects of biology. The prime interest was fisheries, as exemplified by the *Michael Sars* expeditions (1900 to 1910) under the leadership of John Murray and Johan Hjort. Initially, these cruises were in the Norwegian Sea, but eventually expanded to the whole North Atlantic. This would prove to be the last major expedition before the Great War and, although concentrating on fisheries, they described all the associated fauna sampled (Murray and Hjort 1912). The work of the *Michael Sars* awakened the scientific community to the variety and potential of both benthic and pelagic fish. Meanwhile, in 1904–1905, the German zoologist *Franz Doflein* conducted the first Japanese deep-sea biological investigation in Sagami Bay—again concerned with fisheries.

1.2.3 1920s and 1930s

Scientific and technological advances: The interwar years were dominated by economic concerns, and scientific exploration was mainly directed towards physical and chemical oceanography (summarised by Wüst 1964). From 1925 to 1927, the SS *Meteor* under Wüst conducted fourteen latitudinal cross sections of the Atlantic determining the water masses throughout the Eastern and Western Atlantic, giving a three-dimensional visualisation of the structure of the Atlantic Ocean (Wüst 1950, 1964). A major series of expeditions by RRS *Discovery I* and *Discovery II* were undertaken in the late 1920s and 1930s to Antarctic waters (Rice 1986). The aim was to understand the decline in whale populations and describe shallow and deep, benthic and pelagic faunas of Antarctic waters, which were published in thirty-eight volumes of *Discovery Reports* 1929 to 1980. In the Indian Ocean, the HEMS *Mabahis* sampled deep-water faunas for the first time in this region (Sewell 1934; Aleem and Marcos 1984). Although patented in 1913, the echo sounder was only installed on ships after 1924, providing continuous depth recording, which greatly increased data on deep-sea bathymetry. Deep-water echosounding was proven in the 1940s. Sonar methodologies continued to develop during and after the Second World War. In the 1960s, the US Navy developed SASS (Sonar Array Sounding System), which was further developed into a commercial version in 1977 and forms the basis of the data on deep-ocean swath bathymetry that are still being collected today (Figure 1). These methods have also evolved into techniques for fish tracking in fisheries exploration and exploitation.

The first human-occupied vehicle (HOV) designed specifically for studies on marine biology (mostly fisheries investigations) was named the *Nishimura-shiki Mame Sensui-tei ichi-go* and was developed in 1929 in Japan. This vehicle had sampling gear, a diesel engine, lights, two view ports, and an underwater telephone system with a maximum depth of 200 m (Fujikura et al. 2010). By the early 1930s, interest in a human entering the deep sea led to the development of the *Bathysphere* (Beebe 1935). In 1934 William Beebe and Otis Barton descended a 'half mile down' and gave the first account of the deep pelagic fauna *in situ* and in doing so, heightened public perception and interest in the deep ocean (Chapter 8).

International legislation: In 1930, the Hague conference organised by the League of Nations sought to address national claims over waters adjacent to coastlines but ended with no agreement reached (Schoolmeester and Baker 2009; Lodge et al. 2013). The 1920s and 1930s were an era of scientific interest in the oceans, but there was very little commercial deep-water exploitation. This era was brought to an end by the Second World War, when there were technical and scientific developments that had a long-lasting effect on deep-sea biology (see Chapter 3).

1.2.4 1940s to 1960

Scientific and technological advances: The major event in deep-sea biology in the early 1950s was the circum-global Danish *Galathea* expedition that demonstrated that living invertebrates were found in the deepest trenches of the ocean (Bruun et al. 1956). This period saw the development of two major aspects of deep-sea exploitation: radioactive waste

disposal (Chapter 10) and deep-sea fisheries (Chapter 4).

Radioactive waste disposal: The development and subsequent use of the atomic bombs dropped on Hiroshima and Nagasaki presented the allied nations with the problem of disposal of high-level radioactive waste as a product of the development of the bomb. As early as 1946, the United States disposed of its accumulated nuclear waste to the west of the Farallone Islands off California at both relatively shallow and bathyal depths. Disposal of low-level nuclear waste continued off both the West and East Coasts of the United States until 1979, when the United States was the first nation to stop nuclear waste disposal in deep water (Johnson et al. 1984; Koslow 2007). From 1949 to 1982, other nuclear nations including the UK, France, Germany, the Netherlands, Sweden, Belgium, and Switzerland continued to dispose of up to 142,000 tonnes of low-level radioactive waste in the North-East Atlantic (Glover and Smith 2003; Thiel 2003). In addition, the United States had accidentally lost the nuclear submarines USS *Thresher* (1963) and USS *Scorpion* (1968) (see Chapter 10).

Fisheries: Although the first large-scale deep-sea fishery catches, of just a few species, are from 1956 by the Soviets (Victorero et al. 2018), deep-sea fisheries really accelerated during the 1960s (see Chapter 4).

International legislation: States' interest in proclaiming jurisdiction over natural resources on continental shelves increased since 1930. The first United Nations Convention on the Law Of the Sea (UNCLOS I) was held in 1958 and resulted in four conventions, one each on the Territorial Sea and Contiguous Zone, the Continental Shelf, the High Seas, and Fishing and the Conservation of Living Resources of the High Seas (Rothwell and Stephens 2016).

1.2.5 1960s

Scientific and technological advances: In the 1960s, there was a paradigm shift in deep-sea biology. To that date, all sampling had been qualitative or semi-quantitative at best. The development of the box dredge, Reineck box core, and later of the US Naval Electronics Laboratory box corer allowed quantitative sampling of the deep-sea sedimentary environments (Sanders et al. 1965; Hessler and Jumars 1974).

Using careful elutriation samples from bathyal and abyssal depths yielded significant numbers of small-sized species and, together, contributed to the development of the concept of high species-diversity at the deep seabed (Sanders et al. 1965). The determination of asymptotic species-diversity has been a continuous theme in deep-sea biology research ever since.

The 1960s also saw HOVs starting to make their mark. In January 1960, Piccard and Walsh had used the Bathyscaphe *Trieste* to sink to the bottom of the Marianas Trench, a feat not repeated until 2012, when James Cameron took *Deep-Sea Challenger* to the same depths, and, in 2019, Victor Vescovo made the deepest-ever dive in the Marianas Trench. These landmark descents continue to excite interest in the public. Of more immediate benefit to deep-sea research was the commission of Deep-Sea Research Vehicle (DSRV) *Alvin* on 5 June 1964. This was the first free manoeuvrable HOV for deep-sea investigations and has been the workhorse for deep-sea research ever since. The success of DSRV *Alvin* led to development of *Cyana* and *Nautile* (France), *Shinkai* (Japan), and the *MIRS I* and *II* (Russia), with the capacity to take three humans down to depths between 3000 m and 6500 m, and collect samples with a variety of instruments and sensors.

Although not immediately beneficial to deep-sea science, the 1960s saw the development of 'dynamic positioning' (DP) primarily for the positioning of semi-submersible oil rigs. In offshore oceanography, position keeping was maintained by highly skilled deck officers on research vessels. The real benefit of DP in deep-sea science came later in the twentieth century, when it was paired with Differential Global Positioning System (GPS) (available from 1980s) to allow extremely precise positioning and station keeping for the use of submersibles and remotely operated vehicles (ROVs).

Possibly the most iconic output from the 1960s was the map of the bathymetry of the world's ocean produced by Marie Tharp and Bruce Heezen. For the first time, it was easy for scientists and lay persons to visualise that the deep sea was not a single large basin, but a series of basins separated by sills or mountain ranges, with parts of the ocean plunging to great depths (Figure 1). It is from this image, the development of swath bathymetry, as well as harnessing the power of satellites, that the continually increasing resolution of seabed bathymetry has grown to the knowledge we have today.

Deep-sea fisheries: Commercial deep-sea fishing (Chapter 4) at depths greater than 400 m started in the late 1950s, increased during the 1960s, and has continued accelerating to this day (Watson and Morato 2013). Its history is a tale of misunderstanding, greed, weak management, and the inability to understand that a deep-sea fish population needs time to replace itself (Koslow 2007). Deep-sea fishing is not confined to fish. Red corals of the genera *Corallium*, *Lepidisis*, *Keriatosis*, and *Gerardia* and black corals have been collected from shallow water for centuries for jewellery, and deep-water coral fishing peaked in the 1970s and early 1980s (Koslow 2007). Fishing not only reduces the target fish populations and other species through discards, but trawling, in particular, can have profound physical impacts on the seabed and benthic ecosystems such as deep-water corals or sponge beds. Visual observations of the damaged seafloor have been more common since the advent of submersibles and especially ROVs (Benn et al. 2010; Figure 3A, B), with recent studies showing that seabed morphology can be modified by intense trawling activity (Puig et al. 2012).

International legislation: The 1960s were important years for international Law of the Sea that saw growing interest in deep-seabed minerals. In 1967, the Maltese Ambassador Arvid Pardo gave a speech to the United Nations (UN), in which he called for the international seabed to be declared the common heritage of mankind to allow developing states to share the riches of the deep oceans. By the end of the decade, to avoid escalating international tensions, a call is made for the development of an effective international regime over the seabed beyond a clearly defined national jurisdiction. This will lead to the UNCLOS III in the 1970s (Schoolmeester and Baker 2009). UNCLOS II was held in 1960, with the aim of deciding the breadth of the territorial sea and fishery limits. However, no agreements were reached (Shoolmeester and Baker 2009). The four conventions adopted at UNCLOS I (noted above) entered into force during the 1960s.

Figure 3 Major activities and challenges faced by deep-sea ecosystems. A: Cold-water coral reef with *Solenosmilia variabilis* as the main framework builder, Coral Seamount, South West Indian Ridge (Photo: AD Rogers NERC / IUCN Seamounts Project). B: coral rumble as a result of commercial trawling over a cold-water coral reef off New Zealand (Photo: NIWA, NZ). C: cold-water corals and crab entangled with fishing ropes (Photo: P. Tyler, NERC cruise JC36); D: marine litter on the deep seabed in the NE Atlantic (Photo: P. Tyler, NERC cruise JC10); E: shipwrecks may or may not be colonised by deep-sea fauna (Photo: P. Tyler, NERC cruise JC10); F: *Bathystylodactylus* sp. prawn on the abyssal seabed of the Clarion Clipperton Zone showing manganese nodules (Photo: D. Jones, NOCS, cruise JC120/NERC, UK). G: seafloor massive sulphides are formed at hydrothermal vents, here showing vent chimneys with alvinocarid shrimp (Photo: SEHAMA 2002 cruise, PDCTM 1999/MAR/15281, Uni. Azores; PT); H: a cobalt-rich crust seamount in the Pacific showing corals and an echinoid (Photo: NIWA, NZ).

1.2.6   1970s

Scientific and technological advances: The abiding discovery of the 1970s was hydrothermal vents in the Galapagos Rift in the Pacific Ocean (Corliss et al. 1979) and their unique ecosystems (Grassle 1985). Interest by scientists and the media was intense because of the unexpected high biomass, low diversity, and brightly coloured fauna at hydrothermal vents. While the world marvelled at the animals, the geologists were concentrating on the mineral deposits that 4 decades later would initiate the possibility of commercially mining these deposits (Chapter 5). The developments of molecular techniques led also to major changes in the classification of life. In 1977, Woese and Fox separated the Archaea from the Bacteria, based on molecular analyses of their ribosomal RNA. At the time, the Archaea domain was believed to be composed by extremophile microorganisms, such as methanogens, halophilic and hyperthermophilic microbes, but the group includes today microorganisms both from extreme and nonextreme environments (Woese et al. 1990).

Oil Exploration and exploitation: One of the great engineering achievements of the twentieth century was the development of offshore oil platforms, initially for exploration and subsequently for production of oil and gas (Chapter 6). The first offshore platforms (1955) were only in 30 m of water, but by 1970 the engineers had developed a variety of platforms, some of which could drill into the seabed (albeit experimentally) at 1400 m depth. From that date, there has been a steady migration of oil platforms into deeper water (Merrie et al. 2014) with increased risks to deep-sea ecosystems (Bernardino and Sumida 2017). To date, the deepest water depth in which drilling has occurred is 3400 m depth off Uruguay, in the lower part of the bathyal zone. The most active development has been in the Gulf of Mexico, where major reserves are being accessed in waters as deep as 3000 m (Cordes et al. 2016). Although the transport of oil had resulted in significant pollution in shallow water due to wreckage, the oil industry had a responsible track record on containing contamination from oil or drilling muds as exploration proceeded into deeper water (Ramirez-Llodra et

al. 2011). This paradigm was shattered with the massive oil spill from the *Deepwater Horizon* rig in the Gulf of Mexico in 2014. The use of fossil fuels is also raising concerns of additional climate-change impacts to marine ecosystems, which need to be considered in terms of oil-industry impacts (Chapter 6).

International legislation: The third conference on the Law of the Sea (UNCLOS III) was held from 1973 to 1982, and resulted in what is often referred to as the constitution for our oceans, the UN Convention on the Law of the Sea. A number of other important ocean treaties were adopted in the 1970s. The Convention on the Prevention of Marine Pollution by Dumping of Wastes and Other Matter (London Convention), adopted in 1972 and entered into force in 1975, banned the routine dumping of many waste types from ships. The international Convention for the Prevention of Pollution from Ships (MARPOL), adopted in 1973 and entered into force in 1983, regulates marine pollutants from vessel operations or accidents. Both Conventions and their subsequent Protocols are administrated through the International Maritime Organization (IMO).

1.2.7   1980s

Scientific and technological advances: The 1980s saw a rapid expansion of knowledge about the deep sea as more countries became involved in modern deep-water oceanography. The discovery of chemosynthetically-driven hydrothermal-vent communities in the late 1970s was followed by the discovery of cold seeps along the Florida Escarpment (Paull et al. 1984). Understanding how microbial chemoautotrophy supported oases of life on the deep seafloor was one of the great biological discoveries of that decade (Cavanaugh et al. 1981; Felbeck et al. 1981).

To this point it had always been assumed that the energy requirements of the deep-sea fauna were satisfied by the slow sinking of particles from surface production. Visual observation of the seafloor (Lampitt 1985; Hecker 1990) and sediment traps (Deuser and Ross 1980) demonstrated that the particles sank at a rate of 100 m d^{-1}, reaching the seabed in 30 days and reflecting the seasonal pattern of production at the surface (Billett et al. 1983). This seasonal injection of energy drove physiological processes, including reproduction in the previously considered monotonous deep benthos, on a seasonal basis (Tyler et al. 1982).

At the same time there was interest in the sinking of large food packages such as fish, whale, and kelp falls and what contribution they made to deep-sea food webs (Stockton and DeLaca 1982). The first whale falls were found serendipitously during a submersible dive off California. Investigations of the whale carcasses revealed that they provided not only a large carbon source for scavengers but, with time, established their own chemosynthetically driven community at the seabed (Smith et al. 1989). The whale-fall discovery led also to the description of a new genus of tube worm, *Osedax*, that generates its energy heterotrophically (rather than through symbiosis with chemosynthetic bacteria as all other siboglinids) by relying on symbiotic bacteria for the digestion of whale proteins and lipids (Rouse et al. 2004)

The last major scientific discovery of this exciting decade was the observation and analysis of benthic storms (Hollister and McCave 1984). These high-energy, unexpected events occur at abyssal depths as a function of reversal of deep-water flow that accelerates, resuspends sediment, declines, and deposits sediment load on the local fauna (Thistle et al. 1991).

A technological development that would bring the deep sea into everyone's office was the development of the ROV linked to the digital environment. To this point, videos taken by HOV were viewed once back on the mother ship. With the scientific development of ROV *Jason* (Ballard 1993), and subsequent ROVs in later years, video images could be transferred up the optical tether to the ship in real time.

Dumping of waste and sewage sludge: All kinds of litter have been thrown overboard from ships for centuries, as the sea, *in toto*, was being seen as a dumping ground for waste, which became 'out of sight, out of mind' (Chapter 10). A classic example is Deepwater dumpsite (DWD)106 at 2500 m depth in the North-West Atlantic. Between 8 and 9 million tonnes of industrial and sewage waste were dumped between 1986 and 1992, modifying the stable isotope composition of the food chain at the seafloor (Van Dover et al. 1992). It is not just the smothering effect of the particles sinking, but associated contamination can be with trace metals, chlorofluorocarbons, and persistent organics. None of these additional risks had been determined before disposal (Thiel 2003). Although natural 'smothering' occurs from particle flux sinking and the effect of benthic storms, responses by organisms have been adaptive but are unlikely to survive the significantly higher levels of smothering caused by sediment suspension and settlement from mining. The late 1980s also saw the beginnings of disposal of terrestrial mine tailings into deep waters, and this practice continues to grow (Vare et al. 2018).

Growth in deep-water fisheries: The 1980s saw an increase in deep-sea fisheries both in terms of fish species being targeted and regions fished, with large unreported catches between 1985 and 2010 and an estimated 600,000 tonnes of fish being caught in the mid-1970s, late 1980s, and early 2000s (Victorero et al. 2018).

International legislation: Following a decade of negotiations, the UNCLOS was adopted in 1982, though it took another 12 years for it to enter into force.

1.2.8 1990s

Scientific and technological advances: The 1990s saw the establishment of molecular biology as a significant tool in analysing deep-sea biological questions. Population genetics on deep-sea organisms had been carried out in the 1970s (Ayala and Valentine 1974), but modern techniques give a much greater refinement of evolutionary processes, population analysis, connectivity, and biodiversity. Molecular techniques allowed the determination of cryptic species, in essence increasing known biodiversity in the deep sea. Early in the decade, Grassle and Maciolek (1992) had published their seminal paper suggesting the deep sea may contain up to 10 million species. This triggered an international debate, and molecular biology was going to contribute significantly to that debate. Another aspect was the relationship among known species. Worm-like organisms discovered in the early part of the twentieth century were classified as the new phylum Pogonophora (Ivanov 1955). With the discovery of hydrothermal vents, the large, exotic tube worms present in high densities in these ecosystems were classified as yet another new phylum, the Vestimentifera (Jones 1985), related to the pogonophorans. Lastly, the 'bone-eating' *Osedax* species, found on whale bones in 2004, showed great similarities. However, subsequent morphological and molecular analyses and revisions of these groups have shown that the pogonophorans and vestimentiferans are not separate phyla, but a family, the Siboglinidae, in the Class Polychaeta (Phylum Annelida) (Rouse et al. 2004; Hilario et al. 2011).

This decade saw a huge increase in understanding microbial processes in the deep sea. Bacterial activity was believed to be very slow in deep water, with the possible exception of inside the guts of benthic species. Microbial activity was recognised as a significant process affecting the flux of material to the deep seafloor, as well as the identification of prokaryotic life at great depth below the actual seafloor (Huber et al. 2007; Marlow et al. 2014; Corinaldes 2015).

Lost 'hard substrata': One of the fiercest environmental and legal debates in the 1990s was the '*Brent Spar*' saga. The *Brent Spar* was a North Sea tanker loading buoy decommissioned in 1991. Shell, with UK government approval, had decided to sink the *Brent Spar* in 2500 m of water over the North Fenni Ridge in the North Atlantic, rather than take it ashore for break up because of the risks of chemical contamination. After many months of arguments, in 1995 the idea was abandoned, and the buoy is now part of the ferry terminal in Stavanger. On reflection there are no shortage of hard metallic substrata in the sea. Ships sink, the best known being the *Titanic*, but during both world wars, hundreds of thousands of tonnes of shipping was sunk, especially in the North Atlantic. Thiel (2003) estimates 65,000 tonnes of shipping sinks to the deep-sea floor each year. To this can be added to the 1390 tonnes of shipping (of 20 million tonnes total) lost each year during storms. The effect of this metal arriving at the seafloor has never been assessed, and impact on deep-sea biology is unknown. Very few invertebrates are seen on the *Titanic*, although it would appear much of the wood has been removed by wood borer fauna (Chapter 10).

Radioactive disposal: By 1983, the dumping nations participated in the London (Dumping) Convention, at which they agreed to stop the disposal of all radioactive waste. None of the signatories admitted to dumping high-level radioactive waste. This may have closed the chapter on nuclear waste dumping in deep ocean waters, but in 1993, the Russians admitted that up to seventeen nuclear reactors had been 'disposed of' in both shallow and deep waters round Novaya Zemlya in the Barents Sea. To this may be added the accidental loss of the nuclear submarine *Konsomolets* that sank to 1500 m in the Barents Sea in 1989 and still has its nuclear reactor and two nuclear warheads on board. In August 2000, the nuclear submarine *Kursk* was also lost in the Barents Sea. The active disposal of radioactive waste appears to be well regulated, but there is very little information on the effect of this dumping on the deep-sea benthos (Chapter 10).

International legislation: The apparent glacial pace of legislation to protect and manage deep-sea ecosystems has been caused by lack of data, lethargy, and blatant national self-interests. But progress has been made, and this is accelerating. In 1994, after 4 years of informal negotiations to redefine the expression of the common heritage principle, UNCLOS was finally agreed and entered into force. UNCLOS remains one of the most important treaties for the oceans, not least because it defines which maritime zones fall under national jurisdiction and which ones are to be managed by the international community as a whole. Importantly, for the first time, UNCLOS established a framework for the Area (UNCLOS, Part XI). Indeed, Part XI of UNCLOS established the International Seabed Authority (ISA), a dedicated organisation responsible for the regulation and management of deep seabed mining activities, as well as ensuring that the marine environment is protected from any harmful effects that may arise during mining activities, including the exploration and exploitation phases (Jaeckel 2015; UNCLOS 1982—articles 137, 145, 153).

In 1996, the London Protocol to the London Convention was adopted, introducing stricter anti-dumping rules than the Convention had set out. The Protocol entered into force in 2006. The Convention for the Protection of the Marine

Environment of the North-West Atlantic (OSPAR) was adopted in 1992 and entered into force in 1998. It created the OSPAR Commission to administer the Convention and develop policy and international agreements. Despite the growing number of treaties, a key lacuna remains. Submarine or deep-sea tailing disposal associated with both land-based and seabed mining do not fall under the current jurisdiction of the LC/LP (Ramirez-Llodra et al. 2015; Vare et al. 2018), but UNCLOS deals with any harm arising from seabed mining, including tailings.

The 1990s were a key decade for international environmental agreements, several of which were opened for signature at the 1992 UN Earth Summit in Rio. Among these, was the UN Convention on Biological Diversity (CBD), which entered into force in 1993.

1.2.9 2000s

Scientific and technological advances: The single most significant marine research programme of the 2000s was the Census of Marine Life. The aim of this programme was to discover how many species were living in the ocean, how abundant they were, and how they were distributed. The programme was divided into fourteen field programmes, including ecosystems from the intertidal to the abyss, and considering the past, present and future status of life in the oceans (McIntyre 2010; Snelgrove 2010). The deep-sea ecosystems were investigated in five projects: COMARGE (continental margins), CeDAMar (abyssal plains), MAR-ECO (Mid-Atlantic Ridge), CenSeam (seamounts), and ChEss (chemosynthetic-based ecosystems). Each project was responsible for developing the science that underpinned the exploration of biodiversity in the deep-water environments they were examining. The Census projects were supported by advances in technology on board a very wide series of research cruises. The use of ROV technology became an integral part of the sampling activity on most of these cruises. Many of the deep-sea cruises produced vast quantities of image data, and methods for more rapid analysis were developed to speed up what was a slow visual process. These methods have expanded further in the 2010s, to include the use of

artificial intelligence. The new knowledge, results, and data were synthesised in a 2-year effort (SYNDEEP), and data were made available via different open-source databases, such as ChEssBase, SeamountsOnline, and the Biogeographic Information System (OBIS). SYNDEEP led to the creation of the International Network for Scientific Investigation of Deep-sea Ecosystems (INDEEP) in 2010. INDEEP is a global collaborative scientific network dedicated to the acquisition of data, synthesis of knowledge, and communication of findings on the biology and ecology of our global deep ocean, in order to inform its management and ensure its long-term health.

Climate change: Climate change is the most concerning, globally impacting, environmental change at present (Chapter 9). Intuitively, one may have thought that the deep sea might be immune from its effects, but Roemmich and Wunsch (1984) provided the first evidence that significant warming was occurring in an ocean-wide band from 700 to 3000 m in the deep North Atlantic. The deep ocean absorbs considerable amounts of heat and carbon dioxide, acting as an important buffer to climate change (Chapters 8 and 9), but as a result, imposing stress in the form of warming, ocean acidification, deoxygenation, and altered food inputs to vulnerable ecosystems (Levin and LeBris 2015; Sweetman et al. 2017). Danovaro et al. (2004) provided evidence of the effects of water warming on nematodes, at 1540 m depth in the Mediterranean. To emphasise that the deep sea is not isolated from surface waters, Smith et al. (2008) proposed a model by which reduced surface production under warming conditions would change from diatom dominated to pico-plankton dominated, with a net flux of surface production declining from 5 to 1 per cent. The increasing acidity, particularly at bathyal depths, has impacts on the calcium carbonate compensation depth and the calcium shells and skeletons of deep-water species, including habitat-building species such as scleractinian corals. Oxygen loss and oxygen minimum zone expansion is another major climate consequence affecting the bathyal deep sea in particular. There is a general perception that threats to deep-ocean biodiversity and the future services they may provide will be compromised owing to climate change affecting the integrity and function of deep-sea systems (Levin and LeBris 2015, Chapter 9; Sweetman et al. 2017).

International legislation: In 2008, COP9 (Conference of the Parties) to the CBD adopted a series of criteria to identify Ecologically or Biologically Significant Areas (ESBAs) in need of protection in open ocean waters and deep-sea environments (COP9 Decision IX/20). In parallel, the concept of Vulnerable Marine Ecosystems (VMEs) resulted from discussions at the UN General Assembly (UNGA) and was further established under UNGA resolution 61/05 in 2006 (Rogers and Gianni 2010). While the ESBAs have no legal status as protected areas, the VMEs are identified by Regional Fisheries Management Organizations under the FAO as areas where deep-sea fishing is prohibited. International climate negotiations began with the adoption of the UN Framework Convention on Climate Change (UNFCCC) in 1992, with the goal of stabilising atmospheric concentrations of greenhouse gases to avoid dangerous anthropogenic interference with the climate system. This Convention entered into force in 1994 and in 2019 had 197 parties. In 1997, the Kyoto Protocol industrialised countries and those in transition to a market economy agreed to reduce their emissions of six greenhouse gases to 5 per cent below 1990 levels by 2008–2012 (Earth Negotiation Bulletin Vol. 12 no 626). After many years of negotiation, the Paris Agreement was reached in Paris during COP21 in 2015. Its primary aim is to strengthen the global response to the threat of climate change by keeping a global temperature rise this century well below 2 degrees Celsius. It aims to bring all nations together in combating climate change and to adapt to its effects, and generates assistance to developing countries to meet their obligations.

1.2.10   2010s

Scientific and technological advances: As the result of novel sequencing techniques, advanced bioinformatics, and increasing computing power, this decade has seen an unprecedented use of molecular techniques to answer questions in deep-sea biology. Foremost amongst these has been the explosion in understanding taxonomy, especially those taxa, such as sponges and nematodes, that have a chal-

lenging morphological taxonomy. The development of metagenomics and environmental sequencing such as 'e-DNA' analyses is also providing novel tools to understand diversity on global scales. Molecular techniques are also greatly contributing to our understanding of gene flow and population connectivity, which is essential knowledge for sound ecosystem management, including the development of spatial planning, such as the establishment of networks of Marine Protected Areas (e.g. Wedding et al. 2013). Understanding larval ecology and dispersal potential is currently one of the great challenges of deep-sea research. How do propagules from one population disperse and find new suitable habitats? These processes are particularly important for species in fragmented ecosystems, such as hydrothermal vents, cold seeps, seamounts, or cold-water corals, but it applies to all deep-sea ecosystems, including abyssal plains. Molecular biology linked to larval dispersal modelling is making headway in this research (Hilario et al. 2015; Baco et al. 2016) and is helping to find the proverbial needle in a million haystacks.

Digital technology has been whole-heartedly embraced to assist in analysing deep-sea biology. Swath bathymetry is providing high-resolution maps of the seabed, particularly when conducted from autonomous underwater vehicles (AUVs) or near-seabed towed systems such as Ocean Floor Observation and Bathymetry System (or OFOBS) (Clark et al. 2016b). The recent development of artificial intelligence and analytical methods such as deep learning and machine learning are greatly speeding up our capacity to analyse very large datasets, including the great amounts of digital images produced by ROV, AUV, submersible, and towed-camera dives. Modern studies with enhanced technology are also enabling the expansion of outreach via programmes such as the National Oceanic and Atmospheric Administration's *Okeanus Explorer*, the Ocean Exploration Trust *Nautilus* and the Schmidt Ocean Institute *Falkor*, and the plans for the new REV *Ocean* vessel (start operating in 2022), which allows real-time viewing of deep-sea expeditions from land—including science laboratories, schools, and even mobile phones. Video images can be transferred up the optical tether to the ship and transmitted by satellite onto the Internet.

Plastic and other marine litter: Litter is a constant in terrestrial environments and along the seashore in many parts of the ocean (Chapter 10). Marine litter was mostly 'out of sight, out of mind' for most of society, until Sir David Attenborough criticised the amount of litter, especially plastic, and its effect in the ocean, catching the public's imagination like no other pollution. Plastic is the most abundant litter type in the oceans (Ramirez-Llodra et al. 2011; Pham et al. 2014). Plastic is not confined to the surface, and it was not until the regular surveying of the deep seabed by ROVs and digital imagery that the extent of litter and general debris was discovered. Much of the deep seabed has little or no litter, but on those areas near large cities or industrial areas, litter is much more common (Mordecai et al. 2011; Pham et al. 2014) and in semi-closed seas like the Mediterranean, marine litter is widespread (Ramirez-Llodra et al. 2013). Plastic has a long life, and as it breaks down, it forms smaller and smaller particles that result in micro- and nano-plastics accumulating on the sediment and, inter alia, can clog the feeding apparatus of benthic species. Large pieces of plastic such as bags, found even as deep as the Mariana Trench, can cause suffocation and entanglement of fauna and can trap sediment that eventually deoxygenates, creating a reducing habitat on the seabed (Figure 3C–D). Shipwrecks, both from accidents and deliberate sinking (Figure 3E), can have a dual effect on the ecosystem, providing hard substratum at the same time that it has a physical impact on the habitat and can result in pollution (Chapter 10).

Deep-sea mining: The idea of mining metalliferous deposits in the deep ocean is not new (Chapter 5). The prospects for mining manganese nodules from abyssal plains (Figure 3F), particularly in the Pacific Ocean, were raised in the early 1970s (Gerard 1976; Jumars 1981). However, the technological challenge was great, and the required minerals were more easily available on land. Almost simultaneously was the discovery of hydrothermal vents, but the tremendous excitement of these new environments precluded any prospect of mining. Four decades later, the demand for strategic minerals is strong, and interest in mining polymetallic nodules, massive sulphides, and cobalt crusts from the deep sea (Figure 3F–H) has risen up the economic agenda rapidly. This will be examined in detail in Chapter 5, but society has learned a lesson from the past. The past paradigm was exploitation, followed by scientific study and then by legislation. At last this paradigm is breaking so there is international management of exploitation based on scientific data such as that delivered by the Deep-Ocean Stewardship Initiative (DOSI). Scientists have the opportunity to analyse potential mining sites and predict what the outcome of mining on the local deep-sea fauna will be (Glover et al. 2016; Van Dover et al. 2018; Thompson et al. 2018).

Marine genetic resources: Potential wide-reaching benefits of genetic diversity exist in the deep ocean, and there is huge growth and international debate in this arena (Arrieta et al. 2010; Harden-Davies 2016). The immense biodiversity in the ocean, including the estimated 2.2 million species of marine animals and up to a trillion different types of microorganisms, have environmental, economic, societal, and scientific value. For deep-sea ecologists, there is the fundamental scientific discovery and the increasing evidence in support of the importance of this diversity in providing essential ecosystem services that sustain the planet. Marine genetic resources are also sources of commercial bioproducts and support whole industries including fisheries, medical, pharmaceutical, and industrial applications. Research into marine natural products has developed rapidly since it began in the late 1940s, and approximately 30,000 of the 1 million natural products described are from marine origins (Martins et al. 2014), with approximately 600 of these deriving from the deep ocean (Skropeta 2011), and research into deep-sea natural products is increasing (Skropeta 2008; Skropeta and Wei 2014; Harden-Davies 2016). The number of deep-sea natural products described between 2009 to 2013 grew by 188, including compounds derived from depths greater than 5000 m (Skropeta and Wei 2014). This growth area for business, and hot topic of international legislation, is discussed in detail in Chapter 7.

International legislation: The current decade has seen intense developments and negotiations related to deep-sea mining in the Area. A total of twenty-nine exploration contracts have been signed with the ISA, with eight between 2001 and 2006, and twenty-one in this decade. In parallel, the ISA has been developing its Mining Code, to set the rules and regulations for the exploitation phase. Between 2010 and 2013, the ISA established regulations for the prospecting and exploration for deep-sea minerals in the Area, including for polymetallic nodules, polymetallic sulphides, and cobalt-rich ferromanganese crusts. The ISA has developed a regional environmental management plan (or REMP) for the Clarion-Clipperton Zone (nodules) in the Pacific, including protected Areas of Particular Environmental Interest (or APEIs). In parallel, the IMO has recognised the gap in regulations of submarine tailing disposal, and an expert group has been created to provide independent advice on the impacts of wastes and other matter in the marine environment from mining operations including marine mineral mining. IMO also addresses inputs to the ocean related to climate geoengineering, an issue of growing importance. The accelerated depletion of deep-sea fishing stocks and strong campaigning by groups such as the Deep-Sea Conservation Coalition have resulted in the adoption of stronger resolutions by the UN General Assembly (e.g. UNGA resolution 66/68 on Oceans and Sustainable Fisheries), and the EU adopted a new and stronger regulation on deep-sea fisheries in EU waters. Finally, in the second half of this decade, UNGA negotiations have started to develop an international legally binding instrument under UNCLOS on the conservation and sustainable use of marine biological diversity of ABNJ, also known as the Biodiversity Beyond National Jurisdiction process (Chapter 7).

1.3 And the future?

The growing human population, depletion of land and coastal resources, poor management, limited robust international legislation, and accelerating technological developments are all strong drivers for the intensification in the uses of the deep ocean and its resources. If it is financially beneficial to exploit the natural capital of the deep sea, it will

eventually happen (as discussed in Chapter 2). In most cases, these uses can affect ecosystems that are poorly understood, not only in terms of their biodiversity and environmental drivers, but importantly also in relation to the ecosystem functions they sustain and thus the derived ecosystem services. There is an urgent need to ensure that Blue Growth takes place within a sustainable framework that ensures healthy oceans and thus a healthy planet for future generations (Chapter 11). To achieve the UN Sustainable Development Goals, it is imperative to conduct both basic and cutting-edge research and innovation that will inform authorities and policymakers responsible for the development of robust management plans. The momentum created during the Census decade and followed into INDEEP led to the foundation of DOSI in 2013, which seeks to integrate science, technology, policy, law, and economics to advise on ecosystem-based management of resource use in the deep ocean and strategies to maintain the integrity of deep-ocean ecosystems within and beyond national jurisdiction. DOSI is highly active in contributing to the current development of legal instruments related to the use of the deep sea and its resources, such as the ISA Mining Code, the development of a new international legally binding instrument for the conservation and sustainable use of marine biological diversity of ABNJ (BBNJ process), the UNFCCC where DOSI highlights climate change in the deep ocean, and the UN Decade of Ocean Science for Sustainable Development (2021–2030) (Chapter 11). Over the coming couple of decades, there will be further comprehensive, joined-up legislation informed by science, industry, NGOs, economists, policy experts, and other stakeholders. How this new legislation will be effectively implemented is currently cause for concern but will undoubtedly be thoroughly addressed during the legislative processes now underway at the UN and within nations. The new legislation will hope to address all current and future exploitation, including fisheries, mineral extraction, deep-sea dumping, litter, and marine genetic resources. Capacity development and technology transfer are essential elements of these new rules and regulations, as are considerations of cumulative effects of climate change.

Acknowledgements

ERLL would like to acknowledge the support from the Norwegian Institute for Water Research (NIVA) and REV Ocean. MB would like to acknowledge support from Arcadia- a charitable fund of Lisbet Rausing and Peter Baldwin. We would like to thank the external reviewers for their comments, which have helped to improve this chapter.

References

Aguzzi, J. and Company J. B. (2013). Chronobiology of Deep-water Decapod Crustaceans on Continental Margins. *Advances in Marine Biology*, 58, 155–225.

Aleem, A. A. and Marcos, S. A. (1984). John Murray/ Mabahiss Expedition versus the International Indian Ocean Expedition (IIOE) in Retrospect. *Deep-Sea Research*, 31, 583–8.

Anderson, T. R. and Rice, A. L. (2006). Deserts on the Seafloor: Edward Forbes and His Azoic Hypothesis of a Lifeless Deep Ocean. *Endeavour*, 30, 131–7.

Angel, M. V. (2003). The Pelagic Environment of the Open Ocean, in P. A. Tyler (ed.) *Ecosystems of the World*. Amsterdam, Netherlands: Elsevier, pp. 39–80.

Arrieta, J. M., Arnaud-Haond, S., and Duarte, C. M. (2010). What Lies Underneath: Conserving the Oceans' Genetic Resources. *Proceedings of the National Academy of Sciences USA*, 107, 18318–24.

Ayala, F. J. and Valentine, J. W. (1974). Genetic Variability in the Cosmopolitan Deep-Water Ophiuran *Ophiomusium lymani*. *Marine Biology*, 27, 51–7.

Baco, A. R., Etter, R. J., Ribeira, P. A., et al. (2016). A Synthesis of Genetic Connectivity in Deep-sea Fauna and Implications for Marine Reserve Design. *Molecular Ecology*, 25, 3276–98.

Ballard, R. D. (1993). The MEDEA/JASON Remotely Operated Vehicle System. *Deep-Sea Research*, 40, 1673–87.

Beebe, W. (1935). *Half Mile Down*. London: John Lane the Bodley Head.

Benn, A. R., Weaver, P. P., Billett, D. S. M., et al. (2010). Human Activities on the Deep Seafloor in the North East Atlantic: An Assessment of Spatial Extent. *PloS ONE*, 5 (9), e12730.

Bernardino, A. F., Smith, C. R., Baco, A., Atlamira, I., and Sumida, P. Y. G. (2010). Macrofaunal Succession in Sediments Around Kelp and Wood Falls in the Deep NE Pacific and Community Overlap with Other Reducing Habitats. *Deep-Sea Research I*, 57, 708–23.

Bernardino, A. F. and Sumida P. Y. G. (2017). Deep Risks from Offshore Development. *Science*, 358 (6361), 312.

Billett, D. S. M., Lampitt, R. S., Rice, A. L., and Mantoura, R. F. C. (1983). Seasonal Sedimentation of Phytoplankton to the Deep-sea Benthos. *Nature*, 302, 520–2.

Bruun, A. F., Greve, S., Mielche, H., Sparck, R., and Bruun A. F. (eds.) (1956). *The Galathea Deep Sea Expedition 1950–1952: Described by Members of the Expedition*. London: Allen and Unwin

Cavanaugh, C. M., Gardiner, S. L., Jones, M. L., Jannasch, H. W., and Waterbury, J. B. (1981). Prokaryotic Cells in Hydrothermal Vent Tube Worm *Riftia pachyptila* Jones: Possible Chemoautotrophic Symbionts. *Science*, 213, 340–2.

Clark, M., Bagley, N., and Harley B. (2016b). Trawls, in M. Clark, M. Consalvey, and A. A. Rowden (eds.) *Biological Sampling in the Deep Sea*. Chichester: John Wiley & Sons, pp. 126–58.

Clark, M., Consalvey, M., and Rowden A. A. (2016a). *Biological Sampling in the Deep Sea*. Chichester: John Wiley & Sons,

Cordes, E. E., Jones, D. O. B., Schlacher, T. A., et al. (2016). Environmental Impacts of the Deep-Water Oil and Gas Industry: A Review to Guide Management Strategies. *Frontiers in Environmental Science*, 4, 58.

Corinaldes, C. (2015). New Perspectives in Benthic Deep-sea Ecology. *Frontiers in Marine Science*, 2, 17.

Corliss, J. B., Dymond, J., Gordon, L. I., et al. (1979). Submarine Thermal Springs on the Galapagos Rift. *Science*, 203, 1073–83.

Danovaro, R., Dell'Anno, A., and Pusceddu, A. (2004). Biodiversity Response to Climate Change in a Warm Deep Sea. *Ecology Letters*, 7, 821–8.

Deuser, W. G. and Ross, E. H. (1980). Seasonal Change in the Flux of Organic Carbon to the Deep Sargasso Sea. *Nature*, 283, 364–5.

Earth Negotiation Bulletin 12 No. 626 (2015). https://enb.iisd.org/download/pdf/enb12626e.pdf.

Felbeck, H., Chidress, J. J., and Somero, G. N. (1981). Calvin-Benson Cycle and Sulfide Oxidation Enzymes in Animals from Sulfide-rich Habitats. *Nature*, 293, 291–3.

Fernandez-Arcaya, U., Ramirez-Llodra, E., Aguzzi, J., et al. (2017). Ecological Role of Submarine Canyons and Need for Canyon Conservation: A Review. *Frontiers of Marine Science*, 4, 5.

Fujikura, K., Lindsay, D., Kitazato, H., Nishida, S., and Shirayama, Y. (2010). Marine Biodiversity in Japanese Waters. *PloS ONE*, 5(8), e11836.

Gage, J. and May, R. M. (1993). A Dip into the Deep Seas. *Nature*, 365, 609–10.

Gage, J. D. (2003). Food Inputs, Utilization, Carbon Flow and Energetics, in P. A. Tyler (ed.) *Ecosystems of the Deep Sea*. Rotterdam, Netherlands: Elsevier, pp. 313–80.

Gage, J. D. and Tyler, P. A. (1991). *Deep-sea Biology: A Natural History of Organisms at the Deep-sea Floor*. Cambridge: Cambridge University Press.

Gerard, R. (1976). Environmental Effects of Deep-sea Mining. *Marine Technological Society Journal*, 10(7), 7–16.

Glover, A. G. and Smith, C. R. (2003). The Deep-sea Floor Ecosystem: Current Status and Prospects of Anthropogenic Change by the Year 2025. *Environmental Conservation*, 30, 219–41.

Glover, A. G., Wiklund, H., Rabone, M., et al. (2016). Abyssal Fauna of the UK-1 Polymetallic Nodule Exploration Claim, Clarion-Clipperton Zone, Central Pacific Ocean: Echinodermata. *Biodiversity Data Journal*, 4, e7251.

Grassle, J. F. (1985). Hydrothermal Vent Animals: Distribution and Biology. *Science*, 229, 713–17.

Grassle, J. F. and Maciolek, N. J. (1992). Deep-sea Species Richness: Regional and Local Diversity Estimates from Quantitative Bottom Samples. *American Naturalist*, 139, 313–41.

Harden-Davies, H. (2016). Deep-sea Genetic Resources: New Frontiers for Science and Stewardship in Areas Beyond National Jurisdiction. *Deep-Sea Research II*. 137. doi:10.1016/j.dsr2.2016.05.005.

Harris, P. T. and Whiteway, T. (2011). Global Distribution of Large Submarine Canyons: Geomorphic Differences Between Active and Passive Continental Margins. *Marine Geology*, 285, 69–86.

Hecker, B. (1990). Photographic Evidence for the Rapid Flux of Particles to the Sea Floor and Their Transport Down the Continental Slope. *Deep-Sea Research*, 37, 1773–82.

Herring, P. J. (2002). *The Biology of the Deep Ocean*. Oxford: Oxford University Press

Hessler, R. R. and Jumars, P. A. (1974). Abyssal Community Analysis from Replicate Box Cores in the Central North Pacific. *Deep-Sea Research*, 21, 185–209.

Hilário, A., Capa, M., Dahlgren, T. G., et al. (2011). New Perspectives on the Ecology and Evolution of Siboglinid Tubeworms, *PLoS ONE*, 6(2), e16309.

Hilário, A., Metaxas, A., Gaudron, S., et al. (2015). Estimating Dispersal Distance in the Deep Sea: Challenges and Applications to Marine Reserves. *Frontiers in Marine Science*, 2, 6. doi: 10.3389/fmars.2015.00006.

Hollister, C. D. and McCave, I. N. (1984). Sedimentation Under Deep-sea Storms. *Nature*, 309, 220–5.

Huber, J., Mark Welch, D. B., Morrison, H. G., et al. (2007). Microbial Population Structures in the Deep Marine Biosphere. *Science*, 318, 97–100.

Ivanov A. V. (1955). On the Assignment of the Class Pogonophora to a Separate Phylum of Deuterostomia Brachiata A. Ivanov., Phyl. Nov. *Systematic Zoology*, 4, 177–8.

Jaeckel, A. (2015). An Environmental Management Strategy for the International Seabed Authority: The Legal Basis. *The International Journal of Marine and Coastal Law*, 30, 1–27.

Johnson, R. G., Kahn, M., and Robbins, C. (1984). *United States Practices and Policies for Ocean Disposal of Radioactive Wastes, 1946–1984*. EPA 530/1-84-017. Washington, DC: US Environmental Protection Agency,

Jones, M. L. (1985). On the Vestimentifera, New Phylum: Six New Species, and Other Taxa, from Hydrothermal Vents and Elsewhere. *Biological Society of Washington Bulletin*, 6, 117–58.

Jumars, P. A. (1981). Limits to Predicting and Detecting Benthic Community Responses to Manganese Nodule Mining. *Marine Mining*, 3, 213–29.

Kidd, R. B. and Huggett, Q. J. (1981). Rock Debris on Abyssal Plains in the Northeast Atlantic: A Comparison of Epibenthic Sledge Hauls and Photographic Surveys. *Oceanologica Acta*, 4, 99–104.

Koslow, T. (2007). *The Silent Deep: The Discovery, Ecology, and Conservation of the Deep Sea*. Chicago: University of Chicago Press.

Lampitt, R. L. (1985). Evidence for the Seasonal Deposition of Detritus on the Seafloor and Its Subsequent Resuspension. *Deep-Sea Research*, 32A, 885–97.

Levin, L. A. (2003). Oxygen Minimum Zone Benthos: Adaptation and Community Response to Hypoxia. *Oceanography and Marine Biology: An Annual Review*, 41, 1–45.

Levin, L. A. and Le Bris, N. (2015). The Deep Ocean Under Climate Change. *Science*, 350, 766–8.

Levin, L. A. and Sibuet, M. (2012). Understanding Continental Margin Biodiversity: A New Imperative. *Annual Review of Marine Science*, 4, 79–112.

Lodge, M., Lily, H., and Symonds, P. (2013). Legal Rights to Deep Sea Minerals, in E. Baker and Y. Beaudoin (eds.) *Deep Sea Minerals and the Green Economy*. Noumea, New Caledonia: Secretariat of the Pacific Community, pp. 11–24.

Marlow, J. J., Steele, J. A., Ziebis, W., et al. (2014). Carbonate-hosted Methanotrophy Represents an Unrecognized Methane Sink in the Deep Sea. *Nature Communications*, 5, 5094. doi:10.1038/ncomms6094.

Martins, A., Vieira, H., Gaspar, H., and Santos, S. (2014). Marketed Marine Natural Products in the Pharmaceutical and Cosmeceutical Industries: Tips for Success. *Marine Drugs*, 12(2), 1066–101.

McIntyre, A. (2010). *Life in the World's Oceans: Diversity, Distribution, and Abundance*. West Sussex: Wiley-Blackwell.

Menot, L., Galeron, J., Olu, K., et al. (2010). Spatial Heterogeneity of Macrofaunal Communities in and Near a Giant Pockmark Area in the Deep Gulf of Guinea. *Marine Ecology*, 31, 78–93.

Merrie, A., Dunn, D. C., Metian, M., et al. (2014). An Ocean of Surprises—Trends in Human Use, Unexpected Dynamics and Governance Challenges in Areas Beyond National Jurisdiction. *Global Environmental Change*, 27, 19–31.

Moore, P. G. (1999). Fisheries Exploitation and Marine Habitat Conservation: A Strategy for Natural Coexistence. *Marine and Freshwater Ecosystems*, 9, 585–91.

Mora, C., Tittensor, D. P., Adl, S., Simpson, A. G. B., and Worm, B. (2011). How Many Species Are There on Earth and in the Ocean? *PLoS ONE, 9 (8)*, e10001127.

Mordecai, G., Tyler, P. A., Masson, D. G., and Huvenne, V. A. I. (2011). Litter in Submarine Canyons Off the West Coast of Portugal. *Deep-Sea Research II*, 58, 2489–96.

Murray, J. and Hjort, J. (1912). *The Depths of the Ocean.* London: MacMillan and Co.

Paull, C.K., Hecker, B., Commeau, R. et al. (1984). Biological Communities at the Florida Escarpment Resemble Hydrothermal Vent Taxa. *Science*, 226, 965–7.

Pham, C. K., Ramirez-Llodra, E., Alt, C. H. S. et al. (2014). Marine Litter Distribution and Density in European Seas from the Shelves to Deep Basins. *PLoS ONE*, 9, 12.

Puig, P., Canals, M., Company, J. B., et al. (2012). Ploughing the Deep Sea Floor. *Nature*, 489, 286–90.

Ramirez Llodra, E., Brandt, A., Danovaro, R., et al. (2010). Deep, Diverse and Definitely Different: Unique Attributes of the World's Largest Ecosystem. *Biogeosciences*, 7, 2851–99.

Ramirez-Llodra, E., De Mol, B., Company, J. B., Coll, M., and Sardà, F. (2013). Effects of Natural and Anthropogenic Processes in the Distribution of Marine Litter in the Deep Mediterranean Sea. *Progress in Oceanography*, 118, 273–87.

Ramirez-Llodra, E., Trannum, H. C., Evenset, A., et al. (2015). Submarine and Deep-sea Mine Tailing Placements: A Review of Current Practices, Environmental Issues, Natural Analogs and Knowledge Gaps in Norway and Internationally. *Marine Pollution Bulletin*, 97, 13–35.

Ramirez-Llodra, E., Tyler, P. A., Baker, M. C., et al. (2011). Man and the Last Great Wilderness: Human Impact on the Deep Sea. *PLoS ONE*, 6(8), e22588.

Reeburgh, W. S. (2007). Oceanic Methane Biogeochemistry. *Chemical Reviews*, 107, 486–513.

Rex, M. A. and Etter, R. J. (2010). *Deep-Sea Biodiversity: Pattern and Scale.* Cambridge, MA: Harvard University Press

Rex, M. A., Stuart, C. T., Hessler, R. R., et al. (1993). Global-scale Latitudinal Patterns of Species Diversity in the Deep-sea Benthos. *Nature*, 365, 636–9.

Rice, A. L. (1986). *British Oceanographic Research Vessels 1800–1950.* London: The Ray Society.

Ritt, B., Pierre, C., Gauthier, C., Wenzhoefer, F., and Boetius, A. (2011). Diversity and Distribution of Cold-Seep Fauna Associated with Different Geological and Environmental Settings at Mud Volcanoes and Pockmarks of the Nile Deep-Sea Fan. *Marine Biology*, 158, 1187–210.

Roemmich, D. and Wunsch, C. (1984). Apparent Seasonal Changes in the Climatic State of the Deep North Atlantic. *Nature*, 307, 447–50.

Rogers, A. D. (2018) The Biology of Seamounts: 25 Years On. *Advances in Marine Biology*, 79, 137–224.

Rogers, A. D. and Gianni, M. (2010). The Implementation of UNGA Resolutions 61/105 and 64/72 in the Management of Deep-Sea Fisheries on the High Seas, in a report for the Deep-Sea Conservation Coalition. London: International Programme on the State of the Ocean, p. 97.

Rothwell, D. R. and Stephens, T. (2016). *The International Law of the Sea.* Sydney, Australia: Bloomsbury Publishing.

Rouse, G.W., Goffredi, S. K., and Vrijenhoek, R. C. (2004). *Osedax*: Bone-Eating Marine Worms with Dwarf Males. *Science*, 305, 668–70.

Sanders, H. L., Hessler, R. R., and Hampson, G. R. (1965). An Introduction to the Study of the Deep-sea Benthic Faunal Assemblages Along the Gay Head-Bermuda Transect. *Deep-Sea Research*, 12, 845–67.

Schoolmeester, T. and Baker, E. (eds.) (2009). *Continental Shelf—The Last Maritime Zone.* Arendal, Norway: GRID-Arendal.

Scott, K., Harsany, P., and Lyndon, A. R. (2018). Understanding the Effects of Electromagnetic Field Emissions from Marine Renewable Energy Devices (MREDS) on the Commercially Important Crab, *Cancer pagurus* (L.). *Marine Pollution Bulletin*, 131, 560–8.

Sewell, R. B. S. (1934). The John Murray Expedition to the Red Sea, *Nature*, January 20, 86–9.

Skropeta, D. (2011). Exploring Marine Resources for New Pharmaceutical Applications, in W. Gullett, C. Schofield, and J. Vince (eds.), *Marine Resources Management.* Chatswood, NSW, Australia: LexisNexis Butterworths, p. 211–24.

Skropeta, D. (2008). Deep-sea Natural Products. *Natural Product Reports*, 25(6), 1131–166.

Skropeta, D. and Liangqian W. (2014) Recent Advances in Deep-sea Natural Products. *Natural Product Reports*, 31(8), 999–1025.

Smith, C. R. and Baco, A. R. (2003). The Ecology of Whale Falls at the Deep-sea Floor. *Oceanography and Marine Biology: An Annual Review*, 41, 311–54.

Smith, C. R., Kukert, H., Wheatcroft, R. A., Jumars, P. A. and Deming, J. W. (1989). Vent Fauna on Whale Remains. *Nature*, 341, 27–8.

Smith, C. R., Mincks, S., and DeMaster, D. J. (2008). The FOODBANCS Project: Introduction and Sinking Fluxes of Organic Carbon, Chlorophyll-A and Phytodetritus on the Western Antarctic Peninsula Continental Shelf. *Deep-Sea Research II*, 55, 2404–14.

Snelgrove, P. V. R. (2010). *Discoveries of the Census of Marine Life: Making Ocean Life Count.* Cambridge: Cambridge University Press.

Stockton, W. L. and DeLaca, T. E. (1982). Food Falls in the Deep Sea: Occurrence, Quality, and Significance. *Deep-Sea Research*, 29, 157–69.

Sweetman, A. K., Thurber, A. R., Smith, C. R., et al. (2017). Major Impacts of Climate Change on Deep-sea Benthic Ecosystems. *Elementa*, 5. https://www.elementascience.org/article/10.1525/elementa.203/.

Thiel, H. (2003). Anthropogenic Impacts on the Deep Sea, in P. A. Tyler (ed.) *Ecosystems of the World*. Amsterdam, Netherlands: Elsevier, pp. 427–71.

Thistle, D. (2003). The Deep-sea Floor: An Overview, in P. A. Tyler (ed.) *Ecosystems of the World*. Amsterdam, Netherlands: Elsevier, pp. 5–39.

Thistle, D., Ertman, S. C., and Fauchald, K. (1991). The Fauna of the HEBBLE Site: Patterns in Standing Stock and Sediment-dynamic Effects. *Marine Geology*, 99, 413–22.

Thompson, K. F., Miller, K. A., Currie, D., Johnston, P., and Santillo, D. (2018). Seabed Mining and Approaches to Governance of the Deep Seabed. *Frontiers in Marine Science*, 5, 480.

Tunnicliffe, V., Juniper, S. K., and Sibuet, M. (2003). Reducing Environments of the Deep-sea Floor, in P. A. Tyler (ed.) *Ecosystems of the World: Ecosystems of the Deep Sea*. Amsterdam, Netherlands: Elsevier, pp. 81–110.

Tyler, P.A., ed. (2003). *Ecosystems of the World, Vol. 28. Ecosystems of the Deep Oceans*. Amsterdam, Netherlands: Elsevier.

Tyler, P. A., Grant, A., Pain, S., and Gage, J. D. (1982). Is Annual Reproduction in Deep-sea Echinoderms a Response to Variability in Their Environment? *Nature*, 300, 747–9.

UNCLOS (1982). United Nations Convention on the Law of the Sea. Articles 137, 145, 153. https://www.un.org/depts/los/convention_agreements/texts/unclos/unclos_e.pdf. Accessed 27 June 2019.

Van Dover, C. L. (2000). *The Ecology of Deep-sea Hydrothermal Vents*. Princeton, NJ: Princeton University Press.

Van Dover, C. L., Arnaud-Haond, S., Gianni, M., et al. (2018). Scientific Rationale and International Obligations for Protection of Active Hydrothermal Vent Ecosystems from Deep-sea Mining. *Marine Policy*, 90, 20–8.

Van Dover, C. L., Grassle, J. F., Fry, B., Garit, R. H., and Starczak, V. R. (1992). Stable Isotope Evidence for Entry of Sewage Derived Organic Material into a Deep-sea Food Web. *Nature, London*, 360, 153–6.

Vare, L. L., Baker, M. C., Howe, J. A., et al. (2018). Scientific Considerations for the Assessment and Management of Mine Tailings Disposal in the Deep Sea. *Frontiers in Marine Sciences*, 5(17), 13.

Victorero, L., Watling, L., Deng Palomares, M.L., and Nouvian, C. (2018). Out of Sight, But Within Reach: A Global History of Bottom-Trawled Deep-Sea Fisheries From >400 m Depth. *Frontiers in Marine Science*, 5, doi: 10.3389/fmars.2018.00017.

Watson, R. A. and Morato, T. (2013). Fishing Down the Deep: Accounting for Within-species Changes in Depth of Fishing. *Fisheries. Research*, 140, 63–5.

Webb, K. E., Barnes, D. K. A., and Planke, S. (2009). Pockmarks: Refuges for Marine Benthic Biodiversity. *Limnology and Oceanography*, 54, 5, 1776–88.

Wedding, L. M., Friedlander, A. M., Kittinger J. N., et al. (2013). From Principles to Practice: A Spatial Approach to Systematic Conservation Planning In The Deep Sea. *Proceedings of the Royal Society B Biological Sciences*, 280(1773), 20131684.

Woese, C. R., Kandler, O., and Wheelis, M. L. (1990). Towards a Natural System of Organisms: Proposal for the Domains Archaea, Bacteria, and Eucarya. *Proceedings of the National Academy of Sciences USA*, 87, 4576–9.

Woese, C. R. and Fox, G. E. (1977). Phylogenetic Structure of the Prokaryotic Domain: The Primary Kingdoms. *Proceedings of the National Academy of Sciences USA*, 74, 5088–90.

Wust, G. (1950). Blockdiagramme der Atlantischen Zirkulation auf Grund der Meteor Ergebnisse. *Kieler Meeresforschungen*, 7, 24–34.

Wust, G. (1964). The Major Deep-sea Expeditions and Research Vessels 1873–1960. *Progress in Oceanography*, 2, 1–52.

Wyville Thomson, C. (1873). *The Depths of the Sea*. London: MacMillan and Co.

CHAPTER 2

A primer on the economics of natural capital and its relevance to deep-sea exploitation and conservation

Porter Hoagland, Di Jin, and Stace Beaulieu

2.1 Introduction

This chapter reviews the economic underpinnings of the concept of natural capital, examines how it could be measured for pertinent types of deep-sea natural capital, and develops recommendations for future research and practice to enhance its policy relevance for the high seas.

The deep sea was regarded mostly as remote, inaccessible, even threatening, until mankind began to absorb the revelations of the HMS *Challenger* upon its completion of a 68,000 nautical mile 'grand tour' of the world's oceans in the late nineteenth century. Akin to the major mountain ranges on land, the high seas were considered by many to be an immense obstacle, something that had to be surmounted in order to explore and exploit the wealth of new realms. Legends arose, in distant places such as the Sargasso Sea, where great masses of weeds were believed to entangle ships. A century and a half ago, only Jules Verne seemed to posit the notion of a deep sea that was more paradise than threat, virtually liberated from mankind's foibles. In his *20,000 Leagues under the Seas*, Verne (1871) wrote:

The sea does not belong to despots. Upon its surface men can still exercise unjust laws, fight, tear one another to pieces, and be carried away with terrestrial horrors. But at thirty feet below its level, their reign ceases, their influence is quenched, and their power disappears. Ah! sir, live—live in the bosom of the waters! There only is independence! There I recognize no masters! There I am free!

Coastal fishing and whaling has been practiced for many millennia, but it was not until the seventeenth and eighteenth centuries that fish-trawling technologies evolved and fleets of whaleships were organised, enabling yields of fish and mammals in areas that were deeper and more remote from the coast. In the modern Anthropocene epoch, the deep sea increasingly has become the focus of other, even larger scale, real and would-be industrial exploitations, leading potentially to profound changes in its physical characteristics and to its ecological structures and processes. As humankind is wont to do, the quaint Vernean notion of high-seas freedoms has been overcome by excessive exploitation and insulted by wastes; there is now a need for a conservation of the resources of the high seas.

The deep sea can be conceptualised as a resource area, albeit one that exists on a massive scale, comprising a complex and differentiated 'natural' capital. In relentless and innovative ways, using continuously progressing technologies, humans combine this natural capital with labour and skill (human capital) and machines and infrastructure (manufactured capital) to produce goods and services that yield economic benefits: the carriage of goods by sea; the disposal of wastes; the cabling of communications; the supply of seafood, oil, natural

Porter Hoagland, Di Jin, and Stace Beaulieu, *A Primer on the economics of natural capital and its relevance to deep-sea exploitation and conservation* In: *Natural Capital and Exploitation of the Deep Ocean.* Edited by: Maria Baker, Eva Ramirez-Llodra, and Paul Tyler, Oxford University Press (2020). © Oxford University Press. DOI: 10.1093/oso/9780198841654.003.0002 gas, drugs, and minerals; ecotourism and recreation; and the carrying out of scientific research and education. Simultaneously with this industry, human uses of the deep sea increasingly impose costs on other uses or values through environmental changes and degradations, overexploitations of renewable resources (fish and mammals), modifications of habitats, diminishments of biological diversity, and apparent increases in both anthropogenic and natural hazards.

Reference to the deep sea as a kind of natural capital, and the closely related and linked concept of marine-ecosystem services, may enable ocean users, resource managers, and the public to begin to articulate the scales and scopes of the relevant trade-offs that occur or that increasingly must be made as the deep sea experiences the inexorable expansion of industrial development and the spreading out of all manner of human endeavour. A central issue for humans in the 'great acceleration' comprising the Anthropocene is whether humans can exploit ('use' is much too temperate a term) the deep sea in a way that ensures that future human generations can be at least as well-off as the present; that is, can humans achieve a sustainable development of the deep sea? Answering this question requires that natural capital be evaluated in a way that allows comparisons to—and potentially trade-offs with—other capital assets, including both human and manufactured capital.

In this chapter, the meaning of the natural capital of the deep sea is discussed from an economic viewpoint. In particular, the close relationship of natural capital to the idea of ecosystem services is demonstrated (Armstrong et al. 2012). Called for in 2001 by Kofi Annan, then the Secretary General of the United Nations (UN), the classification and evaluation of ecosystem (goods and) services into a Millennium Ecosystem Assessment (MA) was undertaken across a number of environments to elevate the station of many of the difficult-to-value benefits provided by nature (de Groot et al. 2012).

Significant scholarly and pragmatic efforts have been undertaken to refine the MA framework in large part to standardise the definitions of 'ecosystem services' so that they can be readily and consistently linked to systems of national accounting (Carpenter et al. 2009; Mäler et al. 2009; Ring et al. 2010; Landers and Nahlik 2013; Díaz et al. 2018; CICES 2018). Given that there does not yet exist a system of economic accounts for the high seas, which are defined to be areas beyond national jurisdiction (ABNJ), some context-specific cases of the valuation of specific forms of deep-sea natural capital are focused on here. The most significant future challenge will be to evaluate the contributions of natural capital to the overall social value of the deep sea's capital assets (its 'inclusive wealth'), which would help to enable the assessment of progress towards its sustainable development (e.g. UNEP 2018).

The theory that underpins the valuation of natural capital is reviewed first (Fenichel et al. 2016, 2018). Next, some illustrative—but qualitative—examples of the valuation of natural capital in the deep sea, from the surface to the seabed, are introduced, focusing in on the issues that have emerged and the needs for information and analysis that have become evident. The increasing international focus on emerging institutions for the high seas, the terms of which are elaborated in greater depth in Chapters 3 and 7, is argued to provide an opportunity for taking a natural-capital approach to measure progress towards the sustainable use of the deep sea.

2.2 Human perceptions and uses of the deep sea

The industrial revolution was an early inflection point in the human development of the deep sea. The age of sail ended with some of the first naval battles of the US Civil War, ushering in a new era of ocean crossings by steamship that were faster and more predictable. At the same time, progress in understanding and movements to exploit the high seas began to accelerate.

In 1873–1876, the HMS *Challenger* Expedition sampled manganese nodules from the deep seabed and bristlemouths and lanternfish from the mesopelagic. Although some of these extraordinary fish had been described as early as 1810 by the Niçard naturalist Antoine Risso, the *Challenger* observations demonstrated their global distribution. Less than 2 decades later, the German hydrographer Otto Krümmel was able to piece together a map of the spatial distribution of the Sargasso Sea from hundreds of records of sargassum observed by ships crossing the North Atlantic Ocean, thereby shattering one of Alexander von Humboldt's long-standing conjectures about ocean gyres and their flora and fauna.

In their bathysphere in the early 1930s (Chapter 1), William Beebe and Otis Barton were transmitting in real time to NBC radio their 'new' discoveries of remarkable fish from the ocean's 'twilight zone' off the Bermudian coast. By 1957, Bruce Heezen and Marie Tharp at Columbia University's Lamont-Doherty Geological Observatory had published the first comprehensive map of the world's ocean (see Chapter 1), illustrating in detail the midocean ridges, seamounts, and other features that emerged from systematic sounding profiles and calculations of seabed earthquake epicentres. And, by 1977, scientists at the Woods Hole Oceanographic Institution were reporting on a new discovery of unusual creatures found at hydrothermal vent systems situated at seafloor spreading centres off the Galapagos Islands, published by Corliss et al. (1979).

Today's modern communications and exploration technologies, including radio, radar, sonar, and satellite remote sensing, have made the high seas, including the deepest part of the ocean, much less opaque and much more accessible. Modern data collection, especially oceanographic sampling, now is undertaken worldwide and to full depth, revealing the movements of currents; the distributions of nutrients, carbon species, oxygen, and radioisotopes; temperature and salinity gradients; and complex ecological relationships that link the surface waters to the twilight zone to the deepest parts of the ocean and to the seabed itself. Interpretation and conveyance of the data are facilitated by modern computers and software, mapping, and visualisations, further deepening and broadening human understanding of the deep sea.

In his 1898 poem, 'The Explorer', Rudyard Kipling narrated the experience of an archetypal pioneer, responding to supernatural whisperings, who showed the way for relentless economic development. 'You go up and occupy . . . *ores* you'll find

there', he wrote. As with the development of the American West, the prospects of exploiting the resources of the deep sea, from the fisheries for apex predators at the surface to the oil-laden myctophids (lanternfish) and gonostomids (bristlemouths) of the twilight zone to the DNA sequences embodied in oceanic microbes to the massive sulphides of the deep seabed, now embody a hoped-for 'Blue Economy' for the world.

Modern human uses of the deep sea—if defined broadly to be areas where the ocean is deep and therefore also containing the surface waters of the high seas—span the full complement of marine activities, including shipping, commercial fishing (Chapter 4), offshore hydrocarbon development (Chapter 5), waste disposal, naval surveillance and warfare (Chapter 10), piracy, oceanographic research, siting of wind energy, ecotourism and adventure tourism, the search for and collection of potentially useful marine genetic resources (Chapter 7), and others. Future activities include possible expansion of these existing uses, as well as the potential for deep-seabed mining of massive sulfides (for copper and zinc), manganese nodules (for nickel, copper, and cobalt), or ferromanganese crusts (for nickel and cobalt) (see Chapter 6).

2.3 Natural capital and ecosystem services: stocks and flows

Production theory is central to the modern discipline of economics (e.g. Perloff 2012). Beginning with a classical conceptualisation, factors of production are combined in a production process, comprising a 'technology', to produce goods and services that are subsequently exchanged in 'downstream' markets. Normally the factors are purchased in their own markets upstream, where the product of the prices paid and the quantities bought comprise the costs of production. Early economists, initially contemplating agrarian occupations, classified these factors into land, labour, and capital, where 'land' can now be interpreted more generally to represent a type of 'natural' resource. In this early context, 'capital' referred specifically to manmade factors, including machines, ships, buildings, or other infrastructure.

Prices for the factors of production are referred to as the 'interest' (or just the 'price') on manufactured capital, the 'wage' for labour, and the 'rent' for land (or natural resources). In modern terminology, all three factor classes can be interpreted to be forms of capital: physical, human, and natural. (Sometimes a fourth form, 'social capital', is employed to refer to policies, including cultural norms, best practices, or institutions of governance, which may facilitate production, enforce property rights, or improve the means for supplying outputs through markets.) Still, these capital forms may differ in the nature and scale of investments needed to employ them in production and in their rates of depreciation or depletion.

Natural capital is distinguished from other factors in the sense that it is produced organically by nature. Where natural capital is not limited in

Figure 1 Thomas Henry Huxley presenting the inaugural address at the International Fisheries Exhibition in 1882. "I believe, then, that the cod fishery, the herring fishery, the pilchard fishery, the mackerel fishery, and probably all the great sea fisheries, are inexhaustible; that is to say, that nothing we do seriously affects the number of the fish. And any attempt to regulate these fisheries seems consequently, from the nature of the case, to be useless."

supply, there is no need for an explicit upstream factor market, and no rent (price) emerges. Strictly speaking, natural capital is not a resource when it is unlimited, because, by definition, resources are scarce or rare. Examples that are still relevant, but perhaps not in all contexts, include sunlight, the air we breathe, the ocean as a medium for transportation, the wind, currents, waves, or tidal bores that drive renewable energy generators. Once upon a time, even the fish in the sea were considered to be inexhaustible (Huxley 1882; Figure 1).

Where there are limits to the supply of natural capital, the absence of a factor market—or the existence of an imperfect or flawed market—can result in the canonical social costs (termed 'externalities') arising from pollution or resource overexploitation. In the case where there is no market for natural capital in limited supply, and in the absence of regulation, all factors are likely to be misallocated, as the free factor is regarded by firms as an implicit subsidy. This may be due to the lack of legal and enforceable property rights, the existence of significant costs of transactions, or the absence of effective forms of governance (Coase 1960; Ostrom 1990). This phenomenon is known as what Garrett Hardin called, in due course, the 'tragedy of the unmanaged commons' (Hardin 1998). Relevant examples for the deep sea include overfishing, hypoxia, plastic pollution, the rush to claim marine genetic resources and mineral resources, and the warming, acidification, stratification, and deoxygenation that are results of a changing climate.

The term 'natural capital' sometimes is used interchangeably with the term 'ecosystem service', but the two concepts, although linked, can be distinguished as stocks and flows. Both the physical quantities and economic values are important as support for eventual decisions about trade-offs among potentially conflicting uses of resource areas such as the deep sea. For natural capital, it is most useful to conceptualise it as a stock or an asset, from which flows (subtractions from or additions to the stock) may occur (see the general definition provided in Chapter 1). Renewable resources can involve both biological growth and depletion, whereas nonrenewable resources are confined to depletion. (This distinction may be too simplistic, as additions to a nonrenewable resource base can

occur through new discoveries, research and development, and technological changes.) An illustrative example would be a single-species marine fishery, where the biomass at any point in time is the stock of natural capital (Mäler et al. 2009). (Often, the focus of fishery managers is on the 'spawning stock biomass'.) Flows involve additions through births (or recruits of younger cohorts to a spawning stock), and subtractions involve natural mortality or harvests by humans.

With respect to the value of natural capital and its relationship to that of ecosystem services, resource economist Eli Fenichel and his colleagues (2016, 2018) elaborated a natural-capital valuation framework, following the work of macroeconomist and Nobel laureate Kenneth Arrow and his colleagues (2003). The framework was built by extending the most general case set out by resource economist Jon Conrad and mathematician Colin Clark in 1987 (see the Appendices to this chapter for mathematical details). The resultant 'guiding qualitative framework' developed by Fenichel incorporated an 'economic programme' encompassing exploitation (divestment or drawdown) or conservation (investment or build-up) of a natural resource. As envisioned by Fenichel et al. (2016), in this framework, conservation is an investment, not a cost.

Simply stated, the asset value of a unit of stock of a natural capital is the discounted sum of the value of ecosystem-service flows and capital gains. The former is the per-period resource rent, and the latter is an increase (or a decrease, if there is a loss) in value attributed to an expansion (or contraction) of demand, say through changing consumer tastes or a rising (or falling) economy. The discount factor comprises a social rate of time preference that adjusts up or down, depending upon whether investments are made or draw-downs occur in the natural-capital stock.

For example, in the case of a renewable resource, if exploitation exceeds growth, implying a drawdown in the capital stock, then the effective discount factor would be increased, implying that society has taken on a more myopic (near-term) perspective. In contrast, if growth exceeds exploitation, implying an investment in natural capital, then the effective discount factor would be decreased, implying that society has adopted a more hyperopic (far-term) perspective. The basic relationship can be denoted as follows:

$$\begin{aligned} &natural\,capital\,asset\,value \\ &= \frac{ecosystem\,service\,value + capital\,gain}{time\,preference - (net\,investment)}. \end{aligned}$$

In this relationship, the asset value on the left-hand side constitutes a perpetuity. This value is referred to in the economic literature as an 'accounting price'. The product of this value and the extant physical stock would consist of a measure of relevant natural capital in a region's 'inclusive wealth' (Fenichel et al. 2016). In the numerator on the right-hand side, the ecosystem-service value is the resource rent, representing the marginal (i.e. very small) change in a measure of social welfare, such as utility or profits, with a marginal change in the quantity of the relevant natural capital. In economic parlance, the eco-system-service value is sometimes referred to as a 'dividend', and it constitutes an annual flow. Also in the numerator on the right-hand side, capital gains is an annual flow, comprising an annual change in the accounting price, again perhaps due to changing preferences in favour of (or away from, in the case of a loss) the relevant natural capital.

In the denominator on the right-hand side, the time preference is a social rate of time preference. The choice of time preference is a normative one, where, *ceteris paribus*, a smaller value would lead to a *larger* accounting price for the relevant natural capital, and vice versa. The time preference has two components, one representing impatience about potential future uses by the present generation and the second signifying society's anticipation about the welfare of future generations (Moore and Vining 2018). These two components often work in opposite directions; increases in impatience would reduce future natural-capital accounting prices, and increases in the anticipated welfare of future generations would raise them. In practice, plausible values for the social rate of time preference might range from about 2 to 7 per cent, but increasingly analysts have advocated for the use of a social rate of time preference that declines over time, thereby giving greater weight to values in the far distant future (Arrow et al. 2013).

Also in the denominator on the right-hand side, net (natural-capital) investment is the difference between marginal changes in growth (with marginal changes in the natural capital stock) and marginal changes in exploitation (also with marginal changes in the stock). Where marginal growth exceeds marginal exploitation, investment or build-up of the stock occurs, and where exploitation exceeds growth, divestment or drawdown occurs. In the former case, the effective (net) time preference would become more hyperopic; in the latter, more myopic.

As a perpetuity, this relationship represents the asset value of natural capital extending forever into the future when evaluated at any specific time, now or when taken at some chosen point in the future. Note that changes in investment will affect this asset value; for example, following the above, excessive exploitation (a myopic perspective) would increase the effective discount rate, thereby causing a decline in natural-capital asset value. Similarly, technological change that reduces the cost of exploitation also would affect the asset value. Technological change could lead to increases in both the resource rent (in the numerator) and the level of exploitation (in the denominator). Thus, attention ought to be paid to such changes, such as through a management framework, in order to prevent excessive exploitation. From a pragmatic perspective, the accounting price ought to be updated on a regular basis—but especially when significant changes in the relevant stock, capital gain, or net investment programme occur.

As Fenichel et al. (2016) explain, characterising the accounting price of a stock of natural capital helps to put it on a par with the more routinely valued manufactured and human capitals or other natural capitals for which markets exist or accounting prices have been estimated. In circumstances where decisions must be made concerning the allocation of areas of the deep sea for particular uses, for example, assessments of the opportunity costs of specific allocations may thereby be facilitated.

Much of the economic valuation literature for marine resources focuses on assessing only the eco-system-service component of the value of natural capital, such as resource rents in a fishery, for example, and, as explained in Chapters 4, 5, and 6, very little has been accomplished to date regarding the valuation of either natural capital or ecosystem services in the deep sea (Folkersen et al. 2018). In shallower waters, resource economist Yun and his colleagues (2017) applied a natural-capital valuation framework to a multispecies fishery in the Baltic Sea, however, showing the potential benefits of implementing a programme of ecosystem-based fishery management.

Understanding the benefits that are associated with deep-sea natural capital is fundamental to assessing the opportunity costs of proposed activities that might mitigate those benefits, such as unregulated fishery exploitation, the impacts of deep-seabed mining, or the effects associated with climate changes in the ocean, including temperature increases, acidification, lessened dissolved oxygen levels, or shifts in biological diversity. As one measure of inclusive wealth, the accounting price of natural capital is even more central to understanding how the deep sea can be exploited sustainably.

In what follows, some specific examples of human activities and marine-resource valuations on the high seas are examined. Where feasible, extant estimates of ecosystem-service values, capital gains, and net investments are presented, and, from these, crude but qualitative estimates of natural-capital values are attempted. In so doing, needs for additional information or analyses that would contribute to or help to refine the value of natural capital in the deep sea are identified.

2.4 Qualitative examples of natural-capital accounting for the deep sea

If defined as constituting the ocean only at depths below 200 m, and thereby not containing the surface waters, the deep sea therefore comprises areas that lie mostly—but not exclusively—beyond the geological or juridical outer continental shelves of coastal states. For states with narrow shelves, however, such as is the case for many small island states, the deep sea can extend well within their exclusive economic zones (EEZs), incorporating areas above or on their continental shelves. Regardless, questions surrounding the valuation of the natural capital of the deep sea need not be confined to ABNJ, and, because of physical or ecological linkages, they may sometimes pertain to resources and human

activities from the territorial seas to the high seas and from the continental shelves to the deep seabed. As elaborated in the Technical Abstract to help inform the Biodiversity Beyond National Jurisdiction (BBNJ) negotiations (see Chapter 7) produced by the United Nation's First Global International Marine Assessment: '[t]here is an environmental continuum from land through waters under national jurisdiction to areas beyond national jurisdiction...[and] [t]he transitions between the various vertical layers are gradients, not fixed boundaries. Hence, ecological distinctions among the zones are somewhat blurred across the transitions' (FGIMA 2017, para 7, 29). Further, the deep sea is connected by physical and ecological links to the surface waters.

As a consequence of these physical and ecological connections, only illustrative, and mainly qualitative, examples of deep-sea natural-capital valuation are presented here, representing a small selection of different environments, ranging from the surface waters to the deep seabed (Figure 2). These examples might be relevant for individual states whose ocean jurisdiction comprises parts of the deep sea and who are engaged in modifying national accounts to track changes in natural capital. These examples should also be relevant for the deep-sea share of the high seas, where no system of economic accounts exists but where understanding sustainable development has become critical. Important practical questions concern which types of deep-sea natural capital should be accounted for and at what levels of aggregation should they be assessed. The examples presented here are likely to be those that are most relevant for the deep sea, but further work would be required, including additions, subtractions, aggregations, or disaggregations, to incorporate these examples as components in a hypothetical system of economic and environmental accounts for the high seas.

A proper natural-capital accounting would require information about all of the components of the accounting price, including ecosystem-service flows, capital gains, net investments, and the choice of a social rate of time preference (Fenichel et al. 2016). (For the latter, for heuristic purposes, a social rate of time preference of 3 per cent is employed here.) It is noted further where information

(a) BIOZONES

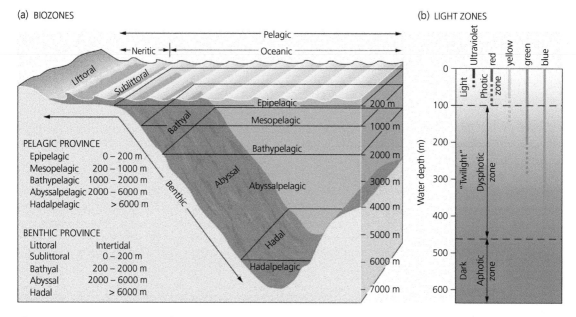

(b) LIGHT ZONES

Figure 2 Conventional vertical zones of the deep sea. Source: Pinet (2009).

is unavailable about one or more of the other components, and some educated guesses are made about the scales or likely qualitative effects of the relevant parameters. Unfortunately, information about capital gains and net investments must await further detailed study for most of the cases presented here, including the development of modelling frameworks and estimates of parameter values, but this information would be fundamental to an accurate assessment of the accounting price of natural capital. More information about the environmental characteristics of the deep-sea natural capital comprising fish stocks, seabed minerals, and oil and natural gas, including their physical flows and gross revenues, can be found in Chapters 4, 5, and 6.

Relatively more work has been undertaken to assess deep-sea ecosystem-service values (Armstrong et al. 2012). Specifically, with respect to ecosystem services, the original broad classification developed by the MA (2005) is referred to here, and ecosystem services are defined broadly to include nonrenewable provisioning services, such as hydrocarbons (Chapter 6) or hard-mineral resources (Chapter 5) or even ocean space for shipping or cabled communications (Chapter 10). Thus, ecosys-

tem services can be classed into one of four general categories, consisting of provisioning, regulating, cultural, and supporting services. Table 1 presents examples of subclasses within each of the four general ecosystem-service categories.

In the MA classification, supporting services are regarded to be upstream of the other three classes, and care must be taken concerning the potential for double counting if the accounting price for the natural capital associated with supporting services is to be evaluated (Fisher et al. 2011). Resource economist Claire Armstrong and her colleagues (2012) provided a comprehensive review of the array of deep-sea ecosystem services, identifying important information needs and methodological issues in valuation, and pointing out that scientists, policymakers, and other stakeholders are at only the early stages of comprehending the full nature of the different kinds of deep-sea supporting services. With respect to mesopelagic fish, it is noted below that supporting service relationships have implications for assessing their ecosystem-service values.

Refinements on this framework have been suggested by many researchers and practitioners (e.g. Haines-Young et al. 2009), and agreement seems to have been reached on version V.1 of a Common

Table 1 Ecosystem-service typology for the deep sea, based upon the MA (2005)

Category	Resource	Ecosystem Service	Description
Provisioning	Ocean space	Shipping	Carriage of goods or humans by sea
		Military Infrastructure	National defence
		Renewable energy	Electricity produced by offshore wind, waves, or currents
		Waste disposal	Plastics, hydrocarbons, and other hazardous chemicals, carbon, and nitrogen
	Living resources	Fisheries	Food products derived from animals and microbes
		Biochemicals and medicines	Pharmaceuticals, nutraceuticals, cosmetics, and other natural products from the sea
		Genetic resources	Genes and genetic information useful for biotechnologies
	Minerals	Hydrocarbons (oil and natural gas)	Fossil fuel exploration, development, and production
		Polymetallic nodules	Hard-mineral exploration, development, and production
		Seafloor massive sulfides	
		Ferromanganese crusts	
Regulating	Seawater and biota	Climate Regulation	Carbon sequestration in the deep sea or on the deep seabed
Cultural	Ocean space, water quality, and biodiversity	Sense of Place	Psychological or emotional connection to a geographic location, such as the deep sea
		Education and communications	Formal education about the ocean, aimed at developing an ocean literacy, and engagement between humans, designed to transfer information about the oceans
		Science	Formal observations, descriptions, hypothesis testing, and development of ocean models
		Recreation and tourism	Direct, but usually nonconsumptive, use of the ocean for pleasure or entertainment
		Cultural heritage	Benefits of using the ocean based upon local ecological knowledge or uses over generations
		Aesthetic values	Passive use of ocean areas based upon their beauty or as a source of artistic inspiration
		Spiritual and religious values	Passive use of the oceans based upon spiritual beliefs or religious feelings
		Existence and stewardship	Nonmarket, nonconsumptive existence, bequest, and option values
Supporting	Seawater and microbes	Primary production	Assimilation of energy and nutrients to produce organic matter, in the deep sea primarily by chemosynthesis
	Seawater and biota	Nutrient cycling	As many as twenty nutrients essential for life, including nitrogen and phosphorus, cycle through ecosystems
	Seawater	Physical properties	Temperature, pressure, salinity, light, sound, and circulation, and pollutant loads
	Biota	Biological diversity	Variability among living organisms including diversity within species, between species, and of ecosystems

International Classification of Ecosystem Services (CICES). These more detailed systems of classification are not needed for an initial illustration of examples for the deep sea, and, in any case, they should be the subject of future research. It will be important as well to match any classification of ecosystem services to a complementary classification of natural capital (Leach et al. 2019), but examples are chosen where the ecosystem services are naturally linked to the relevant natural capital, recognising that a more thorough accounting may be needed to map out cases where multiple ecosystem services are associated with a single type of deep-sea natural capital or where there are linkages among otherwise distinguishable types of deep-sea natural capital.

There are a variety of ways in which humans benefit from the ecosystem services of the deep sea, ranging from the physical consumption of specific resources to the passive appreciation of either individual elements, various combinations, or of the entirety of the relevant environment. Because not all economic benefits are supplied though markets, approaches utilising a variety of economic methodologies may need to be applied to determine the scale of benefits. Again, specific resources or resource areas might provide multiple types of ecosystem services, and, therefore, when considering trade-offs among mutually exclusive human uses of a resource area such as the deep sea, it is important to understand the pattern of existing—and even potential—ways in which humans benefit. As an illustration, the natural capital comprising the stock of fishes of the ocean twilight zone has occasionally been harvested to supply the fishmeal and fish oil markets (a provisioning service); they play an important role in the cycling and export of carbon

from the epipelagic waters to the deeper ocean (a regulating service); they are the subject of marine scientific research programmes (a cultural service); and they are prey for many of the apex predators, including tunas, billfish, sharks, whales, dolphins, seals, and penguins and other seabirds (a supporting service).

Figure 3 outlines a commonly employed typology that links ecosystem services to the ways in which humans can benefit from them (Pascual et al. 2011). In Figure 3, to depict how such a typology can be useful, the services provided by deep-sea hydrothermal vent systems are listed, where examples of individual services are aligned with the types of benefits that extend to humans. The typology separates the total economic value for a resource area into use and nonuse values, where the former implies actual human interactions with the relevant resource and the latter implies either no interactions or only the possibility of interactions in the future.

Use values comprise direct, indirect, and option values. Direct-use values are further subdivided

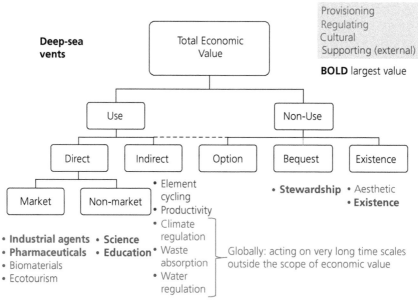

Figure 3 A framework for conceptualizing total economic value, using the example of ecosystem services associated with deep-sea hydrothermal vents systems.

into consumptive (i.e. those that use up a resource) or nonconsumptive values. Nonuse values comprise existence, altruistic, and bequest values. These values often are referred to as 'passive', in that they do not necessarily involve a physical interaction with a resource. In general, provisioning services involve direct uses (or physical exploitation and consumption), regulating services involve indirect uses, and cultural services involve both direct and passive uses. Option value spans the different value types because it comprises economic values associated with using the open ocean in one or more ways in the future. More specifically, option value is the economic value of avoiding an irreversible decision to use the ocean in a certain way, thereby precluding other future uses, either consumptive or passive.

2.4.1 Natural capital of the open-oceans biome

In 2012, Rudolf de Groot and his colleagues presented estimates of the total monetary value for the different biomes or broad resource classes on Earth. Based upon an average of fourteen studies, the monetary value for the 'open-oceans' biome was estimated to be roughly $600/ha ± $500/ha (~95 per cent c.i.) in 2019 US dollars. The underlying studies ranged widely in area coverage and location, from the world ocean to South Africa to Samoa, thereby potentially calling into question their credibility and usefulness as an overall measure. These studies employed a range of valuation techniques from stated preferences (survey-based questioning about willingness to pay for nonuse values) to market prices, and they applied to a wide variety of ecosystem-service classes, including provisioning, regulating, and cultural services. Further, the mean value estimate was applied across geographic regions and depth regimes that were nonuniform with respect to the types of natural capital. Notwithstanding these issues, in a separate study, assuming an open-ocean area of 33.2 billion hectares, Kubiszewski et al. (2017) estimated an open-ocean unit value very similar to that of de Groot et al. (2012) of approximately $800/ha in 2019 dollars. Those authors suggested that the unit value for the open ocean could vary in the future between $580 and $870/ha, depending upon the effectiveness of governance approaches.

de Groot et al. (2012) presented no estimates of capital gains or net investments; nevertheless, there is considerable concern that climate changes could lead to the degradation of ocean quality on a worldwide scale. If so, it is likely that there is now a net disinvestment in the open-ocean biome, tending to cause a decline in its accounting price. At the same time, increases in world population and a growing world economy may imply capital gains, leading to an increase in the open-ocean accounting price. The overall net effects are uncertain, suggesting avenues for future research. Assuming that the two effects roughly balance out, the natural-capital value of the open-ocean biome would be based on the flow of ecosystem services, estimated here on the order of $20,000 ± $16,667/ha.

As argued by the environmental economist Linwood Pendleton and his colleagues (2016), due to the paucity of marine-ecosystem-service valuation studies and their vintages comprising the open-ocean biome data, as well as the nonuniform distributions of ecosystem services over the ocean, open-ocean biome estimates should be approached with great caution. Compilations of total monetary value for the open-ocean biome may best be viewed as a useful means to heighten awareness about the need for further research on and attention to natural-capital accounting and ecosystem-service valuation. While aggregation of natural capital at the open-ocean-biome level may be much too broad to be of value in charting progress towards sustainability, the estimate underlines the importance of identifying and tracking more specific types of natural capital for the deep sea.

2.4.2 Natural capital of the world capture fishery stocks

The stocks of the world's capture fisheries comprise the most well-known form of the ocean's natural capital, and yields from these stocks represent one of its key provisioning services. Provisioning ecosystem services are the 'products obtained from ecosystems, including: "food, genetic resources, biochemicals, natural medicines, pharmaceuticals, or biological materials that serve as sources of energy"' (MA 2005). For the last 3 decades, the world capture fisheries were estimated to have

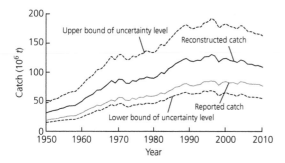

Figure 4 Reconstruction of world fish catch (106 t), showing higher levels of annual catches that those reported by FAO but also showing a more rapid decline in annual catches from their peak in the mid-1990s. Source: Pauly and Zeller (2016).

yielded between 80 and 90 million metric tonnes each year, and year-on-year fluctuations are largely attributed to variability in yields in the Peruvian anchoveta fishery (FAO 2019). Fishery scientists Daniel Pauly and Dirk Zeller have argued that these statistics need to be revised to account for under-reported data, especially discarded catches, and their reconstruction (through only 2010; Figure 4) showed a peak of 130 million metric tonnes in the mid-1990s, followed by a much more rapid decline in landings in the years following that peak (Pauly and Zeller 2016).

Gross revenues (the product of landed quantity times an estimated price) from these fisheries are on the order of $85 billion annually, where the price has been estimated from the values of seafood shipments in international trade. Further, the UN Food and Agriculture Organization (FAO) has estimated that illegal, unreported, and unregulated (so-called IUU) fishing may exploit fisheries yielding another $23 billion in gross revenues (Hudson 2017). According to the FAO, roughly 80 per cent of the world's fisheries are either overexploited (39 per cent) or fully exploited (41 per cent) (FAO, 2016). Although data describing the distribution of IUU fishing across fisheries are limited and problematic, those fisheries that involve such depredations must be considered to be overexploited, as well—even if regulation of legal participants in those fisheries attempts ostensibly to achieve a rational management.

Fisheries economist Ragnar Árnason and his colleagues (2008) have estimated the dissipation of resource rents in the world's fisheries at $50 billion

per year. (Similarly, natural resource economist Chris Costello and his colleagues (2016) estimated that, of 4713 fisheries representing 78 per cent of the world's reported fish catch, $53 billion in profits would be the result of sound management. Such management also would lead to greater yields and very significant biomass increases, constituting investments in the stocks.) The dissipation of rents is a *lost* ecosystem-service value. In addition to this loss, many nations continue to subsidise commercial fishery harvests, especially vessels in their larger scale industrial fleets, and, globally, economist Rashid Sumaila and his colleagues (2019) have estimated that capacity-enhancing subsidies in all commercial fisheries now approach $22 billion a year.

Worldwide, there is incomplete information about either capital gains or net investment in commercial fisheries. Only a few of the world's fisheries are managed for optimal yields, although Costello and his colleagues (2008) have found evidence that quota-based management has been expanding. Considering the large percentage of the world's overexploited fisheries, the very significant subsidisation, and the presence of IUU fisheries, a net drawdown in the capital stocks can be assumed, clearly implying a myopic approach to the conservation of natural capital in the world's commercial fisheries. Again, growth in population and the world's economy suggests increasing capital gains, thereby offsetting the effects of the disinvestment to an unknown extent.

The disinvestment in fisheries leading to the dissipation of resource rents implies that the accounting price of the world's commercial fish stocks is now very small. If the fisheries were managed more appropriately, for example, by reducing subsidies, investing in stocks by reducing excess fishing effort, and allowing resource rents to become fully realised, the natural capital of the world's commercial fisheries could be on the order of $1.7 trillion. The combined surface area of the world's oceans is approximately 36 billion hectares, and the high seas represent about 64 per cent of that area, or 23 billion hectares. If applied uniformly over the surface area of the ocean, this suggests that current drawdown policies result in a lost natural-capital value of $47/ha.

Most of the world's commercial fisheries are prosecuted within EEZs, however, and consequently they are not equivalent to the commercial fisheries that are limited to the deep sea, although overlaps exist where shelves are narrow or where fish straddle or migrate across boundaries. Enric Sala, from the National Geographic Society, and his collaborators (2018) have estimated the spread of aggregate 'profits' to the high-seas fishing fleets in 2016, ranging from losses of -$0.36 billion to gains of $1.43 billion. These authors also calculate very significant subsidies to the high-seas fleets of individual nations, especially Japan, Spain, China, South Korea, the United States, and Taiwan.

Assuming that these profits approximate resource rents, in the absence of specific information about either gains or investments in capital, the natural-capital value of the deep-sea fisheries is on the order of $18 billion, ranging from -$12 to $48 billion, using a time preference of 3 per cent. If applied uniformly over the actively fished surface area of the high seas (~11 billion ha; Sala et al. 2018), this suggests a natural-capital value of $1.63/ha ± $2.73/ha, with the lower 25 per cent of that interval constituting an actual net loss of natural-capital value.

2.4.3 Natural capital of the ocean twilight zone's fish stocks

As prey for apex predators, in the surface waters, including finfish, mammals, and birds, the fish biomass in the ocean twilight zone comprises an important supporting service. Supporting services are 'those environmental features and ecological elements that contribute to provisioning, regulating, and cultural ecosystem services' (MA 2005). The ecological linkages between twilight zone fish and their predators are not well constrained, however, thereby limiting our understanding of the scales of both the supporting services and their associated natural-capital values.

Historically, twilight zone fish have been the targets of infrequent commercial fisheries in only a few locations (Priede 2017). Consequently, there has been no serious human exploitation to date, and stocks should be at or near their carrying capacities. Note, however, that the growth rate of the twilight zone fish stocks may be influenced by the stocks of

predators; where the predators are exploited, perhaps to excess, a drawing down of the predators may constitute a *de facto* investment in the twilight zone fish as prey (Koehn et al. 2017). Although the scale of such an investment awaits a deeper understanding of the ecological relationships, from a qualitative perspective, such an indirect investment would imply a lowering of the discount rate and therefore an increase in the accounting price of natural capital for twilight zone fish in their supporting service role.

In the ocean twilight zone, supporting services comprise not only the fish biomass but also environmental features, such as light levels, water temperatures, and concentrations of nutrients, as well as the ecological structures and processes, involving bacteria and other microbes, zooplankton, krill, shrimps, siphonophores, salps, jellies, cephalopods, echinoderms, and organisms from other taxonomic groups. By definition, supporting services are not used directly by humans, implying that the assignment of economic values to those services often can be problematic. Where commercial uses are linked to supporting services, such as in the case of the predator-prey relationships among commercial fisheries for the apex predators and twilight zone fish, in principle, the resource rents from the commercial fisheries could be imputed to the twilight zone fish in their supporting service role (e.g. Hoagland et al. 2013). In undertaking such a valuation, it is important to avoid the 'double counting' of economic values; for example, it would be incorrect to sum resource rents in the surface fisheries in addition to the imputed value of supporting services that have been assessed from those same rents.

An important insight is gained from the idea of imputing rents from a commercial fishery to the supporting service, however. In particular, the dissipation of resource rents through the overfishing of higher trophic levels, as described above, would imply that society imputes little to no value for twilight zone organisms in their supporting service role. Thus, the sustainable conservation of apex predators is of critical importance to characterising the economic value of the ocean twilight zone as a supporting service, even if doing so results in a natural drawdown of twilight zone fish stocks.

2.4.4 Natural capital of the ocean's biological carbon pump

The biological carbon pump (BCP) is an important factor in the cycling of organic carbon in the deep sea via marine organisms; it represents possibly the most important type of regulating service. Regulating ecosystem services are the 'benefits obtained from the regulation of ecosystem processes, including, for example, the regulation of climate, water, and some human diseases' (MA 2005). Mediated by the BCP, large amounts of carbon are exported to the deep sea where carbon may be sequestered for centuries or even millennia. Moreover, approximately 1 per cent of the total particulate organic carbon produced by phytoplankton at the surface is exported to the ocean floor where it is considered to be sequestered permanently. Without the BCP, atmospheric CO_2 could be as much as 200 ppm higher than it is today, making the ocean's carbon sequestration a critical ecosystem service.

There exist a wide range of estimates of ocean twilight zone ocean carbon sequestration, from 4 to 12 billion metric tonnes per year (Buesseler et al. 2007). Using a midrange estimate of a 'carbon price' of $30 per metric tonne of CO_2 (~$110/t of carbon), this regulating service benefits the world on the order of $400–$1300 billion each year. The carbon price is expected to increase over time, implying a capital gain for the ocean as a means of exporting and sequestering carbon. Due to the uncertainties involved in oceanic export and sequestering of carbon, the net capital investment is uncertain. For example, does the likely increase in the supply of CO_2 to the ocean affect its environmental characteristics in a way that slows the rate of sequestration, say due to a shift in phytoplankton species or surface warming, deoxygenation, or acidification? With a more accurate estimate of ocean carbon sequestration, society could make better-informed decisions regarding the mitigation of CO_2 emissions.

Importantly, the ocean's carbon sequestration is threatened potentially by the effects of climate change, leading to phenomena such as thermal stratification and acidification that could affect the flux of carbon to deep water in as yet unpredictable ways (IPCC 2019). Further, the possibility of commercial harvests of twilight zone fish could affect the functioning of the BCP in some locations. Using a social rate of time preference of 3 per cent, the asset value of this service would range from $13 to $43 trillion. While recognising that ocean carbon sequestration may not be uniformly distributed over the world ocean, when applied over the areal extent of the high seas, the estimated natural-capital value of this regulating service ranges between $565/ha to $1870/ha.

2.4.5 Natural capital of deep-seabed minerals

Interest in mining minerals from the deep sea occurs in EEZs and in the 'Area', defined as the deep seabed beyond national jurisdiction, with an overview in Chapter 5. If it were to take place, the recovery of minerals from the deep seabed would be a provisioning ecosystem service. Importantly, governments and private firms with interests in looking for and recovering minerals from the deep seabed in the Area are under a legal obligation to 'to prevent, reduce, and control pollution and other hazards to the marine environment' (ISA 2019). For the Area, the International Seabed Authority (ISA) has jurisdiction over prospecting, exploration, and mining, and it has been developing regulations that are designed to protect the environment of the deep sea.

The ISA has entered into twenty-nine exploration contracts with enterprises that are investigating the nature of three types of resources in areas of the deep seabed: manganese nodules (in the Central Indian Ocean [one contract] and in the Pacific Ocean in a region located between the Clarion and Clipperton Fracture Zones [sixteen contracts]); seafloor massive sulfides (in the Indian Ocean along the Southwest and Central Indian Ridges and in the Atlantic Ocean along the Mid-Atlantic Ridge [seven contracts]); and cobalt-rich ferromanganese crusts (on seamounts in the Western Pacific Ocean [five contracts]). Most of the contracts are with governments or consortia comprised of mining companies and governments. These contracts last 15 years, and there exists the possibility of 5-year extensions. In 2017, seven contractors were given 5-year extensions, expiring in 2022.

The large number of exploration contracts suggests to many observers that deep-seabed mining is imminent, but, to date, only two of the governments or consortia have decided to move forward to enter into a contract for commercial recovery, and neither one has begun to recover minerals on a commercial scale. Many hypotheses have been put forward about why there has been no actual mining. A critical impediment is that some of the ISA's regulations for commercial recovery are still in the process of being promulgated, imposing a legal uncertainty about the likely scales of mining costs (Koschinsky et al. 2018). Further, the technology for commercial recovery has never been implemented commercially, imposing another layer of costly uncertainty. Importantly, some observers have argued that there exists a disadvantage to being a first-mover because of this uncertainty; most enterprises would prefer to watch and learn from the mistakes of the early entrants, and this strategic behaviour puts all entrants on hold. The industrial organisation economist Jim Broadus (1987) likened this to a slow-speed bicycle race, such as the Marymoor Crawl (2014), where the participants are required to stay upright in approximately the same place for an undetermined period of time before the race begins.

Broadus's insight relates to circumstances where a choice of whether or not to invest in deep-seabed mining can be characterised as irreversible (or reversible only at great cost) in the sense that the equipment and materials used for mining are firm- or industry-specific and not easily retrieved and used for other purposes. The investment expenditures would therefore comprise sunk costs, and they can be steep—on the order of at least $1 billion. Importantly, potential seabed miners may choose to delay in order to consider carefully the timing and scale of their investments.

Specifically, in the face of uncertainty and irreversibility, there may be value to a deep-seabed mining firm for the option of delaying entry into commercial recovery. Option value can exist even for risk-neutral investors, such as the nations that participate alone or as members of a seabed-mining consortium. Specifically, option value is defined as the difference in uncertain net benefits between two development strategies: invest immediately or wait until new information becomes available (Arrow and Fisher 1974).

When a deep-seabed mining firm makes an irreversible investment, it exercises ('kills') its option to

invest and thereby gives up the possibility of waiting for the arrival of new information that might bear on the desirability or timing of the investment. Thus, the traditional rule of making an investment solely upon its expected net present value should be modified (Pindyck 1991). With irreversibility and uncertainty, an investor should invest if the benefits of the project are at least as large as the investment costs plus the value of keeping the investment option alive.

According to this rule, investment costs are fixed, but the realisation of benefits will depend upon the timing of the investment. Option value is affected by the discount rate, any expected rate of appreciation in benefits, the spread of possible benefits, and costs.

The existence of option value depends upon the nature of the relevant uncertainties and the opportunities for gaining information to reduce them (Freeman 1984). In general, information about uncertain benefits and costs can be gained by waiting. If uncertainty is due to a lack of information about the benefits of deep-seabed mining, then waiting (and carrying out environmental or technological or resource or market analyses) could resolve the uncertainty. In this case, it is the waiting strategy that creates option value.

A key feature here is that the project value may change over time. An investor can maximise a project's net present value by choosing the optimal time at which to invest. Importantly, a deep-seabed-mining firm needs to consider option value only if the project's benefits are appreciating, for example, due to a rising price or declining costs. Note that, even if the current benefits are less than costs, implying that net present value would be negative, because benefits are growing, the option value would be positive. Option value will be larger with larger rates of benefit growth or with greater uncertainty.

For any economic appraisal of a mining investment, including deep-seabed mining, a major source of uncertainty concerns minerals prices (Auger and Guzmán 2010). The markets for the minerals that are the main targets for recovery show episodic—but fleeting—movements into price ranges that might be regarded as commercially viable (e.g. Hoagland 1993). Working against commercialisation,

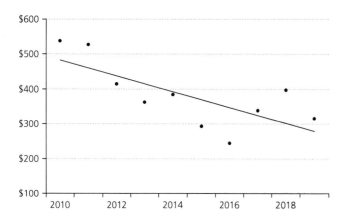

Figure 5 Ten-year, grade-weighted manganese nodule price ($/t) and linear trend: 2010-2019. Grades comprise: Ni (1.3%); Cu (1.1%); Co (0.2%); Mn (27.0%). Prices have been adjusted to 2019 US dollars using the US Consumer Price Index. Source: USGS (2019) and the London Metal Exchange (2019) for metal prices.

a grade-weighted manganese nodule 'price' using annual average London Metal Exchange prices and metal grades for an average nodule, exhibited a declining trend over the last decade (Figure 5). Further, an analysis for the European Union of supply and demand in the relevant metals markets found that '. . . polymetallic sulphides are expected to show the highest commercial viability, whereas nodules and crust [sic] are only marginally or not commercially feasible' (ECORYS 2014). Some analysts have countered that an increasing trend in the nodule price would be evident, if prices over longer periods were to be included. Upon close inspection, however, such evidence depends upon relatively short-lived market upticks.

All of the relevant metals across all of the resource classes, including nickel, copper, zinc, cobalt, and possibly manganese, involve markets characterised by existing onshore mines that are capable of expanding production in reaction to market signals (USGS 2019), and they involve significant metal recycling. Further, metal consumers respond to price increases by looking for substitutes and cutting back (Koschinsky et al. 2018). The market for cobalt is a contemporary example, as mine production in the Congo ramped up in response to price increases in the past year, and as substitutes for cobalt in lithium ion batteries (especially nickel for cobalt) have now become a reality (e.g. Lee et al. 2018).

Several models of a manganese nodule mining operation have appeared in the published and grey literatures, and all reach similar conclusions about likely positive profitability. Most employ simula-

tion approaches, including Monte Carlo analysis, to accommodate stochasticity in important model parameters, such as metal grades and volumes, operating costs, or mineral prices (Van Nijen et al. 2018). Descriptions of the distributions of prices as inputs into simulation models are based typically upon naïve models of historical prices, however, but these may be supplemented by subjective expert forecasts of future price behaviour (Roth and Royo 2018).

As one example of a mining operation model, the resource economists Jeffrey Wakefield and Kelley Myers estimated the net benefits (as measured by rents and royalties) of a manganese nodule mining operation in the EEZ of the Cook Islands to be on the order of $500 million, bounded by a range of $150 million to $1 billion. The area to be mined is 2.71 million hectares, implying an ecosystem-service value of $185/ha within a range of $55/ha to $370/ha. Other models are perceived to be more problematic in terms of their use to estimate resource rents, as royalties were calculated on the basis of a small fixed proportion of gross—not net—revenues.

Given the reservation of some individual nations or industrial consortia moving forward to undertake deep-seabed mining, it is reasonable to conclude that these entities perceive a significant option value and that its scale is at least as large as the net benefits estimated by discounted cash-flow approaches. In the presence of option value, the ecosystem-service dividend would be zero or negative. Capital gains might be forecasted using predictions of a manganese nodule price, and recent arguments for rising minerals prices have been based upon forecasts of the production of batteries for electric cars, for example. Again, the grade-weighted 'price' of manganese nodules has been declining over the last decade, however. If mining starts, the activity will result in a drawdown of stocks, suggesting a net disinvestment, although further prospecting and exploration could lead to an increase in the resource stock. Absent proof that mining can be net beneficial through actual extractive activity, at least for some of the resources, the accounting price of deep-seabed resources as a form of natural capital could be regarded as merely trivial or approaching zero.

2.4.6 Natural capital of the cultural aspects of the deep sea

For conservation of the natural capital of the deep sea, a critical need is the development of a deeper understanding of its physical processes and ecological relationships. Importantly, refinement of natural-capital values hinges on such an understanding, and oceanographic research is therefore a relevant and important cultural ecosystem service. Cultural ecosystem services have been defined as the 'non-material benefits people obtain from ecosystems through spiritual enrichment, cognitive development, reflection, recreation, and aesthetic experience' (MA 2005), and they are explored further in Chapter 8. For example, a reduction in the uncertainty surrounding the functioning of the BCP as a regulating service would lead to improvements in the skill of integrated assessment models in assessing future economic damages from climate change. In this case, the natural capital relevant to the service provided by the BCP comprises an oceanic physical-ecological system connected from the surface to the deep sea: blooms of phytoplankton in the epipelagic surface waters, stocks of zooplankton that feed on phytoplankton, stocks of mesopelagic fish that feed on the zooplankton, the bacteria that help to break down animal excrement and dead flora and fauna, and other important components of the ocean's carbon cycle. Advances in knowledge about the workings of this system may lead to increases in its accounting price.

Jin et al. (2019) developed a Bayesian decision framework for estimating the value of oceanographic research in reducing the uncertainty surrounding the estimate of ocean carbon sequestration. This estimate depends upon both predictive skill and the scale of future climate-caused economic damages. The authors' estimate of the value of marine science directed at ocean sequestration is on the order of tens of billions of dollars each year. The estimated value is positively correlated with the level of uncertainty, but the authors found that increases in research investments to enhance predictive skill exhibited diminishing returns.

Because improvements in predictive skill are likely to take place over time as observations are compiled and interpreted, an extension of the framework would model the benefits of increases in prediction accuracy as research advances and more knowledge is gained. Further, the potential economic damages from ocean acidification have not yet been incorporated into estimates of the carbon price, and these may be especially important to take into consideration with respect to potential climate-change effects in the deep sea (see Chapter 9).

2.4.7 Natural capital of the passive use of deep-sea hydrothermal vents

Globally, active deep-ocean hydrothermal vent fields represent only a small percentage of the Earth's seabed area, yet they encompass extraordinary, uncommon, and as yet incompletely understood ecosystems. Their seabed locations comprise environments characterised by very high pressures, cold ambient temperatures that interact with high-temperature fluid releases, and the absence of light. The primary productivity generated by these systems relies upon chemosynthesis, contrasting sharply with the productivity driven by photosynthesis in the overlying waters. Because these systems are isolated and rare, some have been identified as ecologically or biologically significant marine areas, proposed as world heritage sites, and incorporated already into marine protected areas within national jurisdictions (Beaulieu et al. 2013; Bax et al. 2016; Freestone et al. 2016).

Based upon their rarity, natural scientists have argued that hydrothermal vent systems must have great value per unit area (Van Dover et al. 2018), but careful assessments of their economic value as natural capital will require further research (Le et al. 2016). The recovery of hard minerals (mainly copper and zinc) from vent fields is the foremost potential provisioning service, and, although there has been some prospecting and limited testing of recovery technologies, there has been no commercial exploitation to date. It is still unclear that the mineral occurrences at vent fields are of such a grade or size to support significant seabed-mining production.

Seabed mining is expected to degrade the environment significantly, but some observers suggest that biological recovery could occur naturally, based upon the sporadic shutting down and opening up of unexploited vent systems. Some marine genetic resources associated with vent systems have been identified and commercialised already, constituting another provisioning service. Exploitation of genetic resources would occur on a small scale, and it appears unlikely to result in any significant degradation of the localised environment.

Other known ecosystem services associated with vent systems include the regulating, supporting, and cultural types. Biotic and abiotic processes help to cycle basic elements in these systems, providing an important regulating service within the local vent field environment. Primary productivity at vent systems also has been shown to help support ecological processes in nearfield and overlying ocean environments (Resing et al. 2015; Levin et al. 2016; Le Bris et al. 2019), but much more research is needed to characterise the scales and importance of these services. Vent systems likely act on time scales much too long to contribute to economic value as regulating or supporting services.

Until these other services can be better characterised or realised, cultural services will be of overriding importance, with the generation of knowledge through scientific research being the most prominent. During the last 4 decades, substantial efforts have been undertaken to advance the science of hydrothermal vent systems, involving disciplines ranging from biology and ecology to geochemistry and geophysics. In turn, the scientific knowledge of vent systems has inspired nonconsumptive, passive values, including K-12 education programmes, documentary videos, and the work of fine artists, contributing to what the deep-sea ecologists Cindy

Van Dover et al. (2018) have described as a 'marine wonderment industry'.

Consequently, assessing the passive use (i.e. non-use) values generated by the cultural ecosystem services associated with hydrothermal vent systems is central to understanding the value of natural capital for these environments, a conclusion expressed for deep-sea ecosystems in general in Chapter 8. Ultimately, this will require survey-based, stated-preference approaches, such as those obtainable through contingent valuation or choice experiment methodologies (Aanesen and Armstrong 2019). Moreover, significant efforts will need to be undertaken to increase the literacy of the public about these systems so that the trade-offs explored through these methods can be regarded as credible for assigning economic values.

Without the collection of primary data about human preferences, policy analysts must rely upon the cruder approach of transferring estimates of benefits from other contexts, making qualitative arguments about the utility of such transfers to valuing hydrothermal vent systems. Benefit transfer involves the application of valuation data from other contexts, relying significantly upon expert judgement about the relevance of the disparate contexts (Bateman et al. 2010). Such an approach was taken to value the passive use of the SuSu Knolls hydrothermal vent field (Batker and Schmidt 2015). This benefit transfer estimate was employed to explore trade-offs associated with the loss of such values as a potential consequence of the exploitation of minerals proposed by Nautilus Minerals for the Solwara I project.

For the Solwara I study, Batker and Schmidt (2015) argued that estimates of passive values from other unique or sensitive ecosystems should be considered valid for benefits transfer. The study relied upon estimated passive-use values ranging from $1 to $896/ha/year associated with the Intag Cloud Forest in Ecuador. (This forest environment also was subject to potential loss as a consequence of the opening of a copper mine.) The analysts argued that, as a productive and biologically diverse ecosystem, albeit one much more accessible to humans than a deep-sea vent field would be, the estimate of passive benefits associated with the Intag Cloud Forest constituted a conservative estimate for SuSu

Knolls, one that likely overestimated the latter's passive-use value.

A worrisome aspect of this particular application of benefit transfer is that it comprised *serial* transfers of benefits of differing types, deriving from a range of contexts and spanning studies carried out as much as 3 decades earlier. Table 2 depicts the relationships among these studies. The Intag Cloud Forest study relied upon two estimates: (i) a 1984 study of recreation (hiking, a direct-use value) in a temperate, deciduous forest, the Ramsey's Draft Wilderness in the George Washington National Forest in the Commonwealth of Virginia, USA, and (ii) a compilation of three other studies (Gössling 1999). The latter compilation was based upon: (i) a 1988 study of tourism (direct use) in Korup National Park, Cameroon; (ii) a 1994 study of existence value (passive use) for *all forests* in Mexico; and (iii) a 1995 study of combined tourism and option values (direct and passive uses) for the Borivli National Park in Mumbai, India. Because the transferred values were described as conservative (believed to overestimate likely passive values for hydrothermal vent fields), other analysts have rejected the approach, arguing that such estimates were much too high, and relying on yet other studies providing even lower economic values (e.g. Wakefield and Myers 2018 relying on Asquieth et al. 2008).

It is important to begin to consider the full range of likely ecosystem-service values for remote, difficult-to-access deep-sea environments, such as hydrothermal vent fields. In doing so, it will be critical to avoid ad hoc methods of benefits transfer as these are likely to result in ad hoc decisions about the net benefits of allowing particular consumptive or irreversible uses to occur. This point is driven home by Maja Folkersen et al. (2018), who found a critical knowledge gap relating to the economic value of the deep sea. In particular, Folkersen and her colleagues found that knowledge of ecological structures and processes influences estimates of the economic value of deep-sea natural capital, suggesting the need for further research on the science of ocean ecosystems, the education of the public about these systems, and the development of estimates about the values of ecosystem services.

2.5 Emerging institutions for deep-sea governance

Arguments for developing the concepts of natural capital and ecosystem services were put forward specifically for the purpose of making better decisions about allocating natural resources. In particular,

Table 2 Serial benefit transfers for valuing the passive use of the SuSu Knolls hydrothermal vent field

Citation	Date	Location	Use/Nonuse	Area (ha)	Surplus ($/ha/y)
(1) Batker and Schmidt (2015)	2015	Intag Province (Cloud Forest), Ecuador Benefit transfers from earlier studies (1.1) and (1.2)	Aesthetics and recreation values (direct, nonconsumptive use)	5350	$5– $893
(1.1) Prince and Ahmed (1988)	1984	Ramsey's Draft Wilderness (temperate deciduous forest) Virginia, USA	Recreation value (direct, nonconsumptive use)	2638	$115
(1.2) Gössling (1999)	1999	Benefit transfers from earlier studies (1.2.1), (1.2.2), and (1.2.3)	Passive value (nonuse)	--	$0–$29
(1.2.1) Ruitenbeck (1992)	1988	Korup National Park (rainforest) Cameroon	Tourism value (direct, nonconsumptive use)	35,100	$1
(1.2.2) Adger et al. (1995)	1994	Mexican forests (all) (temperate and tropical, deciduous and coniferous forests)	Existence value (nonuse)	50,000,000	$0–$18
(1.2.3) Hadker et al. (1997)	1995	Borivli (Sanjay Gandhi) National Park Mumbai, India	Tourism and option values ('maintain and preserve' the park)	10,309	$950

assigning economic values to direct and indirect uses of natural resources allows decision-makers to attempt to understand the opportunity costs of proposed actions when the capacity to use (or passively appreciate) particular resources is lost, either temporarily or forever. There are many instances, particularly for the deep sea, where it can be challenging to estimate such values. While crude attempts at valuation are fairly criticised, they comprise steps in the right direction for making allocation decisions. With appropriate resources, such efforts can be refined and improved, thereby bettering human welfare. The consequences of a failure to attempt even rough estimates inevitably leads to a situation where unvalued natural capital or ecosystem services are treated by those making critical allocation decisions as if they have no value at all.

For the deep sea, there are many examples where such allocation decisions are required, including those relating to the regulation of existing high-seas fisheries; the prospects of harvesting fisheries that are as yet unexploited, such as those for lanternfish in the twilight zone; the recovery of minerals from the deep seabed, including massive sulphides from hydrothermal systems; and the establishment of institutions for allowing access to marine genetic resources on the high seas. Historically and continuing through to the present, these decisions have been made poorly, leading to the overexploitation of fishery resources, the selection of royalty terms that are unlikely to capture resource rents on ocean minerals fully, and a rush to discover and patent the genetic resources of the ocean.

As laid out in Chapter 3, this situation now is changing; the international community has begun to negotiate an international legally binding instrument to govern the conservation and sustainable use of marine biodiversity in ABNJ. If successful in its goals, that instrument would provide a means for allowing access to marine genetic resources, including the fair distribution of any realised benefits to mankind; require assessments of impacts for ocean uses that could lead to significant environmental effects; and establish tools for the management of designated marine areas (explained further in Chapter 7). It is too early to predict what the specific elements of each of these institutions might look like, but it is unclear that they would incorporate

explicit requirements for the development of estimates of the value of natural capital to assist in measuring the progress towards sustainable development. Without the latter, a stated goal of the sustainable use of marine biodiversity rings hollow.

There is now an opportunity, however, at this early stage in the design of international institutions for governance of the deep sea, for specific provisions to be incorporated in these institutions calling for assessments of natural-capital accounting prices. Even if the BBNJ instrument were to call for such assessments, however, the realisation of this opportunity faces significant hurdles. First, the deep sea, where it occurs beyond national jurisdictions, has no system of accounting for 'inclusive wealth' comprising the value of human, manufactured, and natural capital precisely because it is beyond the purview of any one nation. Second, until fairly recently, many nations had not been measuring or incorporating the accounting value of natural capital in their own national accounts, and efforts to do so are still nascent, covering in many cases only subsets of natural-capital stocks (Edens 2013; UNEP 2018). Third, efforts to develop hierarchical classifications of natural capital to link with existing ecosystem-service classifications are still under development (Leach et al. 2019), and this is especially true for the important deep-sea natural capital.

These hurdles naturally help to characterise appropriate focuses for future research and practice. A future research agenda would involve refining estimates of accounting prices for the most important types of deep-sea natural capital, locating these within linked classifications of ecosystem services and natural capital, and the design and implementation of a system of economic and environmental accounts for the deep sea comprising the high seas.

Sustainable use implies an objective of equity, especially across generations. As laid out by Arrow and outlined in the Appendix to this chapter, an objective of sustainable use does not necessarily require optimising economic benefits. For a sustainable development path through time, what is required instead is that the welfare of human society must be nondecreasing with time. Simply put, sustainable development requires that the sum of the values of all forms of capital (the products of accounting prices and capital stocks) must not be decreasing (IHDP 2014).

The BBNJ deliberations are focused also on a goal of conservation, which can be interpreted to imply an objective of minimising the unnecessary or wasteful use of scarce resources (Scott 1955). The joint goals of conservation and fairness that have been put forward by the BBNJ process often can find themselves in conflict, necessitating trade-offs that require a loosening of either one goal or the other. While strict adherence to the conservationist goal of minimising waste can be consistent with sustainability, the result that sustainable development requires that the value of inclusive wealth—from all forms of capital—be nondecreasing with time may allow significant flexibility in balancing the two. Only with significant and continuing analytical efforts to characterise ecosystem-service dividends, capital gains, net investments, and the resulting accounting prices for natural capital can such a balance be attempted for the deep sea.

2.6 Conclusions

> History was part of the baggage we threw overboard when we launched ourselves into the New World. We threw it away because it recalled old tyrannies, old limitations, galling obligations, bloody memories. Plunging into the future through a landscape that had no history, we did both the country and ourselves some harm along with some good. Neither the country nor the society we built out of it can be healthy until we stop raiding and running, and learn to be quiet part of the time, and acquire the sense not of ownership but of belonging . . . Only in the act of submission is the sense of place realised and a sustainable relationship between people and earth established.
> **Wallace Stegner (1992)**

Humans are still in the earliest stages of thinking about how best to conserve the natural capital of the deep sea. Much of it is still unexplored, and it may take decades—if not much longer—to understand in sufficient detail the physical characteristics and dynamic features of the system. As human understanding grows, and as the outlines of a hoped-for 'Blue Economy' begin to emerge, it will be critical to

account for the value of the deep sea's natural capital.

A central issue relates to the need for financial resources to accomplish the goals of conservation and sustainable use of the deep sea. Financial resources are needed for administration, convening of interested parties, public outreach, environmental assessments, monitoring and enforcement, scientific research, and support for management decisions. Little attention has been paid to this issue in international discussions relating to any of the deep-sea institutions, including the BBNJ, the implementation of the targets of Sustainable Development Goal 14, the updating of the World Ocean Assessment, or preliminary discussions about the forthcoming Decade of Ocean Science for Sustainable Development (2021–2030). Few nations are likely to make more than minor financial contributions to deep-sea conservation out of their foreign aid budgets or other funds, especially given pre-existing commitments under other conventions and agreements.

An obvious source of financial resources comprises the dividends from natural capital: the potential resource rents that would be generated from the human uses of the resources of the deep sea, including most prominently the prosecution of commercial fisheries, the recovery of minerals and hydrocarbons from the deep seabed, and the discovery and commercialisation of marine genetic resources. Unfortunately, for the reasons discussed earlier and elsewhere in this volume, none of these potential financial resources are likely to be realised without very significant changes to existing practices and policies. Thus, as in Stegner's thesis about how a sense of place comes about, the world is still raiding and running.

The deep sea is remote and costly to access. Most of humanity may have only an indirect connection to it during their lifetimes. What may be most critical—but also most difficult to value in economic terms—are the cultural ecosystem services, including science, education, and the building up of an ocean literacy. All three can be conceptualised as (cultural) supporting services for a fourth, sense of place. Given the countless overwhelming policy priorities at domestic and international levels, including the threat of nuclear war; armed conflict;

terrorism; climate change; poverty; degree of food security; spread of disease; human diaspora; natural catastrophe; air, water, and land-based pollution; slavery; gender discrimination; among many others, little can be accomplished unless humans are able to develop a profound emotional and psychological connection with the deep sea. Nonetheless, the emergence of a sense of place is the *sine qua non* for future human investments in conservation and sustainable use of the deep sea's natural capital.

Acknowledgements

The authors thank the editors and two peer reviewers for their guidance and very useful insights and suggestions. Any errors remain the responsibility of the authors. PH and DJ thank the Audacious Project, a collaborative endeavour housed at TED, for funding through the WHOI Ocean Twilight Zone Program. PH and SB thank The Joint Initiative Awards Fund of the Andrew W. Mellon Foundation for funding through a Woods Hole Oceanographic Institution Interdisciplinary Study Award.

References

Aanesen, M. and Armstrong, C. W. (2019). Trading Off Co-produced Marine Ecosystem Services: Natural Resource Industries Versus Other Use and Non-Use Ecosystem Service Values, *Frontiers in Marine Science*, 6, 102. https://doi.org/10.3389/fmars.2019.00102.

Adger, W. N., Brown, K., Cervigini, R., and Moran, D. (1995). Total Economic Value of Forests in Mexico. *Ambio*, 24, 286–96.

Armstrong, C. W., Foley, N. S., Tinch, R., and van den Hove, S. (2012). Services From the Deep: Steps Towards Valuation Of Deep Sea Goods and Services. *Ecosystem Services*, 2, 2–13.

Árnason, R., Kelleher, K., and Willman, R. (2008). *The Sunken Billions: The Economic Justification for Fisheries Reform*. Washington, DC: The World Bank.

Arrow, K. J., Cropper, M., Gollier, C., et al. (2013). Determining Benefits and Costs for Future Generations. *Science*, 341, 349–50.

Arrow, K. J., Dasgupta, P, and Mäler, K. (2003). Evaluating Projects and Assessing Sustainable Development in Imperfect Economies. *Environmental and Resource Economics*, 26, 647–85.

Arrow, K. J. and Fisher, A. C. (1974). Environmental Preservation, Uncertainty, and Irreversibility. *Quarterly Journal of Economics*, 88, 312–19.

Asquieth, N. M., Vargas, M. T., and Wunder, S. (2008). Selling Two Environmental Services: In-kind Payments for Bird Habitat and Watershed Protection in Los Negros, Bolivia. *Ecological Economics*, 65, 675–84.

Auger, F. and Guzmán, J. I. (2010). How Rational Are Investment Decisions in the Copper Industry? *Resources Policy*, 35, 292–300.

Bateman, I. J., Mace, G. M., Fezzi, C., Atkinson, G., and Turner, K. (2010). Economic Analysis for Ecosystem Service Assessments. *Environmental and Resource Economics*, 37, 211–32.

Batker, D. and Schmidt, R. 2015. *Environmental and Social Benchmarking Analysis of Nautilus Minerals Inc. Solwara 1 Project. SL01-NMN-XEE-RPT-0180-001*. Tacoma, WA: Earth Economics.

Bax, N. J., Cleary, J., Donnelly, B., et al. (2016). Results of Efforts by the Convention on Biological Diversity to Describe Ecologically or Biologically Significant Marine Areas. *Conservation Biology*, 30, 571–81.

Beaulieu, S. E., Baker, E. T., German, C. R., and Maffei, A. (2013). An Authoritative Global Database for Active Submarine Hydrothermal Vent Fields. *Geochemistry-Geophysics-Geosystems*, 14, 1–14.

Broadus, J. M. (1987). Seabed Materials. *Science*, 235, 853–60.

Buesseler, K. O., Lamborg, C. H., Boyd, P. W., et al. (2007). Revisiting Carbon Flux Through the Ocean's Twilight Zone. *Science*, 316, 567–70. https://doi.org/10.1126/science.1137959.

Carpenter, S. R., Mooney, H. A., Agard, J., et al. (2009). Science for Managing Ecosystem Services: Beyond the Millennium Ecosystem Assessment. *Proceedings of the National Academy of Sciences USA*, 106, 1305–12. https://doi.org/10.1073/pnas.0808772106.

Clark, C. W. and Munro, G. R. (1975). The Economics of Fishing and Modern Capital Theory: A Simplified Approach. *Journal of Environmental Economics and Management*, 2, 92–106.

Coase, R. H. 1960. The Problem of Social Cost, *Journal of Law and Economics*, 3, 1–44.

Common International Classification of Ecosystem Services (CICES) (2018). *CICES V.1*. Brussels, Belgium: Programme on Natural Systems and Vulnerability, European Environment Agency, European Union. https://cices.eu/.

Conrad, J. M. and Clark, C. W. (1987). *Natural Resource Economics: Notes and Problems*. New York: Cambridge University Press.

Corliss, J. B., Dymond, J., Gordon, L. I., et al. (1979). Submarine Thermal Springs on the Galápagos Rift. *Science*, 203 (4385), 1073–83.

Costello, C., Gaines, S. D., and Lynham, J. (2008). Can Catch Shares Prevent Fisheries Collapse? *Science*, 321, 1678–81.

Costello, C., Ovando, D., Clavelle, T., et al. (2016). Global Fishery Prospects Under Contrasting Management Regimes. *Proceedings of the National Academy of Sciences USA*, 113(18), 5125–9.

de Groot, R., Brander, L., van der Ploeg, S., et al. (2012). Global Estimates of the Value of Ecosystems and Their Services in Monetary Units. *Ecosystem Services*, 1, 50–6.

Díaz, S., Pascual, U., Stenseke, M., et al. (2018). Assessing Nature's Contributions to People. *Science*, 359, 270–2.

ECORYS (2014). *Study To Investigate State of Knowledge of Deep Sea Mining. Final Report Annex 3: Supply and Demand. Director General of Maritime Affairs and Fisheries.* Rotterdam, Netherlands, and Brussels, Belgium: ECORYS Nederland BV (15 October).

Edens, B. (2013). Depletion: Bridging the Gap Between Theory and Practice. *Environment and Resource Economics*, 54, 419–41. https://doi.org/10.1007/s10640-012-9601-3.

Fenichel, E., Abbott, J. K., Bayham, J., et al. (2016). Measuring the Value of Groundwater and Other Forms of Natural Capital. *Proceedings of the National Academy of Sciences USA*, 113, 2382–7.

Fenichel, E. P., Abbott, J. K., and Yun, S. D. (2018). The Nature of Natural Capital and Ecosystem Income, in V. K. Smith, P. Dasgupta, and S. Pattanayak (eds.) *Handbook of Environmental Economics*. Amsterdam, Netherlands: North Holland, pp. 85–142.

First Global International Marine Assessment (FGIMA) (2017). *The Conservation and Sustainable Use of Marine Biological Diversity of Areas Beyond National Jurisdiction.* Technical Abstract. New York: United Nations.

Fisher, B., Polasky, S., and Sterner, T. (2011). Conservation and Human Welfare: Economic Analysis of Ecosystem Services. *Environmental & Resource Economics*, 48, 151–9.

Folkersen, M. V., Fleming, C. M., and Hasan, S. (2018). The Economic Value of the Deep Sea: A Systematic Review and Meta-analysis. *Marine Policy*, 94, 71–80.

Food and Agriculture Organization (FAO) (2016). *The State of World Fisheries and Aquaculture 2016: Contributing to Food Security and Nutrition For All.* Rome: FAO.

Food and Agriculture Organization (FAO) (2019). *The State of World Fisheries and Aquaculture (SOFIA).* Rome: FAO Fisheries and Aquaculture Department (27 February). http://www.fao.org/fishery/.

Freeman, A. M. (1984). The Quasi-option Value of Irreversible Development. *Journal of Environmental Economics and Management*, 11, 292–5.

Freestone, D., Laffoley, D., Douvere, F., and Badman, T. (2016). World Heritage in the High Seas: An Idea Whose Time Has Come. World Heritage Reports 44. Paris: UN Educational, Scientific and Cultural Organization. https://unesdoc.unesco.org/ark:/48223/pf0000245467.

Gössling, S. (1999). Ecotourism: A Means to Safeguard Biodiversity and Ecosystem Functions. *Ecological Economics*, 29, 303–20.

Hadker, N., Sharma, S., Davis, A., and Muraleedharan, T. R. (1997). Willingness-to-Pay for Borivli National Park: Evidence from a Contingent Valuation. *Ecological Economics*, 21, 105–22.

Haines-Young, R., Potschin, M., de Groot, R., Kienast, F., and Bolliger, J. (2009). *Towards a Common International Classification of Ecosystem Services (CICES) for Integrated Environmental an Economic Accounting.* Draft report to the European Environment Agency. Nottingham: Centre for Environmental Management, School of Geography, University of Nottingham.

Hardin, G. (1998). Extensions of the 'Tragedy of the Commons'. *Science*, 280, 682–3.

Hoagland, P. (1993). Manganese Nodule Price Trends: Dim Prospects for the Commercialization of Deep Seabed Mining. *Resources Policy*, 19, 287–98.

Hoagland, P., Kite-Powell, H. L., Jin, D., and Solow, A. R. (2013). Supply-side Approaches to the Economic Valuation of Coastal and Marine Habitat in the Red Sea. *Journal of the King Saud University-Science*, 25, 217–28.

Hudson, A. (2017). Restoring and Protecting the World's Large Marine Ecosystems: An Engine for Job Creation and Sustainable Economic Development. *Environmental Development*, 22, 150–5.

Huxley, T. H. (1882). Inaugural Address to the Fisheries Exhibition, London, in C. Blindermand and D. Joyce (eds.) *The Huxley File.* Worcester, MA: Clark University.

Intergovernmental Panel on Climate Change (IPCC) (2019). 'Summary for Policymakers', in H-O. Pörtner, D. C. Roberts, V. Masson-Delmotte, et al. (eds.) *IPCC Special Report on the Ocean and Cryosphere in a Changing Climate.* Monaco: IPCC 51st Session, Working Groups I and II (24 September), pp. 5–54.`

International Human Dimensions Program on Global Environmental Change (IHDP) (2014). *Inclusive Wealth Report 2014.* Delhi, India: United Nations University and United Nations Environment Programme.

International Seabed Authority (ISA) (2019). Biodiversity. Kingston, Jamaica. https://www.isa.org.jm/biodiversity-0.

Jin, D., Hoagland, P., and Buesseler, K. (2019). *The Value of Scientific Research on the Ocean's Biological Carbon Pump.* Mimeo. Woods Hole, MA: Marine Policy Center, Woods Hole Oceanographic Institution.

Kipling, R. (1898). The Explorer, in *Rudyard Kipling's Verse, Inclusive Edition, 1885–1918.* Garden City, NY: Doubleday, Page & Co (1922).

Koehn, L. E., Essington, T. E., Marshall, K. N., et al. (2017). Trade-Offs Between Forage Fish Fisheries and Their

Predators in the California Current. *ICES Journal of Marine Science*, 74, 2448–58. doi:10.1093/icesjms/fsx072.

Koschinsky, A., Heinrich, L., Boehnke, K., et al. (2018). Deep-sea Mining: Integrated Research on Environmental, Legal, Economic, and Societal Implications. *Integrated Environmental Assessment and Management*, 14, 672–91.

Kubiszewski, I., Costanza, R., Anderson, S., and Sutton, P. (2017). The Future Value of Ecosystem Services: Global Scenarios and National Implications. *Ecosystem Services*, 26, 289–301.

Landers, D. H. and Nahlik, A. M. (2013). *Final Ecosystem Goods and Services Classification System (FEGS-CS)*. EPA/600/R-13/ORD-004914. Washington, DC: Office of Research and Development, US Environmental Protection Agency.

Le Bris, N., Yücel, M., Das, A., et al. (2019). Hydrothermal Energy Transfer and Organic Carbon Production at the Deep Seafloor. *Frontiers in Marine Science*. https://doi.org/10.3389/fmars.2018.00531.

Le, J. T., Levin, L. A., and Carson R. T. (2016). Incorporating Ecosystem Services Into Environmental Management of Deep Seabed Mining. *Deep-Sea Research II*, 137, 486–503

Leach, K., Grigg, A., O'Connor, B., et al. (2019). A Common Framework of Natural Capital Assets for Use in Public and Private Sector Decision Making. *Ecosystem Services*, 36, 100899.

Lee, J., Kitchaev, D. A., Kwon, D-H, et al. (2018). Reversible Mn2 /Mn4 Double Redox in Lithium-excess Cathode Materials. *Nature*, 556, 185. https://doi.org/10.1038/s41586-018-0015-4.

Levin, L. A., Baco, A. R., Bowden, D. A., et al. (2016). Hydrothermal Vents and Methane Seeps: Rethinking the Sphere of Influence. *Frontiers in Marine Science*, 3, 72. https://doi.org/10.3389/fmars.2016.00072.

London Metal Excahnge (LME) (2019). Non-ferrous: LME Official Prices, US$ per Tonne. https://www.lme.com/en-GB/Metals/Non-ferrous#tabIndex=0.

Mäler, K.-G., Aniyar, S., and Jansson, Å. (2009). Accounting for Ecosystems. *Environment and Resource Economics*, 42, 39–51. https://doi.org/10.1007/s10640-008-9234-8.

Marymoor Crawl (2014). https://vimeo.com/101885379.

Millennium Ecosystem Assessment (MA) (2005). https://www.millenniumassessment.org/en/index.html.

Moore, M. A. and Vining, A. R. (2018). *The Social Rate of Time Preference and the Social Discount Rate*. Arlington, VA: Mercatus Symposium, Mercatus Center at George Mason University (November).

Ostrom, E. (1990). *Governing the Commons: The Evolution of Institutions for Collective Action*. Cambridge: Cambridge University Press.

Pascual, U., Muradian, R., Brander, L., et al. (2011). The Economics of Valuing Ecosystem Services and Biodiversity, in P. Kumar (ed.) *The Economics of Ecosystems and Biodiversity: Ecological and Economic Foundations*. London: Routledge, pp. 183–256. https://doi.org/10.4324/9781849775489.

Pauly, D. and Zeller D. (2016). Catch Reconstructions Reveal That Global Marine Fisheries Catches Are Higher Than Reported and Declining. *Nature Communications*, 7, 10244.

Pendleton, L. H., Thébaud, O., Mongruel, R. C., and Levre, H. (2016). Has the Value of Global Marine and Coastal Ecosystem Services Changed? *Marine Policy*, 64, 156–8. http://dx.doi.org/10.1016/j.marpol.2015.11.018.

Perloff, J. M. (2012). *Microeconomics*, 6th edn. Boston: Addison-Wesley.

Pindyck, R. S. (1991). Irreversibility, Uncertainty, and Investment. *Journal of Economic Literature*, 29, 1110–48.

Pinet, P. R. (2009). *Invitation to Oceanography*, 5th edn. Boston: Jones and Bartlett Publishers Inc.

Priede, I .G. (2017). *Deep-Sea Fishes: Biology, Diversity, Ecology and Fisheries*. New York: Cambridge University Press.

Prince, R. and Ahmed, E. (1984). Estimating Individual Recreation Benefits Under Congestion and Uncertainty. *Journal of Leisure Research*, 21, 61–76.

Resing, J. A., Sedwick, P. N., German, C. R., et al. (2015). Basin-scale Transport of Hydrothermal Dissolved Metals Across the South Pacific Ocean. *Nature*, 523, 200–3. https://doi.org/10.1038/nature14577.

Ring, I., Hansjürgens, B., Elmqvist, T., Wittmer, H., and Sukhdev, P. (2010). Challenges in Framing the Economics of Ecosystems and Biodiversity: The TEEB Initiative. *Current Opinions in Environmental Sustainability*, 2, 15–26.

Roth, R. and Royo, C. M. (2018). Update on Financial Payment Systems: Seabed Mining for Polymetallic Nodules. PowerPoint presentation. Kingston, Jamaica: International Seabed Authority (July 16).

Ruitenbeek, H. J. (1992). The Rainforest Supply Price: A Tool for Evaluating Rainforest Conservation Expenditures. *Ecolological Economics*, 6, 57–78.

Sala, E., Mayorga, J., Costello, C., et al. (2018). The Economics of Fishing the High Seas. *Science Advances*, 4 (6), eaat2504.

Scott, A. (1955). *Natural Resources: The Economics of Conservation*. Toronto, Canada: University of Toronto Press.

Stegner, W. (1992). *The Sense of Place*. New York: Random House Inc.

Sumaila, U. R., Ebrahim, N., Schuhbauer, A., et al. (2019). Updated Estimates and Analysis of Global Fisheries Subsidies. *Marine Policy*, 109, 103695.

UN Environment Programme (UNEP). 2018. *Inclusive Wealth Report 2018*. Nairobi, Kenya: UNEP.

US Geological Survey (USGS) (2019). *Mineral Commodity Summaries 2019*. Reston, VA: USGS.

Van Dover, C. L., Arnaud-Haond, S., Gianni, M., et al. (2018). Scientific Rationale and International Obligations for Protection of Active Hydrothermal Vent Ecosystems from Deep-sea Mining. *Marine Policy*, 90, 20–8.

Van Nijen, K., Van Passel, S., and Squires, D. (2018). A Stochastic Techno-economic Assessment of Seabed Mining of Polymetallic Nodules in the Clarion Clipperton Fracture Zone. *Marine Policy*, 95,133–41.

Verne, J. (1871). *Vingt Mille Lieues sous les Mers*. Paris: J. Hetzel et Cie. https://gallica.bnf.fr/ark:/12148/btv1b8600258f/f1.item.

Wakefield, J. R. and Myers, K. (2018). Social Cost Benefit Analysis for Deep Sea Minerals Mining. *Marine Policy*, 95, 346–55. http://dx.doi.org/10.1016/j.marpol.2016.06.018i.

Yun, S. D., Hutniczak, B., Abbott, J. K., and Fenichel, E. P. (2017). Ecosystem-based Management and the Wealth of Ecosystems. *Proceedings of the National Academy of Sciences USA*, 114, 6539–44.

Appendix

A1 Theoretical framework for sustainable development

According to Arrow et al. (2003), the wealth of an economy is the worth of its capital assets, including manufactured, human, and natural capital. The value function of an economy (V) describes social welfare as a function of capital stocks ($\mathbf{K}$) and a resource allocation mechanism (a):

$$V(\mathbf{K}_t, \alpha, t) = \int_t^\infty U(C_\tau) e^{-\delta(\tau - t)} d\tau, \qquad (1)$$

where t is time, $\mathbf{K}$ is a vector with elements (K_{it}) denoting a comprehensive list of capital assets, U is a social utility function, C is consumption, and δ is the discount rate. The resource allocation mechanism (a) is an economic programme $\{C_\tau, \mathbf{R}_\tau, \mathbf{K}_\tau\}_t^\infty$ describing relationships among consumption, resource flows ($\mathbf{R}_\tau$), and capital stocks from time t to the indefinite future. In other words, the value function (V) considers benefits to both the current and future generations.

The accounting price (p_{it}) of the ith capital stock (K_{it}) is:

$$p_{it} = \frac{\partial V(\mathbf{K}_t, \alpha, t)}{\partial K_{it}}. \qquad (2)$$

The accounting price of capital asset i consists of the present discounted value of the perturbations to U resulting from a marginal increase in the quantity of asset i.

The economic programme $\{C_\tau, \mathbf{R}_\tau, \mathbf{K}_\tau\}_0^\infty$ is on a *sustainable development path* at t if:

$$\frac{dV_t}{dt} \geq 0. \qquad (3)$$

Note that this framework does not require optimality (i.e. an economic programme to maximise V).

Instead, there may be any number of technological and ecologically feasible economic programmes that satisfy the sustainability condition (3).

The wealth of the economy is a linear combination of its capital stocks, and the weights are the accounting prices of corresponding stocks:

$$V_t = \sum_i p_{it} K_{it}. \qquad (4)$$

A2 Accounting price for global public goods

Countries interact with each other not only through trade but also through transnational externalities (e.g. CO_2 in the atmosphere or the extraction of high-seas resources). Let G_t be the stock of a global public good at t. G is an argument in the value function of every country. Let V_j be the value function of country j, we have:

$$V_{jt} = V_j\left(\mathbf{K}_{jt}, \alpha, G_t\right), \qquad (5)$$

where $\mathbf{K}_{jt}$ represents the stock of country j's assets. The accounting price for asset i in country j is:

$$p_{ijt} = \frac{\partial V_j\left(\mathbf{K}_{jt}, \alpha, G_t\right)}{\partial K_{ijt}}. \qquad (6)$$

For a global public good G, the accounting price is:

$$g_{jt} = \frac{\partial V_j\left(\mathbf{K}_{jt}, \alpha, G_t\right)}{\partial G_t}. \qquad (7)$$

It is possible that G is both an economic 'good' for some countries ($g_{jt} > 0$) and an economic 'bad' for others ($g_{jt} < 0$). The accounting prices would be affected by international resource allocation mechanisms (Arrow et al. 2003).

A3 Accounting price for natural capital

Perhaps the most challenging task in social and environmental accounting is to estimate the accounting price (p) of natural capital (i.e. the unit price of natural capital). Here, we describe the results of Fenichel et al. (2016), using the well-known bioeconomic model. Note, however, that these authors derived their formula for the accounting price without the optimality requirement, and the accounting price formula remains valid even when the economic programme is nonoptimal.

A3.1 The classical bioeconomic model

Conrad and Clark (1987) describe the 'most general case' of a commercial fisheries bioeconomic model as:

$$\max \int_0^T U\big(S(t), H(t)\big) e^{-\delta t} dt \qquad (8)$$

$$\text{subject to } \dot{S} = F\big(S(t)\big) - H(t), \qquad (9)$$

where U is a utility function, S is a fish stock, F is the stock growth function, H is harvest, and δ is a discount rate. The control variable is H. Note that, following Clark and Munro (1975), in most bioeconomic models, the utility function is specified as the revenue from harvest minus the cost of fishing.

The current-value Hamiltonian is:

$$H = U(S, H) + p\big[F(S) - H\big], \qquad (10)$$

where p is the shadow value of S (i.e. the accounting price). The first-order conditions include:

$$H_H = 0 \qquad (11)$$

$$\dot{p} - \delta p + H_S = 0. \qquad (12)$$

From (10), (11), and (12), we have:

$$p = U_H \qquad (13)$$

$$\dot{p} - \delta p + U_S + pF_S = 0. \qquad (14)$$

Rearrange (14):

$$p = \frac{U_S + \dot{p}}{\delta - F_S}. \qquad (15)$$

Equation (15) is similar to the Fenichel et al. (2018) accounting price formula, except that the denominator does not include the term H_S.

A3.2 The Fenichel et al. (2018) framework

The Fenichel et al. (2018) formula can be obtained by modifying the classical bioeconomic model as follows:

$$\max \int_0^T U(S(t), x(S(t))) e^{-\delta t} dt \qquad (16)$$

$$\text{subject to } \dot{S} = F\big(S(t)\big) - H(S(t), x(S(t))), \qquad (17)$$

where x is the 'economic programme', which, according to Yun et al. (2017), comprises decisions about 'time at sea and gear choices'. Thus, harvest is now a function of both S and x. Here, x is the control.

The first-order conditions require:

$$p = \frac{U_x}{H_x} \qquad (18)$$

$$\dot{p} - \delta p + U_S + p(F_S - H_S) = 0. \qquad (19)$$

Rearrange (19) to obtain:

$$p = \frac{U_S + \dot{p}}{\delta - (F_S - H_S)}, \qquad (20)$$

which is equation (S8) in Fenichel et al. (2016).

CHAPTER 3

The legal framework for resource management in the deep sea

Aline Jaeckel, Kristina Gjerde, and Duncan Currie

3.1 Introduction

The deep oceans and their protection, management, research, and resources are governed by a range of legal instruments, summarised in this chapter. The legal frameworks relevant to the deep oceans span from fisheries (Chapter 4) to mineral mining (Chapter 5), hydrocarbons (Chapter 6), and marine scientific research (Chapters 7 to 9), and pollution (Chapter 10).

The framework convention is the 1982 *UN Convention on the Law of the Sea* (UNCLOS). Negotiated in the 1970s and early 80s, it sought to settle 'all issues relating to the law of the sea' (UNCLOS, preamble), though some matters remained for future development. UNCLOS outlines the jurisdictional limits, rights and obligations of states with respect to maritime activities and the marine environment. Responsibility for developing specific rules and regulations is generally allocated to sector-focused institutions and bodies. As such, a plethora of legal instruments exists, and more are being developed, for the governance and management of our oceans. A recent example is the development of a new instrument to regulate the conservation and sustain-able use of marine biodiversity in areas beyond national jurisdiction (ABNJ) (see Chapter 7). Treaty negotiations are currently underway to develop rules and fill the legal and policy gaps relating to the conservation and sustainable use of marine biodiversity left by UNCLOS (Wright et al. 2018). In addition to treaty law, the legal framework for the oceans is also informed by international case law, as well as existing and emerging customary international law principles and approaches, such as the duty to prevent harm to the marine environment, the precautionary principle, and the ecosystem approach (Tanaka 2015), though implementing them in practice is not always straight forward.

The content of the legal framework for ocean resource management differs, depending on where the resources occur. UNCLOS has, for better or for worse, divided the oceans into different jurisdictional zones, starting from the baseline around a coast and heading seawards: territorial sea, exclusive economic zone, continental shelf, high seas, and the international seabed legally known as the 'Area' (Figure 1).

Natural resources in the first three maritime zones—the territorial sea, the exclusive economic zone, and the continental shelf—are subject to the jurisdiction of the coastal state. The other maritime zones—the high seas and the international seabed 'Area'—are known as ABNJ. Much though by no means all of the deep ocean lies in ABNJ with the natural resources being governed by international law (Figure 2).

Problematically, however, current international law has proven inadequate to protect valuable resources in the deep oceans from human-made stressors, such as overfishing and pollution (Norse et al. 2012; Van Cauwenberghe et al. 2013; Van Dover et al. 2014; Taylor et al. 2016). As our understanding of the importance of the deep oceans increases, and discoveries of valuable deep-sea

Aline Jaeckel, Kristina Gjerde, and Duncan Currie, *The legal framework for resource management in the deep sea* In: *Natural Capital and Exploitation of the Deep Ocean.* Edited by: Maria Baker, Eva Ramirez-Llodra, and Paul Tyler, Oxford University Press (2020). © Oxford University Press.
DOI: 10.1093/oso/9780198841654.003.0003

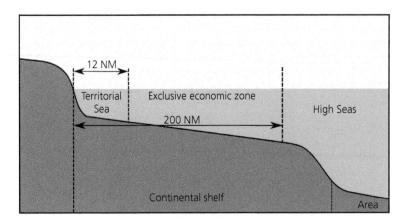

Figure 1 Maritime zones. © Aline Jaeckel

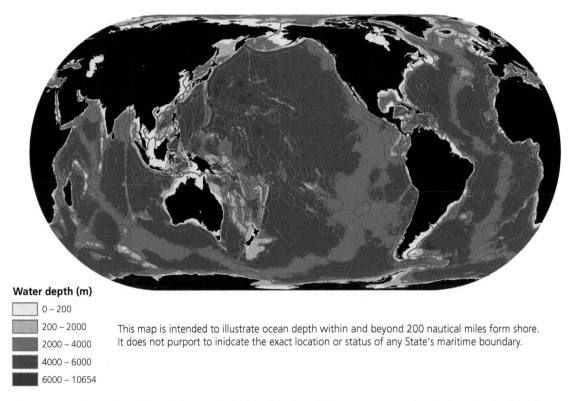

Water depth (m)

- 0 – 200
- 200 – 2000
- 2000 – 4000
- 4000 – 6000
- 6000 – 10654

This map is intended to illustrate ocean depth within and beyond 200 nautical miles form shore. It does not purport to inidcate the exact location or status of any State's maritime boundary.

Figure 2 Ocean depth within and beyond 200 nautical miles from shore. © Les Watling. Image courtesy of Les Watling, University of Hawaii at Manoa (Dec 2019)

resources accelerate, the deficiencies of the legal framework become more and more conspicuous and concerning.

This chapter outlines and discusses the legal framework applicable to natural resources and activities in the deep oceans. It starts by summaris-

ing the legal regime for natural resources in areas under national jurisdiction, including fisheries and hydrocarbons (Section 3.2). It then outlines the specific legal regimes under international law for deep-sea fisheries (Section 3.3.1), prevention of pollution and other environmental harm (Section 3.3.2), deep-seabed mining (Section 3.3.3), marine scientific research (Section 3.3.4), and some remaining gaps in the international legal framework (Section 3.3.5). Next, the chapter offers some brief discussion about the role of scientists in decision-making over marine resource management (Section 3.4). Concluding remarks are offered in Section 3.5.

3.2 National law

Marine resources in areas under national jurisdiction are regulated and managed primarily by the coastal state. All natural resources in the territorial sea, out to 12 nautical miles from the baseline, which is usually the low water line, are subject to regulation and management by the coastal state (UNCLOS, article 2). The same is true for the exclusive economic zone (UNCLOS, article 56), the water column up to 200 nautical miles from the baseline, although here other states have some rights with respect to navigation, overflight, and laying submarine cables (UNCLOS, article 58). Coastal states also have jurisdiction over marine scientific research (UNCLOS, article 56(1)(b)(ii); see Section 3.3.4).

The continental shelf, used here as a legal term and not to be confused with the physical continental shelf around coasts (Chapter 1), describes the seabed and subsoil adjacent to a state to a distance of 200 nautical miles from the baseline. All states have a legal continental shelf, regardless of the topography of the ocean floor beyond their coastlines. If a state's outer continental margin extends further than 200 nautical miles, it is entitled to an extended continental shelf, the outer limits of which are based on a complex formula set out in article 76 of UNCLOS, to a maximum of 350 nautical miles from the baseline or 100 nautical miles from the 2500 metre isobath. The outer limits of the extended continental shelf are approved by the UN Commission on the Limits of the Continental Shelf. The natural resources of the continental shelf, mainly hydrocarbons, minerals, and sedentary species, are also regulated and managed by the coastal state (UNCLOS, article 77).

As a result of these maritime zones and associated rights and obligations of the coastal states,

some of the deep oceans fall in areas under national jurisdiction, especially around islands and regions with narrow continental shelves. Indeed, the extension of coastal states' sovereign rights to the exclusive economic zone and continental shelf has significantly reduced the size of the high seas and Area, bringing more natural resources under the control of coastal states.

Importantly, the protection of the marine environment in areas under national jurisdiction is also regulated by the coastal state (UNCLOS, 56(1)(b)(iii)), although UNCLOS sets out general obligations for marine environmental protection that apply to *all* areas of the oceans (Nordquist et al. 1991). Amongst these is the unequivocal obligation to protect and preserve the marine environment (UNCLOS, article 192) including in the context of exploiting marine resources (UNCLOS, article 193). Moreover, states must 'prevent, reduce and control pollution of the marine environment from any source' subject to their capabilities (UNCLOS, article 194). Particular attention is to be paid to 'rare or fragile ecosystems as well as the habitat of depleted, threatened or endangered species and other forms of marine life' (UNCLOS, article 194(5)). States also have to take 'all measures necessary to ensure that activities under their jurisdiction or control are so conducted as not to cause damage by pollution to other States and their environment' or to ABNJ (UNCLOS, article 194(2)).

The following paragraphs provide a very brief overview of the rights and obligations of states over living and nonliving resources in areas under national jurisdiction, focusing on the exclusive economic zone and continental shelf, as these can capture deep-sea resources depending on the ocean topography adjacent to states.

The management of nonliving resources, including authorising prospecting and exploration for or exploitation of oil, gas, and minerals, falls to the coastal state (UNCLOS, articles 77, 81, 246). However, coastal states are required to have laws and regulations to prevent, reduce, and control pollution from seabed activities, which must be no less effective than relevant international rules, standards, and recommended practices and procedures (UNCLOS, article 208). This would include measures for mineral mining adopted by the International Seabed Authority. It would also include international measures for hydrocarbon production, although no such measures exist: a clear gap in the international law framework (Rochette et al. 2014).

Two exceptions apply to seabed resources on the extended continental shelf (i.e. the continental shelf beyond 200 nautical miles). First, while a coastal state may withhold consent for commercially driven research for natural resources on its continental shelf, it may not do so on the extended continental shelf unless the coastal state is itself exploring or exploiting the seabed resources of the extended continental shelf (UNCLOS, article 246(6)). Here again, the Convention seeks to balance rights of coastal states with the interests of other states to exploit the ocean's natural resources. Second, a percentage of revenue from the extraction of hydrocarbons or minerals on the extended continental shelf must be transferred to the International Seabed Authority, to be shared with other states (UNCLOS, article 82). The criteria for distribution of revenues to states remain undeveloped.

Living resources in the exclusive economic zone and the continental shelf also fall under the jurisdiction of the coastal state, although international law sets framework conditions. Specifically, coastal states have discretion in setting allowable catch limits for fisheries, taking into account best scientific evidence (UNCLOS, article 61). However, they are required to conserve and manage fish stocks to prevent overexploitation while also ensuring optimum utilisation (UNCLOS, article 61, 62). Indeed, the notion of optimum utilisation extends beyond the coastal state. If a coastal state does not have the capacity to harvest the entire allowable catch, it must allow other states to fish the surplus (UNCLOS, article 62). However, coastal states have broad discretion in setting the conditions for fishing in their exclusive economic zones, including licensing of fishing vessels and equipment and regulating seasons and areas of fishing (UNCLOS, article 62(4)). In other words, states can prevent access to the surplus (if any) and thereby implement 'a conservation-friendly approach' (Matz-Lück and Fuchs 2015).

Aside from fisheries, coastal states have rights over marine genetic resources in their exclusive economic zones and on their continental shelf (UNCLOS, articles 56, 77). In many cases, access to marine genetic resources is facilitated via marine scientific research, which in turn requires consent from the coastal state, as discussed below in the section on marine scientific research (Section 3.3.4). The sharing of benefits derived from marine genetic resources is currently not regulated by law of the sea but instead by the 1992 *Convention on Biological Diversity* and its 2010 *Nagoya Protocol on Access to Genetic Resources and the Fair and Equitable Sharing of Benefits Arising from their Utilization*. In brief, a bilateral agreement is required between the provider and the user accessing genetic resources, which must be based on prior informed consent and include mutually agreed terms (Nagoya Protocol, articles 5, 6; CBD, article 15).

3.3 International law

3.3.1 Deep-sea fishing

Deep-sea fishing (Chapter 4) within a state's exclusive economic zone is regulated by the coastal state as set out in Section 3.2, while fisheries beyond national jurisdiction, often referred to as high-seas fisheries, are subject to an inadequate international legal framework. In brief, all states have a right to fish on the high seas, including the deep ocean (UNCLOS, article 116), but they also have a corresponding duty to conserve living resources (UNCLOS, article 117) and to cooperate with other states in the conservation and management of high-seas fisheries (UNCLOS, article 118).

Cooperation is facilitated through regional fisheries management organisations (RFMOs) (see also chapter 4 on deep-sea fisheries). Indeed, the 1995 *United Nations Fish Stocks Agreement* (UNFSA) imposes an obligation on contracting parties to cooperate with and through RFMOs (UNFSA, article 8(3)) and places an obligation on states to establish RFMOs for straddling and highly migratory fish stocks where they do not exist (UNFSA, article 8(5)). If an RFMO or arrangement exists for a particular region or species, states can only access the relevant fisheries if they are a member of the RFMO or arrangement or agree to apply its conservation and management measures (UNFSA, article 8(4)).

Nevertheless, deep-sea bottom fisheries were allowed to advance without RFMOs in place (Wright et al. 2015) and were previously considered not to be economically significant enough to require regulation.

Article 5 of UNFSA introduces specific requirements relevant to deep-sea fishing. Parties are to adopt measures, based on the best scientific evidence available, to ensure long-term sustainability of straddling fish stocks and highly migratory fish stocks and promote the objective of their optimum utilisation. Parties must also apply the precautionary approach (see also UNFSA, article 6), minimise discards and bycatch, protect biodiversity, take measures to prevent or eliminate overfishing, collect and share complete and accurate data, and 'implement and enforce conservation and management measures through effective monitoring, control and surveillance' (UNFSA, article 5). Article 6 requires states to determine stock reference points and action that must be taken if those points are exceeded. However, UNFSA still relies on the concept of maximum sustainable yield, as qualified by relevant environmental and economic factors (UNFSA, article 5(b); UNCLOS, articles 61(3), 119), which is a problematic concept not least because of its focus on an isolated single species population, inconsistent with the ecosystem approach (Legovic et al. 2010; Ghosh and Kar 2013).

RFMOs have a range of functions, including agreeing 'conservation and management measures to ensure the long-term sustainability' of straddling and highly migratory fish stocks, allocating allowable catch quotas, reviewing the status of stocks, distributing best scientific evidence relevant to their fisheries, and establishing monitoring, control, surveillance, and enforcement mechanisms (UNFSA, article 10). RFMOs and other regional bodies which have the legal competence to regulate deep-sea fishing include the Commission for the Conservation of Antarctic Marine Living Resources (CCAMLR), the North Atlantic Fisheries Organization (NAFO), the Northeast Atlantic Fisheries Commission (NEAFC), the Southeast Atlantic Fisheries Organization (SEAFO), the South Pacific Regional Fisheries Management Organization (SPRFMO), the Southern Indian Ocean Fisheries Agreement (SIOFA), the North Pacific Fisheries Commission (NPFC) and the General Fisheries Commission of the Mediterranean (GFCM).

Substantive requirements for bottom-fishing RFMOs are set out in United Nations General Assembly Resolutions. In 2006, the UN General Assembly (UNGA) adopted Resolution 61/105 (2006) to ensure the long-term sustainability of deep-sea fish stocks and protection of vulnerable marine ecosystems (VMEs) from the significant adverse impacts of bottom fisheries. The resolution called for impact assessments to determine whether individual bottom-fishing activities would have significant adverse impacts on VMEs, and to ensure that activities are either managed to prevent these impacts or not authorised to proceed (UNGA 2006, para 83). The resolution also called upon RFMOs to establish 'move-on' rules requiring vessels to cease bottom fishing in areas where VMEs are encountered, and to report the encounter so that appropriate measures can be adopted, and closure of certain areas to bottom fishing (UNGA 2006, para 83).

Details for operationalising these measures were adopted by the Food and Agriculture Organization in its 2008 *International Guidelines for the Management of Deep-sea Fisheries in the High Seas*. Later UN resolutions added other specific requirements, such as calling upon states and RFMOs not to authorise vessels to fish until the measures are adopted and implemented (UNGA 2009, para 120). Following workshops on bottom fisheries, the United Nations called for improved and public assessments (UNGA 2011, para 129) and assessment of cumulative impacts and improved implementation of thresholds and move-on rules (UNGA 2016, para 180). Despite these resolutions, the implementation of the measures in RFMOs has been patchy (Gianni et al. 2011, 2016; see also Chapter 4). A further UN General Assembly workshop is scheduled for 2020.

The effectiveness of fisheries governance by RFMOs has been criticised (Gjerde et al. 2013) because consensus decision-making can mean that necessary decisions are not taken (Molenaar 2004), scientific advice is too often not followed, and sometimes states that do participate can opt out of decisions by lodging a reservation. Moreover, a lack of institutional capacity makes it difficult for RFMOs to take an ecosystem approach instead of a single-stock focus.

More detailed accounts on the international legal regime for deep-sea fishing are available in (Wright et al. 2015; Caddell 2018; Korseberg 2018; Oanta 2018).

3.3.2 Pollution

Marine pollution (Chapter 10) affecting the deep sea is regulated by a diverse array of agreements and bodies. UNCLOS provides the general framework setting forth minimum obligations to be elaborated at the national, regional, and global levels. The challenge is that much of the pollution stems from diffuse land-based sources and activities that are more challenging to regulate than ships at sea. As will be seen, the transboundary and synergistic nature of pollution requires sustained domestic environmental programmes, as well as ambitious bilateral, regional, and global cooperative efforts (Osborn 2015).

As noted, UNCLOS obliges states to take all measures necessary to prevent, reduce, and control pollution from any source, using the best practicable means and in accordance with national capacities (UNCLOS article 194). These measures are to include, amongst others, those designed to minimise to the fullest possible extent:

(a) the release of toxic, harmful, or noxious substances from land-based sources, from or through the atmosphere, or by dumping;
(b) pollution from vessels;
(c) pollution from installations and devices used in exploration or exploitation of natural resources of the seabed and subsoil; and
(d) pollution from other installations and devices operating in the marine environment.

UNCLOS also imposes a duty on states to think before they act by requiring them, inter alia, to ensure that measures to control pollution do not simply transfer damage or hazards from areas to another or transform one type of pollution to another (UNCLOS, article 195).

Other general measures in UNCLOS requiring scientific consideration include a duty on states to monitor the risks or effects of pollution, to keep under surveillance any permitted activities to determine whether such activities are likely to cause pol-lution, and to publish and make available such reports (UNCLOS articles 204, 205). In addition, states are to assess the potential effects of any planned activity under their jurisdiction or control if there are reasonable grounds for believing such activity may cause substantial pollution or significant and harmful changes to the marine environment (UNCLOS, article 206). As there are no accepted definitions of what is considered to be significant or harmful changes, these obligations are difficult to implement in practice, and require further elaboration. For this purpose, UNCLOS does require states to cooperate to establish appropriate scientific criteria for the formulation and elaboration of rules, standards, and recommended practices and procedures for the prevention, reduction, and control of pollution of the marine environment (UNCLOS article 201), and to promote studies, undertake programmes of scientific research, and exchange relevant information and data on pollution of the marine environment (UNCLOS article 200). UNCLOS additionally requires states to provide scientific and technical assistance to developing states including to improve their capacity to prevent and manage pollution and to prepare environmental impact assessments (UNCLOS, article 202). It is hoped that the UN Decade of Ocean Science (2021–2030) can contribute broadly to this under-implemented requirement.

With respect to pollution from land-based sources, the *Global Programme of Action for the Protection of the Marine Environment from Land-based Activities* (GPA), established through the nonbinding 1995 *Washington Declaration*, is the primary global instrument. The GPA does not set global rules or standards but looks instead to the development of national and regional programmes of action. The voluntary nature of the GPA has, according to some (Birnie et al. 2009), undermined its ability to stem problems effectively such as the rising tide of marine litter including plastics and microplastics, eutrophication from human waste, agriculture and burning of fossil fuels, or activities that degrade wetlands, riparian zones, and other coastal and marine environments. The lack of mandatory reporting requirements further hinders the ability to measure progress or to stimulate action to redress lack of progress (Osborn 2015).

In contrast, there are legally binding agreements to deal with specific substances, such as mercury, or broader trade-related issues. Relevant instruments include the 1989 *Convention on the Control of Transboundary Movements of Hazardous Wastes and their Disposal* (Basel Convention), the 1998 *Convention on the Prior Informed Consent Procedure for Certain Hazardous Chemicals and Pesticides in International Trade* (Rotterdam Convention), and the 2001 *Convention on Persistent Organic Pollutants* (Stockholm Convention). The 2013 *Minimata Convention on Mercury* is the most recent and most stringent: it calls for the setting of release limits and the use of best available techniques and best environmental practices in controlling mercury pollution. The Minimata Convention further calls for development of a multipollutant control strategy and international cooperation in the development of harmonised methodologies and the sharing of information regarding mercury pollution (Osborn 2015). This in many ways could serve as a model for what is required for effective pollution control.

Pollution from vessels arises from the discharge of wastes accumulated during ship operations (oily waste, chemicals, sewage, garbage, and air emissions) or the dumping of wastes transported from land to be disposed of at sea. Unlike pollution from land, requirements for polluting activities at sea are quite specific. For dumping, there are two relevant agreements: the 1972 Convention on the Prevention of Marine Pollution by Dumping of Wastes and Other Matter (the London Convention) and the 1996 London Protocol to the London Convention. The original 1972 London Convention takes a permissive approach to regulating the deliberate disposal at sea of wastes or other matter. While matters listed on a 'blacklist' are prohibited, the London Convention allows dumping at sea as long as a permit is obtained from a Contracting Party (London Convention, article IV; VanderZwaag 2015). In contrast, the more recent London Protocol 'is grounded on a precautionary approach to ocean dumping' (VanderZwaag 2015). It only allows the disposal of wastes that are on a global 'accepted list' and only after a detailed waste assessment which considers whether reuse and recycling options are available (London Protocol, annex 2; VanderZwaag 2015). Wastes that can still be dumped include dredged

material (the largest by amount), as well as fish waste, sewage sludge, vessels and platforms, organic materials of natural origin, inert inorganic geological materials, and spoilt cargoes (VanderZwaag 2015).

Operational discharges and other types of impacts from shipping, as well as more general safety aspects, are regulated at the international level by the International Maritime Organization (IMO). Under UNCLOS, coastal states must apply internationally accepted pollution standards set by IMO in their exclusive economic zone (and no higher), whereas they can adopt stricter controls for vessels calling in their home ports or passing through their territorial waters out to 12 nautical miles (UNCLOS, article 211).

One of the key IMO instruments regulating vessel source pollution is the 1973/78 *International Convention on the Prevention of Pollution from Ships* (MARPOL). MARPOL addresses both ship and cargo-generated pollutants in six technical annexes. In short, MARPOL prohibits any discharge of oil, noxious liquid substances, sewage, or garbage at sea unless specific conditions are met including discharge rate and location (Ringbom 2015). To address concerns over cumulative effects of such pollutants in ocean areas with low circulation, the IMO has developed standards for designating 'special areas' where discharges are essentially prohibited. To minimise and monitor the generation of pollution, MARPOL also sets standards for ship construction and monitoring equipment, and authorises sanctions for violations. MARPOL's newest annex VI addresses air pollution from ships including NO_x and SO_x emissions and establishes energy-efficiency standards as an initial step to address greenhouse gas emissions from ships (Ringbom 2015).

More detailed accounts of the international legal regime for marine pollution are available in (Osborn 2015; Ringbom 2015; VanderZwaag 2015; Harrison 2017).

3.3.3 Deep-sea mining

The regulation of ocean mining differs depending on the location of the mineral deposits. Mining oil, gas (Chapter 6), or solid minerals (Chapter 5), including iron sands or diamonds, on the continental shelf is regulated by the coastal state. In contrast, minerals on the deep seabed beyond national jurisdiction are subject to the jurisdiction of the International Seabed Authority (ISA). This section focuses on deep-sea mining on the international seabed, legally known as the 'Area'.

The Area has mineral deposits containing copper, manganese, cobalt, and other metals. The three types of deposits that are currently of most interest are polymetallic nodules that occur on abyssal plains, seafloor massive sulphides at hydrothermal vents, and ferromanganese crusts associated with seamounts (Koschinsky et al. 2018).

The ISA, established by UNCLOS, regulates and controls all exploration for and exploitation of seabed minerals in the Area. However, some of the mandatory requirements under UNCLOS proved to be too controversial for developed states, such as the need to transfer mining technology to an international mining body called the 'Enterprise'. As a result, the particularly controversial aspects of the regime were renegotiated and substantially altered through the 1994 *Agreement Relating to the Implementation of Part XI of the United Nations Convention on the Law of the Sea* (hereafter '1994 Implementing Agreement') (Nandan et al. 2002a). The resulting legal framework is a blend of the original UNCLOS provisions, which were considered a milestone for enabling developing states to share the riches of natural resources, and the 1994 Implementing Agreement, which was driven by industrialised states that sought to place seabed mining on a more commercial footing and minimise notions of wealth redistribution. UNCLOS must be read in conjunction with the 1994 Implementing Agreement, which prevails in case of inconsistencies with Part XI of UNCLOS.

Exploring minerals on the international seabed requires a contract from the ISA (UNCLOS, article 153(3)), which grants exclusive rights to explore minerals in a defined area for 15 years, with extensions possible. Contractors can be public or private entities and need to be sponsored by a state, which in turn can be held liable for environmental harm beyond what is allowed by an exploration contract, unless the state met its due diligence obligation to prevent such harm

(UNCLOS, articles 139, 153(2); Seabed Disputes Chamber 2011).

Interest in mining minerals on the deep seabed has sharply increased in recent years. The ISA has approved twenty-nine exploration contracts, spanning the Pacific Ocean, Indian Ocean, and the Atlantic Ocean (see https://www.isa.org.jm/contractors/exploration-areas). Exploration work has been underway for over 17 years and is regulated by the ISA's exploration regulations for either polymetallic nodules (ISA 2013b), seafloor massive sulphides (ISA 2010), or ferromanganese crusts (ISA 2012). These regulations set out the conditions for obtaining an exploration contract, as well as the obligations of the ISA, sponsoring states, and contractors during the exploration phase. These obligations include the need to apply a precautionary approach and best environmental practices.

The ISA is currently developing the regulatory framework for commercial-scale exploitation of minerals in the Area, which will ultimately enable exploitation to commence. In doing so, the ISA needs to act on behalf of humankind as a whole and give effect to its mandate of administering the Area and its mineral resources as the 'the common heritage of mankind' (UNCLOS, articles 136, 137).

The common heritage principle requires the seabed to be managed in accordance with the following criteria. First, access to minerals is controlled by the ISA (UNCLOS, article 137), and any use of the Area must be for peaceful purposes only (UNCLOS, article 141). Second, the ISA regime must enable the full participation of developing states. To this end, those holding an exploration contract with the ISA have to provide training programmes for scientists and technical personnel from developing states to build scientific capacity in developing states (UNCLOS, articles 143(3), 144, annex III article 15; 1994 Implementing Agreement, annex section 5(1)(c); ISA 2013a). Additionally, developing states have access to so-called 'reserved areas', that is, exploration sites that have been preresearched by other contractors to ensure they have sufficiently large mineral deposits to support a commercial operation (ISA 2013b, regulations 16, 17).

Third, all benefits derived from the Area, and its resources must be shared equitably with both present and future generations (UNCLOS, article 140;

Jaeckel et al. 2016). As such, a proportion of the economic profits derived from mining will have to be shared with the international community via the ISA, likely by way of royalties and/or profit-sharing (UNCLOS, article 140; Lodge et al. 2017). The Area offers crucial ecosystem services to all, with additional economic benefits described in the previous chapter, which links to the next point.

Fourth, the marine environment must be effectively protected from harmful effects of seabed mining for the benefit of present and future generations (UNCLOS, article 145). This is simultaneously the most difficult criterion to realise and the most important one for future generations who may want to benefit from ecosystem services of the deep oceans (Levin et al. 2016; Jaeckel, et al. 2017; Niner et al. 2018). At the same time, sponsoring states and the ISA are both under an obligation to apply the precautionary approach in respect of seabed-mining activities in the Area. The Seabed Disputes Chamber has ruled that the precautionary approach is also an integral part of the general obligation of due diligence of sponsoring States, which is applicable even outside the scope of the Regulations (Seabed Disputes Chamber 2011, para 131).

Marine scientific research plays a key role in identifying environmental standards and indicators, as well as best environmental practice to enable effective environmental management. Importantly, international law confirms the right to conduct marine scientific research freely in the ocean beyond national jurisdiction (UNCLOS, articles 87, 143, 256; ISA 2013b, regulation 1(4)). Nonetheless, potential challenges exist when scientific research by one party affects the mineral exploration activities of another party (Hamann 2015). Moreover, it can be difficult to distinguish exploration activities from marine scientific research, especially when contractors participate in collaborative research projects (Hamann 2015).

The ISA itself is obliged to promote and encourage marine scientific research in the Area, and coordinate and disseminate the results (UNCLOS, article 143(2)). A particular focus is placed on research about the environmental impacts of seabed mining, as well as technology relating to the protection and preservation of the marine environment (1994 Implementing Agreement, annex section

1(5)). Pursuant to this mandate, the ISA has supported several collaborative research projects, such as the KAPLAN project on analysing biodiversity in the Clarion-Clipperton Fracture Zone (Lodge 2008).

More detailed accounts on the international legal regime for deep-seabed mining are available in (Nandan et al. 2002b; Harrison 2017; Jaeckel 2017; Koschinsky et al. 2018).

3.3.4 Marine scientific research

Managing ocean resources starts with understanding the relevant resources and ecosystems and associated communities through marine scientific research. Studying the oceans is one of the aims of the UNCLOS (Preamble). Indeed, states parties to the Convention are required to promote and facilitate marine scientific research (UNCLOS, article 239).

All states have a right to conduct marine scientific research (UNCLOS, article 238) subject to a number of rules, which differ depending on where the research is carried out. Three options exist: (a) research within the coastal waters of a state that is also the flag state of the research vessel, that is, the state in which the vessel is registered; (b) research in the coastal waters of another state; and (c) research on the high seas or the international seabed 'Area'. First, if the research takes place in the coastal waters that are controlled by the same state in which the research vessel is registered (in other words, flag state and coastal state are the same entity) that state sets the rules regarding regulation and authorisation of research.

Second, if the research takes place in the waters under national jurisdiction of another state, it is subject to the rules, regulations, and conditions of, as well as express consent from, the coastal state (UNCLOS, articles 245, 246). However, while scientific research in the territorial sea of a coastal state (12 nautical miles from the baseline) always requires express consent from the coastal state, research in the exclusive economic zone or on the continental shelf of the coastal state shall normally receive consent, if the research is 'pure' research as opposed to applied research (UNCLOS, article 246(3)). Pure research focuses on 'increas[ing] scientific knowledge of the marine environment for the benefit of all mankind' (UNCLOS, article 246(3)), while applied research is generally linked to commercial aims, such as exploration or exploitation of natural resources. However, the distinction between pure and applied research can be unclear in practice.

Researching states have a number of obligations for research in the exclusive economic zone or on the continental shelf of a coastal state, including allowing the coastal state to participate in the research and providing it with the final results and conclusions of the research, as well as access to the research data (UNCLOS, articles 248, 249). Researching states must ensure that the research is conducted in accordance with UNCLOS or face liabilities for contraventions of UNCLOS or for damage caused by pollution arising from marine scientific research (UNCLOS, article 263).

Third, research in the deep ocean that forms part of the high seas or the international seabed Area can be conducted by any state or competent international organisation (UNCLOS, articles 87, 256, 257). States are to promote international cooperation in research, especially research about pollution of the marine environment (UNCLOS, articles 143, 200, 242). In fact, in 2018, the UN General Assembly called upon:

States, individually or in collaboration with each other or with competent international organizations and bodies, to continue to strive to improve understanding and knowledge of the oceans and the deep sea, including, in particular, the extent and vulnerability of deep-sea biodiversity and ecosystems, by increasing their marine scientific research activities in accordance with the Convention. (2018a, para 279)

The ISA can serve as a focal point for deep-sea research (Nandan 2006). Indeed, the ISA can itself conduct research regarding the Area and its resources and it has an obligation to promote and encourage research as well as to coordinate and disseminate research results (UNCLOS, article 143). Importantly, while research on the high seas can be carried out freely, research in the Area, pursuant to UNCLOS article 143(1) must be 'carried out exclusively for peaceful purposes and for the benefit of mankind as a whole [...]'.

The rules on marine scientific research are also directly relevant for bioprospecting of marine genetic resources, which involves the search for and usage of genetic material for commercial purposes (Leary et al. 2009). A key issue under consideration is the extent to which bioprospecting requires separate rules from marine scientific research. UNCLOS does not specifically regulate bioprospecting as it predates the discovery of the economic potential of marine genetic resources (Armas-Pfirter 2009). Nonetheless, some argue that access to marine genetic resources can be largely regulated by existing rules on marine scientific research, given that bioprospecting does not generally involve large-scale extraction of resources but rather small sample sizes of living resources to extract genetic material. However, while research is considered to have intrinsic benefits, the question is whether commercial profits from bioprospecting extracted from marine genetic resources from the high seas and international seabed should be shared internationally. Marine genetic resources extracted from within national waters and seafloor, including the extended continental shelf, are already subject to national regulation under the 2010 *Nagoya Protocol on Access to Genetic Resources and the Fair and Equitable Sharing of Benefits Arising from their Utilization to the Convention on Biological Diversity*. Whether and how benefits might be shared from marine genetic resources derived from areas beyond national boundaries are questions that are now being negotiated by states as part of the new agreement on the conservation and sustainable use of marine biological diversity in ABNJ. These negotiations are summarised in the following section and further discussed in Chapter 7, which provides an in-depth analysis of marine genetic resources and bioprospecting.

More detailed accounts on the international legal regime for marine scientific research are available in (Verlaan 2012; Hubert 2015, 2018; Stephens and Rothwell 2015).

3.3.5 Current gaps in the law

Despite the broad scope of UNCLOS, gaps remain in the legal framework for the deep ocean. The following paragraphs briefly summarise gaps relating to (a) marine biodiversity in ABNJ, and (b) ocean fertilisation.

A key gap in the law is the protection of marine biodiversity in ABNJ (Chapter 6). For example, as outlined in Wright et al. (2018), as of 2019, there is no global framework to establish marine protected areas to safeguard biodiversity, nor does UNCLOS specify procedures or minimum standards for environmental impact assessments or strategic environmental assessments. In addition, there are gaps in the management of high-seas fisheries, the governance of marine activities in ABNJ is highly fragmented, and there are no agreed governance principles for the high seas. Access to marine genetic resources in ABNJ, and any sharing of benefits derived from these resources, is not addressed either.

These gaps have been recognised by the international community following lengthy discussions (UNGA 2015, 2017). Negotiations are being conducted from 2018 to 2020 to develop a new legally binding instrument on the conservation and sustainable use of marine biological diversity in ABNJ. The instrument is focusing on a package of four issues: (a) marine genetic resources, (b) area-based management tools including marine protected areas, (c) environmental impact assessment, and (d) capacity building and technology transfer. It is hoped that the future legal instrument will close many of the current gaps relating to the conservation and sustainable use of marine biodiversity in ABNJ.

A further gap in the law relates to ocean fertilisation, and other forms of geoengineering that may affect the ocean. Ocean fertilisation involves the artificial introduction of iron or other substances into the oceans 'with the principal intention of stimulating primary productivity in the oceans' (2013 Amendments to the London Protocol, Annex 4). Ocean fertilisation is not addressed under UNCLOS beyond the generic requirement of article 195 noted above, which requires states not to transform one type of pollution to another. Nevertheless, the London Convention and Protocol have begun to address ocean fertilisation, starting in 2007 with a Statement of Concern (IMO 2008a) followed by a 2008 resolution in which state parties agreed to a moratorium on ocean fertilisation unless carried out as legitimate marine scientific research (IMO 2008b). The resolution specifically recognizes ocean fertilization as coming under the ambit of the London Convention and Protocol which apply to 'dumping of wastes and the placement of other matter contrary to the objectives of the agreement'. In 2010, state parties followed up with an initial Assessment Framework for Scientific Research Involving Ocean Fertilization (IMO 2010), which required environmental assessment, including risk management and monitoring of any authorised experiments. The Framework required revising or rejecting projects if the risks were deemed unacceptable (IMO 2010, para 4.3). There is no threshold below which experiments are exempt from assessment, meaning the Framework applies regardless of the size or scale of the project.

Parties to the London Protocol put the above on a legally binding footing by adopting the 2013 amendments to the annex of the London Protocol. The amendments, once in force, will prohibit ocean fertilisation unless it is genuine marine scientific research, has been approved by a permit, and has been subject to an environmental impact assessment (EIA), which is not usually a requirement for scientific research (2013 Amendments to the London Protocol, Article 6*bis*). A binding assessment framework to determine whether a particular activity is marine scientific research is set out in annex 5 to the 2013 Amendments. However, this legal framework is not yet in force and will only bind a small group of states (Ginzky and Frost 2014; Scott 2015).

The Parties to the CBD have also sought to address ocean fertilisation and geoengineering more generally and have essentially called for a moratorium on climate-related geoengineering activities, other than legitimate small-scale research. More specifically, in 2010, the CBD Conference of Parties invited states parties to ensure that, in accordance with the precautionary approach, 'no climate-related geo-engineering activities that may affect biodiversity take place, until there is an adequate scientific basis on which to justify such activities' (Conference of the Parties to the Convention on Biological Diversity 2010). Two years later the Conference of the Parties highlighted 'the lack of science-based, global, transparent and effective control and regulatory mechanisms for climate-related geoengineering' (Conference of the Parties to the Convention on Biological Diversity 2012). In summary, while tentative steps have been taken towards restricting climate-related geoengineering that could affect the ocean, including ocean fertilisation, the issue lacks comprehensive regulation.

Global requirements, standards, and procedures for EIA of activities affecting marine biodiversity in ABNJ are also under negotiation as part of the above-mentioned future agreement on marine biodiversity in ABNJ, which is currently being negotiated. A controversial part of the negotiations is whether the EIA provisions should address only activities carried out in ABNJ or those which could affect ABNJ. The outcome will affect which forms of geoengineering might be covered by EIA requirements under the new treaty.

3.4 The role of scientists in ocean governance

Successful diplomacy requires that negotiators have a sound understanding of the relevant science, which itself requires input from scientists who understand and can contribute to the policy process (Moomaw 2018). Indeed, scientists play an important role in informing policy and regulatory development, as well as resource management, through scientific understanding (Marshall et al. 2017).

The role of science and scientists in international law of the sea is deeply rooted in UNCLOS, which requires allowable fish catch to be based on the 'best scientific evidence available' to the states concerned (UNCLOS, article 61(2), 119). Similarly, the UNFSA reinforces the importance of best scientific information, including in relation to precautionary management of fish stocks (UNFSA articles 5(b), 6, 10(f)), while the ISA has recognised the importance of best available scientific and technical information in relation to the potential impacts of deep-seabed mining on VMEs (ISA 2013b, regulation 31(4)).

Scientists influence international law in a number of ways, including through provision of scientific advice to states or to scientific advisory bodies, such as scientific committees of RFMOs. The role of the scientific committee and of its members is typically spelled out in the relevant rules of procedure. Scientists may be contracted directly by a state, RFMO, or the ISA, or they may provide advice for a commercial fishing industry or nongovernmental organisation (NGO) observer. Scientists also contribute to the development of international law and

policy by publishing scientific papers or contributing to international workshops as scientific advisers, such as those commonly held by the ISA, where they can suggest specific environmental management measures. Indeed, in the case of deep-seabed mining, it was the initiative of scientists that eventually led to the adoption of a regional environmental management plan for the Clarion-Clipperton Zone in the central Pacific (Lodge et al. 2014). Scientists can also directly participate in the meetings of some international organisations as accredited observers, such as through the Deep-Ocean Stewardship Initiative, and use this position to provide scientific guidance and commentary to international bodies. This involvement is subject to the rules of procedure of the particular governance body. Involvement of scientists will be particularly important during the UN Decade of Ocean Science for Sustainable Development (2021–2030). Lastly, scientists also contribute to initiatives such as the First Global Integrated Marine Assessment (DOALOS 2015), which examined the state of knowledge of our oceans and the many ways in which we benefit from and affect them. A second assessment is now underway.

Translating scientific advice into law and policy is, of course, not without its challenges. Indeed, too often scientific advice is either not implemented at all or the implementation is severely constrained by political or economic compromises. To revisit the example of the regional environmental management plan for the Clarion-Clipperton Zone, while the final plan established no-mining areas as recommended by scientists, the location of the no-mining areas was changed to avoid overlaps with existing mineral exploration sites. This resulted in 'substantial modifications to the spatial location of the science-based recommendations for the proposed [marine protected area] network' (Wedding et al. 2015), and it compromised the potential effectiveness of the environmental management measure.

Transparency and scientific advice go hand in hand. The annual UN Sustainable Fisheries Resolution in 2018, as in previous years, urges RFMOs to improve transparency and to rely on the best scientific information available (UNGA 2018b, para 166). Similarly, the Rio+20 Outcome document *The Future We Want* resolves to 'promote the science-policy interface through inclusive, evidence-based and transparent scientific assessments' (UNGA 2012, para 76(g)). It also reaffirms the need for effective, transparent, accountable, and democratic institutions (UNGA 2012, para 10) and recognises the importance of science-based fisheries management plans, as well as the contribution of the scientific and technological community to sustainable development (UNGA 2012, paras 48, 168).

Best practice in public participation in environmental decision-making is seen in the European-based 1998 Aarhus Convention. The Almaty Guidelines developed under the Aarhus Convention state that participation of the public 'should be as broad as possible' (UNECE 2005, para 30) and that 'an international forum, or a process within it, should in principle be open to the participation of the public' (UNECE 2005, para 31).

In sum, scientists are involved in developing and implementing international law and policy affecting the deep oceans in a number of ways. Indeed, the importance of scientific advice is underlined in the relevant legal frameworks.

3.5 Conclusion

Human activities affecting the deep oceans are subject to a range of legal rights and obligations, as summarised in this chapter. The vast majority of the deep oceans lie beyond the jurisdiction of coastal states and are governed by international law. UNCLOS sets the legal framework for the protection, use, and exploration of the deep oceans, though it is by no means the only relevant legal instrument. In fact, UNCLOS is supported by a host of other international agreements, including specific agreements on ocean dumping, pollution, and straddling and migratory fisheries.

Despite its goal of settling 'all issues relating to the law of the sea' (UNCLOS, preamble), gaps remained after the entry into force of UNCLOS. Examples include the conservation and sustainable use of marine biodiversity in ABNJ, which is the subject of ongoing negotiations for a new international agreement. A further example is ocean fertilisation through geoengineering, which states parties to the London Protocol on dumping have started to regulate under the auspices of the IMO.

The regulation of individual sectors covered by UNCLOS continues to evolve as well. Deep-sea fisheries are subject to decisions by RFMOs and periodic review by the UN General Assembly, while the regulation of vessel sourced marine pollution is facilitated by the IMO. The rules on deep-seabed mining in the Area continue to be developed by the ISA.

Scientists play an important role not only in the management of the deep oceans but also in the development of rules and regulations pertaining to human activity that affect the deep oceans. They can ensure the legal framework incorporates best scientific advice through serving on scientific advisory bodies to governance institutions, participating in international negotiations as observers, presenting the latest scientific findings at side-events of diplomatic meetings, or acting as advisers to national governments.

Acknowledgements

AJ received funding from the Australian Research Council's Discovery Early Career Researcher Award scheme (grant number: DE190101081). KG would like to thank The Gallifrey Foundation.

References

Armas-Pfirter, F. (2009). How Can Life in the Deep Sea Be Protected? *The International Journal of Marine and Coastal Law*, 24(2), 281–307. doi: 10.1163/157180809X436017.

Birnie, P., Boyle, A., and Redgwell, C. (2009). *International Law and the Environment*. 3rd edn. Oxford: Oxford University Press.

Caddell, R. (2018). International Environmental Governance and the Final Frontier: The Protection of Vulnerable Marine Ecosystems in Deep-Sea Areas beyond National Jurisdiction. *Yearbook of International Environmental Law*, 27(1), 28–63. doi: 10.1093/yiel/yvy002.

Conference of the Parties to the Convention on Biological Diversity (2010). *Biodiversity and Climate Change*. UNEP/CBD/COP/DEC/X/33 (29 October 2010). https://www.cbd.int/doc/decisions/cop-10/cop-10-dec-33-en.pdf.

Conference of the Parties to the Convention on Biological Diversity (2012). *Climate-related Geoengine*. UNEP/CBD/COP/DEC/XI/20 (5 December 2012). https://www.cbd.int/doc/decisions/cop-11/cop-11-dec-20-en.pdf.

DOALOS (2015). *A Regular Process for Global Reporting and Assessment of the State of the Marine Environment, Including Socio-economic Aspects (Regular Process).* https://www.un.org/Depts/los/global_reporting/WOA_RegProcess.htm.

Ghosh, B. and Kar, T. K. (2013). Possible Ecosystem Impacts of Applying Maximum Sustainable Yield Policy in Food Chain Models. *Journal of Theoretical Biology*, 329, 6–14. doi: 10.1016/j.jtbi.2013.03.014.

Gianni, M. et al. (2011). *Unfinished Business: A Review of the Implementation of the Provisions of UNGA Resolutions 61/105 and 64/72 Related to the Management of Bottom Fisheries in Areas Beyond National Jurisdiction.* Deep Sea Conservation Coalition. http://www.savethehighseas.org/wp-content/uploads/2011/09/DSCC_review11.pdf.

Gianni, M. et al. (2016). *How Much Longer Will It Take? A Ten-year Review of the Implementation of United Nations General Assembly Resolutions 61/105, 64/72 and 66/68 on the Management of Bottom Fisheries in Areas Beyond National Jurisdiction.* Deep Sea Conservation Coalition. http://www.savethehighseas.org/publicdocs/DSCC-Review-2016_Launch-29-July.pdf.

Ginzky, H. and Frost, R. (2014). Marine Geo-engineering: Legally Binding Regulation Under the London Protocol. *Carbon & Climate Law Review*, 8(2), 82–96.

Gjerde, K. M. et al. (2013). Ocean in Peril: Reforming the Management of Global Ocean Living Resources in Areas Beyond National Jurisdiction. *Marine Pollution Bulletin*, 74(2), 540–51. doi: 10.1016/j.marpolbul.2013.07.037.

Hamann, K. (2015). Deep Seabed Mining: Regulating Scientific Research in the Area in the Light of Environmental Policy Challenges. Master thesis at Christian-Albrechts-Universität zu Kiel, Germany.

Harrison, J. (2017). *Saving the Oceans Through Law: The International Legal Framework for the Protection of the Marine Environment.* Oxford: Oxford University Press. doi: 10.1093/law/9780198707325.001.0001.

Hubert, A.-M. (2015). Marine Scientific Research and the Protection of the Seas and Oceans, in R. Rayfuse (ed.) *Research Handbook on International Marine Environmental Law.* Cheltenham: Edward Elgar, pp. 313–36.

Hubert, A-M. (2018). Marine Scientific Research, in M. Salomon and T. Markus (eds) *Handbook on Marine Environment Protection.* New York: Springer, pp. 933–51.

IMO (2008a). *Ocean Fertilization: Report of the Legal and Intersessional Correspondence Group on Ocean Fertilization (LICG).* LC 30/4 (25 July 2008). http://ceassessment.org/wp-content/uploads/2014/07/Report-of-the-Legal-and-Intersessional-Correspondence-Group-on-Ocean-Fertilization-LICG-2008.pdf.

IMO (2008b). *Resolution LC-LP.1(2008) on the Regulation of Ocean Fertilization.* Lc-LP.1(2008). https://web.whoi.edu/ocb-fert/wp-content/uploads/sites/100/2017/07/OF_Resolution_at_LC_30_-_LP_3_56339.pdf.

IMO (2010). *Resolution LC-LP.2(2010) on the Assessment Framework for Scientific Research Involving Ocean Fertilization.* LC-LP.2(2010) (14 October 2010). http://www.imo.org/blast/blastDataHelper.asp?data_id=31100&filename=2010resolutiononAFOF.pdf.

ISA (2010). Regulations on Prospecting and Exploration for Polymetallic Sulphides in the Area. ISBA/16/A/12/Rev.1 (7 May 2010). https://www.isa.org.jm/documents/isba16a12-rev-1.

ISA (2012). Regulations on Prospecting and Exploration for Cobalt-rich Ferromanganese Crusts in the Area. ISBA/18/A/11 (22 October 2012). https://www.isa.org.jm/documents/isba18a11.

ISA (2013a). Recommendations for the Guidance of Contractors and Sponsoring States Relating to Training Programmes under Plans of Work for Exploration Issued by the Legal and Technical Commission. ISBA/19/LTC/14 (12 July 2013). https://www.isa.org.jm/documents/isba19ltc14.

ISA (2013b). Regulations on Prospecting and Exploration for Polymetallic Nodules in the Area. ISBA/19/C/17 (22 July 2013). https://www.isa.org.jm/documents/isba19c17.

Jaeckel, A., Ardron, J. A., and Gjerde, K. M. (2016). Sharing Benefits of the Common Heritage of Mankind—Is the Deep Seabed Mining Regime Ready? *Marine Policy*, 70, 198–204. doi: 10.1016/j.marpol.2016.03.009.

Jaeckel, A., Gjerde, K. M., and Ardron, J. A. (2017). Conserving the Common Heritage of Humankind—Options for the Deep-seabed Mining Regime. *Marine Policy*, 78, 150–7. doi: 10.1016/j.marpol.2017.01.019.

Jaeckel, A. L. (2017). *The International Seabed Authority and the Precautionary Principle.* Leiden, Netherlands: Brill Nijhoff.

Korseberg, L. (2018). The Law-making Effects of the FAO Deep-Sea Fisheries Guidelines. *International and Comparative Law Quarterly*, 67(4), 801–32. doi: 10.1017/s0020589318000192.

Koschinsky, A. et al. (2018). Deep-sea Mining: Interdisciplinary Research on Potential Environmental, Legal, Economic, and Societal Implications. *Integrated Environmental Assessment and Management*, 14(6), 672–91 doi: 10.1002/ieam.4071.

Leary, D., Vierros, M., Hamon, G., Arico, S., and Mongale, C. (2009). Marine Genetic Resources: A Review of Scientific and Commercial Interest. *Marine Policy*, 33(2), 183–94. doi: 10.1016/j.marpol.2008.05.010.

Legovic, T., Klanjšcek, J., and Gecek, S. (2010). Maximum Sustainable Yield and Species Extinction in Ecosystems.

Ecological Modelling, 221(12), 1569–74. doi: 10.1016/j.ecolmodel.2010.03.024.

Levin, L. A. et al. (2016). Defining 'Serious Harm' to the Marine Environment in the Context of Deep-seabed Mining. *Marine Policy*, 74, 245–59. doi: 10.1016/j.marpol.2016.09.032.

Lodge, M. W. (2008). Collaborative Marine Scientific Research on the International Seabed. *Journal of Ocean Technology*, 3, 30–6. http://www.isa.org.jm/files/documents/EN/efund/JOT-article.pdf.

Lodge, M., Johnson, D., Le Gurun, G., Wengler, M., Weaver, P., and Gunn, V. (2014). Seabed Mining: International Seabed Authority Environmental Management Plan for the Clarion–Clipperton Zone. A Partnership Approach. *Marine Policy*, 49, 66–72. doi: 10.1016/j.marpol.2014.04.006.

Lodge, M. W., Segerson, K., and Squires, D. (2017). Sharing and Preserving the Resources in the Deep Sea: Challenges for the International Seabed Authority. *International Journal of Marine and Coastal Law*, 32(3), 427–57. doi: 10.1163/15718085-12323047.

Marshall, N. et al. (2017). Empirically Derived Guidance for Social Scientists to Influence Environmental Policy. *PLoS ONE*, 12(3), 1–9. doi: 10.1371/journal.pone.0171950.

Matz-Lück, N. and Fuchs, J. (2015). Marine Living Resource, in D. Rothwell et al. (eds) *The Oxford Handbook of the Law of the Sea*. Oxford: Oxford University Press, pp. 491–515. doi: 10.1093/law/9780198715481.003.0022.

Molenaar, E. J. (2004). Unregulated Deep-sea Fisheries: A Need for a Multi-level Approach. *International Journal of Marine and Coastal Law*. 19(3), 223–46. doi: 10.1163/1571808042886048.

Moomaw, W. R. (2018). Scientist Diplomats or Diplomat Scientists: Who Makes Science Diplomacy Effective? *Global Policy*, 9, 78–80. doi: 10.1111/1758-5899.12520.

Nandan, S. (2006). Administering the Mineral Resources of the Deep Seabed, in D. Freestone, R. Barnes, and D. Ong (eds.) *The Law of the Sea: Progress and Prospects*. Oxford: Oxford University Press, pp. 75–92. doi: 10.1093/acprof:oso/9780199299614.001.0001.

Nandan, S. N., Lodge, M. W., and Rosenne, S. (2002a). *The Development of the Regime for Deep Seabed Mining*. International Seabed Authority. http://www.isa.org.jm/files/documents/EN/Pubs/Regime-ae.pdf.

Nandan, S. N., Lodge, M. W., and Rosenne, S. (2002b). *United Nations Convention on the Law of the Sea, 1982: A Commentary, Volume VI*. Leiden, Netherlands: Martinus Nijhoff Publishers. http://books.google.com/books?id=3HoWAQAAIAAJ&pgis=1.

Niner, H. J. et al. (2018). Deep-Sea Mining with No Net Loss of Biodiversity—An Impossible Aim. *Frontiers in Marine Science*, 5, 53. doi: 10.3389/fmars.2018.00053.

Nordquist, M. H. et al. (1991). *United Nations Convention on the Law of the Sea, 1982: A Commentary, Volume IV*. Leiden, Netherlands: Martinus Nijhoff Publishers. http://books.google.com/books?id=4zS6fOrs2LAC&pgis=1.

Norse, E. A. et al. (2012). Sustainability of Deep-sea Fisheries. *Marine Policy*, 36(2), 307–20. doi: 10.1016/j.marpol.2011.06.008.

Oanta, G. A. (2018). International Organizations and Deep-sea Fisheries: Current Status and Future Prospects. *Marine Policy*, 87, 51–9. doi: 10.1016/j.marpol.2017.09.009.

Osborn, D. (2015). Land-based Pollution and the Marine Environment, in R. Rayfuse (ed.) *Research Handbook on International Environmental Law*. Cheltenham: Edward Elgar, pp. 81–104.

Ringbom, H. (2015). Vessel-source Pollution, in R. Rayfuse (ed.) *Research Handbook on International Environmental Law*. Cheltenham: Edward Elgar, pp. 105–31.

Rochette, J., Wemaëre, M., Chabason, L., and Callet, S. (2014). Seeing Beyond the Horizon for Deepwater Oil and Gas: Strengthening the International Regulation of Offshore Exploration and Exploitation. *IDDRI, Study n° 01/14*, pp. 1–36. https://www.iddri.org/sites/default/files/import/publications/st0114_jr-et-al._offshore-en.pdf.

Scott, K. N. (2015). Geoengineering and the Marine Environment, in R. Rayfuse (ed.) *Research Handbook on International Marine Environmental Law*. Cheltenham: Edward Elgar, pp. 451–72.

Seabed Disputes Chamber (2011). *Responsibilities and Obligations of States Sponsoring Persons and Entities with Respect to Activities in the Area (Advisory Opinion)*. http://www.itlos.org/fileadmin/itlos/documents/cases/case_no_17/adv_op_010211.pdf.

Stephens, T. and Rothwell, D. (2015). Marine Scientific Research, in D. Rothwell et al. (eds) *The Oxford Handbook of the Law of the Sea*. Oxford: Oxford University Press, pp. 559–81. doi: 10.1093/law/9780198715481.003.0025.

Tanaka, Y. (2015). Principles of International Marine Environmental Law, in R. Rayfuse (ed.) *Research Handbook on International Marine Environmental Law*. Cheltenham: Edward Elgar, pp. 31–56. doi: 10.4337/9781781004777.00009.

Taylor, M. L., Gwinnett, C., Robinson, L. F., and Woodall, L. C. (2016). Plastic Microfibre Ingestion by Deep-sea Organisms. *Scientific Reports*, 6, 1–9. doi: 10.1038/srep33997.

UNECE (2005). *Almaty Guidelines: Decision II/4: Promoting the Application of the Principles of the Aarhus Convention in International Forums*. ECE/MP.PP/2005/2/Add.5 (20 June 2005). http://www.unece.org/fileadmin/DAM/env/documents/2005/pp/ece/ece.mp.pp.2005.2.add.5.e.pdf.

UNGA (2006). *Sustainable Fisheries, Including Through the 1995 Agreement for the Implementation of the Provisions of the United Nations Convention on the Law of the Sea of 10 December 1982 Relating to the Conservation and Management of Straddling Fish Stocks and Highly Migratory Fish Stocks, and Related Instruments.* UN Doc A/RES/61/105 (8 December 2006). https://undocs.org/en/A/RES/61/105.

UNGA (2009). *Sustainable Fisheries, Including Through the 1995 Agreement for the Implementation of the Provisions of the United Nations Convention on the Law of the Sea of 10 December 1982 Relating to the Conservation and Management of Straddling Fish Stocks and Highly Migratory Fish Stocks, and Related Instruments.* UN Doc A/RES/64/72 (4 December 2009). https://undocs.org/en/A/RES/64/72.

UNGA (2011). *Sustainable Fisheries, Including Through the 1995 Agreement for the Implementation of the Provisions of the United Nations Convention on the Law of the Sea of 10 December 1982 Relating to the Conservation and Management of Straddling Fish Stocks and Highly Migratory Fish Stocks, and Related Instruments.* UN Doc A/RES/66/68 (6 December 2011). https://undocs.org/en/a/res/66/68.

UNGA (2012). *The Future We Want.* UN Doc A/RES/66/288 (27 July 2012). https://undocs.org/en/A/RES/66/288.

UNGA (2015). *Development of an International Legally Binding Instrument Under the United Nations Convention on the Law of the Sea on the Conservation and Sustainable Use of Marine Biological Diversity Of Areas Beyond National Jurisdiction.* UN Doc A/RES/69/292 (19 June 2015). https://undocs.org/en/a/res/69/292.

UNGA (2016). *Sustainable Fisheries, Including Through the 1995 Agreement for the Implementation of the Provisions of the United Nations Convention on the Law of the Sea of 10 December 1982 Relating to the Conservation and Management of Straddling Fish Stocks and Highly Migratory Fish Stocks, and Related Instruments.* UN Doc A/RES/71/123 (7 December 2016). https://undocs.org/en/A/RES/71/123.

UNGA (2017). *International Legally Binding Instrument Under the United Nations Convention on the Law of the Sea on the Conservation and Sustainable Use of Marine Biological Diversity of Areas Beyond National Jurisdiction.* UN Doc A/RES/72/249 (24 December 2017). https://undocs.org/en/A/RES/72/249.

UNGA (2018a). *Oceans and the Law of the Sea.* UN Doc A/RES/73/124 (11 December 2018).

UNGA (2018b). *Sustainable Fisheries, Including Through the 1995 Agreement for the Implementation of the Provisions of the United Nations Convention on the Law of the Sea of 10 December 1982 Relating to the Conservation and Management of Straddling Fish Stocks and Highly Migratory Fish Stocks, and Related Instruments.* UN Doc A/RES/73/125. https://undocs.org/en/A/RES/73/125.

Van Cauwenberghe, L., Van Reusel, A., Mees, J., and Janssen, C.R. (2013). Microplastic Pollution in Deep-sea Sediments. *Environmental Pollution*, 182, 495–9. doi: 10.1016/j.envpol.2013.08.013.

VanderZwaag, D. L. (2015). The International Control of Ocean Dumping: Navigating from Permissive to Precautionary Shores, in R. Rayfuse (ed.) *Research Handbook on International Environmental Law.* Cheltenham: Edward Elgar, pp. 132–47.

Van Dover, C. L. et al. (2014). Ecological Restoration in the Deep Sea: Desiderata. *Marine Policy*, 44, 98–106. doi: 10.1016/j.marpol.2013.07.006.

Verlaan, P. (2012). Marine Scientific Research: Its Potential Contribution to Achieving Responsible High Seas Governance. *The International Journal of Marine and Coastal Law*, 27(4), 805–12. doi: 10.1163/15718085-12341260.

Wedding, L. M. et al. (2015). Managing Mining of the Deep Seabed. *Science*, 349(6244), 144–5. doi: 10.1126/science.aac6647.

Wright, G., Ardon, J., Gjerde, K., Currie, D., and Rochette, J. (2015). Advancing Marine Biodiversity Protection Through Regional Fisheries Management: A Review of Bottom Fisheries Closures in Areas Beyond National Jurisdiction. *Marine Policy*, 61, 134–48. doi: 10.1016/j.marpol.2015.06.030.

Wright, G. Rochette, J., Gjerde, K., and Seeger, I. (2018). *The Long and Winding Road: Negotiating a Treaty for the Conservation and Sustainable Use of Marine Biodiversity in Areas Beyond National Jurisdiction.* IDDRI Study 8/18. https://www.iddri.org/sites/default/files/PDF/Publications/Catalogue Iddri/Etude/201808-Study_HauteMer-long and winding road.pdf.

International agreements cited

1972 London Convention: Convention on the Prevention of Marine Pollution by Dumping of Wastes and Other Matter (adopted 29 December 1972, entered into force 30 August 1975) 1046 UNTS 138.

1973 MARPOL Convention: International Convention for the Prevention of Pollution from Ships (adopted 2 November 1973, entered into force 2 October 1983) and Its Protocol of 1978 (adopted 17 February 1978, entered into force 1 October 1983) 1340 UNTS 62.

1982 United Nations Convention on the Law of the Sea (adopted 10 December 1982, entered into force 16 November 1994) 1833 UNTS 3.

1989 Basel Convention: Convention on the Control of Transboundary Movements of Hazardous Wastes and their Disposal (adopted 22 March 1989, entered into force 5 May 1992) 1673 UNTS 126.

1992 Convention on Biological Diversity (adopted 5 June 1992, entered into force 29 December 1993) 1760 UNTS 79.

1994 Agreement Relating to the Implementation of Part XI of the United Nations Convention on the Law of the Sea (adopted 28 July 1994, entered into force 28 July 1996) 1836 UNTS 3.

1995 Agreement for the Implementation of the Provisions of the United Nations Convention on the Law of the Sea of 10 December 1982 relating to the Conservation and Management of Straddling Fish Stocks and Highly Migratory Fish Stocks (adopted 4 August 1995, entered into force 11 December 2001) 2167 UNTS 3.

1995 Washington Declaration on Protection of the Marine Environment from Land-based Activities (adopted 1 November 1995) https://wedocs.unep.org/handle/20.500.11822/13421.

1996 London Protocol: Protocol to the Convention on the Prevention of Marine Pollution by Dumping of Wastes and Other Matter (adopted 7 November 1996, entered into force 24 March 2006) 36 ILM 1 (1997).

1998 Rotterdam Convention: Convention on the Prior Informed Consent Procedure for Certain Hazardous Chemicals and Pesticides in International Trade (adopted 10 September 1998, entered into force 24 February 2004) 2244 UNTS 337.

1998 Aarhus Convention: Convention on Access to Information, Public Participation in Decision-making and Access to Justice in Environmental Matters, 25 June 1998, 2161 UNTS 447.

2001 Stockholm Convention: Convention on Persistent Organic Pollutants (adopted 22 May 2001, entered into force 17 May 2004) 2256 UNTS 119.

2010 Nagoya Protocol on Access to Genetic Resources and the Fair and Equitable Sharing of Benefits Arising from Their Utilization to the Convention on Biological Diversity (adopted 29 October 2010, entered into force 12 October 2014) UNEP/CBD/COP/DEC/X/1.

2013 Minimata Convention on Mercury (adopted 10 October 2013, entered into force 16 August 2017) 55 ILM (2016).

2013 Amendments to the London Protocol: Resolution LP.4(8) on the Amendment to the London Protocol to the Regulate the Placement of Matter for Ocean Fertilization and Other Marine Geoengineering Activities (adopted 18 October 2013, not yet in force).

CHAPTER 4

Exploitation of deep-sea fishery resources

Les Watling, Lissette Victorero, Jeffrey C. Drazen, and Matthew Gianni

4.1 The development of deep-sea fisheries

One of the greatest resources of the ocean for humankind are fisheries (by definition including fish and other animals, such as crustaceans and mollusc but not reptiles or mammals), which are estimated to provide nutrition for 3.2 billion people with an industry worth an estimated US$152 billion (FAO 2018). In the deep sea, fisheries are primarily promulgated between 200 and 1800 m depth, with approximately 300 fish species being exploited (Priede 2017), either as target species or as bycatch, the latter being caught unintentionally (Figure 1). While shallow-water industrial fishing and artisanal deep-sea fishing can be traced back hundreds of years, industrial deep-sea fisheries developed primarily after the 1950s as continental shelf stocks declined and became more tightly regulated (Koslow et al. 2000). The venturing into deeper waters was spurred by technological advances, such as the development of large and heavy trawl gear and winches, advanced navigational capabilities, and the capacity to catch, process, and freeze the fish off-shore (Merrett and Haedrich 1997). Additional key factors associated with the commercial exploitation of deep-sea fish stocks were the absence of regulations and lack of knowledge about the biological characteristics of the target species and their surrounding ecosystem.

In the early years of the industry, the first official landings from deep-sea fisheries were relatively small (< 10,000 tonnes) as there were few target species of high value, such as the Greenland halibut and blue ling. Other fish caught were discarded as bycatch. The early leader of deep-sea fishing was the Soviet Union, even developing a research programme specialising in fishing techniques below 500 m (Priede 2017). Over time, the industry continued to expand, and in 1967, one of the first large-scale deep-sea fishing events occurred on the high seas when the Soviet Union trawled the Hawaiian-Emperor seamount chain, obtaining catches of 145,000 tonnes of pelagic armorhead in the first years of the fishery (Clark et al. 2007). Awarded with this initial success, the 'gold rush' for the deep-sea had begun. Fishing vessels repeatedly encountered high biomass, mainly around topographic highs, such as seamounts, of which there are many in the deep sea and along the upper continental slopes at mid to high latitudes. In particular, fish species in the upper bathyal waters were found to be exploitable and marketable, since these species occurred in high abundance and resembled their shelf relatives in terms of size and flesh, while the deeper species displayed bad-tasting watery flesh and unattractive appearance.

However in less than a decade, after the initial success, the pelagic armorhead fishery had collapsed due to overexploitation, and even now, over 50 years later, the stocks have not recovered (Kiyota et al. 2015). This event provided the first glimpse into the boom-and-bust cycle of deep-sea fisheries, a pattern that is now well-recognised. A more modern

Les Watling, Lissette Victorero, Jeffrey C. Drazen, and Matthew Gianni, *Exploitation of deep-sea fishery resources* In: *Natural Capital and Exploitation of the Deep Ocean.* Edited by: Maria Baker, Eva Ramirez-Llodra, and Paul Tyler, Oxford University Press (2020). © Oxford University Press.
DOI: 10.1093/oso/9780198841654.003.0004

Figure 1   Representative images of deep-sea fish and fisheries. A. Blue antimora (*Antimora rostrata*), Kelvin Seamount. B. Blue Ling (*Molva dypterygia*) off Norway. C. Roundnose grenadier (*Coryphaenoides rupestris*) Manning Seamount. D. Giant grenadier (*Albatrossia pectoralis*) Emperor seamounts. E. Pale chimaera (*Hydrolagus pallidus*) Bear seamount. F. False boarfish (*Neocyttus helgae*) Corner Rise seamounts. G. bottom otter trawl catch of grenadiers and thornyheads, miscellaneous other species, from 1200 m, Monterey Bay, California. H. orange roughy (*Hoplostethus atlanticus*) from bottom trawl catch off New Zealand. Photo credits: A, C, E, F, courtesy of NOAA OE/URI/Les Watling. B, courtesy of MAREANO-Institute of Marine Research, Norway. D, courtesy of Schmidt Ocean Institute/Les Watling. G, courtesy of Michelle Kay. H, courtesy of Claire Nouvian/Bloom Association.

example of a boom-and-bust fishery is the round-nose grenadier, which in Norwegian waters lasted only 5 years (2001–2006) before being subjected to a moratorium (ICES 2016). Despite the exploitation levels leading to a rapid depletion of many stocks, the industry kept expanding for several decades by exploring new fishing grounds and targeting new species. This expansion, in both fishing depth and spatial extent, was able to mask a pattern of decline for individual deep-sea fisheries, since catches appeared to either remain stable or increase with time. Moreover, management was continuously lagging behind, and the proposed regulations were based on assumptions from shallow-water fisheries since there was little understanding about the biology and life history of deep-sea species.

4.1.2 Deep-sea fishing methods

Deep-sea fisheries use methods adapted from shallow fisheries, such as gill nets, pots, tangle nets, or more specialised deep-sea fishing techniques, such as bottom longlines and otter trawls, to capture the target species of fish and crustaceans (Clark and Koslow 2007; FAO 2008). Gillnets, usually made from monofilament, use a floating headline anchored at each end and a bottom-weighted line, which can drag along the seafloor. Most swimming fish that are not small enough to swim through the mesh will be caught. Pots are used to catch crustaceans, primarily deep-sea red crabs in the Atlantic, and lithodid crabs along the Aleutian Islands, and some fish species (eels and hagfish). The pots can be large, 2 m or more across, and are weighted to keep them on the bottom. In the crab fishery, generally a number of pots are connected together on one line marked by a surface buoy. Tangle nets are used in limited areas, particularly on the sides of seamounts, to retrieve corals for jewellery making. This gear typically consists of a large iron bar to which are tied large pieces of frayed ropes that would snag coral colonies. On the Emperor seamounts, tangle nets were deployed to obtain *Corallium* species or bamboo corals during the years 1965–1992.

Bottom longlines comprise a long setline anchored to the bottom and are furnished with around 1000 hooks, and weights to keep the line on the bottom. The hooks are baited in an effort to catch scavenging or predaceous fish species (Priede et al. 2010). Longline fisheries target a select number of species, such as the Patagonian toothfish, since most species of deep-sea fish are not attracted to bait (Yeh and Drazen 2011). Trawl gear, such as otter trawls, are the most ubiquitous of all gear used to capture deep-sea fish and crustaceans. An otter trawl consists of a large net connected to metal doors by steel cables, which can weigh up to 5000 kg, with the trawl spanning a total width of 80–200 m (Merrett and Haedrich 1997). The footrope of the trawl is armoured with large bobbins, rollers, or steel discs, which can weigh several tonnes, allowing the trawl to be hauled across rough bottom. When used for midwater trawling, where bottom contact might be intermittent, the footrope is equipped with chains and weights. The top opening of the net is lined with floats to keep the net from collapsing. During bottom trawling operations, it is generally the doors, the cables connecting the doors to the net, and the footrope that drag across the seafloor, often for several kilometres, thus the extent of the seafloor that is damaged by the trawl is generally equal to the distance between the doors and the distance trawled—typically several kilometres.

4.1.3 The footprint of deep-sea fisheries

Deep-sea fishing activities occur within the Exclusive Economic Zone (EEZ) waters of nations and on the high seas over the continental slope, ridges, banks, and seamounts. Throughout the past 65 years, deep-sea fisheries are estimated to have caught approximately 109–142 million tonnes of fish (Table 1) for use as food, fish meal, and oil. The pattern of catches since 1950 are illustrated in Figure 2, which also depicts the amount of unregulated landings and discarded fish. Historically, unregulated landings for deep-sea fish species correspond to fishermen discovering a new species to exploit, which was void of regulations. However, once the industry was

Table 1 World total fisheries catch and deep-sea demersal fishery catch as reported to the Food and Agriculture Organization of the United Nations and as estimated by the Sea Around Us research initiative

	Reported Catch (FAO) t	Total Catch (FAO + SAU) t
Global fisheries landings (or catch)	3,916,784,240	5,183,437,078
Deep-sea fisheries catch >400 m	14,397,146	24,905,883
Portion of global catch	0.37%	0.48%
Deep-sea fisheries catch 200–400 m	109,699,516	141,958,982
Portion of global catch	2.8%	2.7%
Total deep-sea fisheries catch	3.2%	3.2%

Abbreviations: FAO, Food and Agriculture Organization of the United Nations; SAU, Sea Around Us; t, metric tonnes.

established, due to economic incentives, unregulated landings and illegal catches resulted in the total catch exceeding the quota assigned to certain stocks. The high numbers of discards associated with deep-sea fishing in some areas of the world are characteristic of bottom trawling as this gear catches nontarget species, undersized fish, or species that have no quota remaining (Zeller et al. 2017). Within deep-sea fisheries, there is an apparent distinction between the upper depth fisheries (~ 200–400 m) and deeper ones (>400 m) (Figure 2). Catches from shallower depths are about six-times greater than the deeper catches, illustrating that the deeper fisheries are generally less productive (Watling et al. forthcoming).

Globally, the leading trawling nations include Japan, Russia, New Zealand, Iceland, South-Korea, Spain, and France. The largest spatial footprint of

deep-sea fisheries has historically been focused in the NE Atlantic and SW Pacific Food and Agriculture Organization of the United Nations (FAO) fishing regions (Table 2). For shallower deep-sea fisheries, the SE Atlantic and Pacific have also provided catches of high magnitude, while for deeper species the NW Pacific represents an important fishing region (Table 2). Within these general areas, the most problematic fishing in terms of regulations and habitat destruction occurs on the high seas and seamounts, and in deep coral-reef areas along the continental slope.

In the SW Pacific, the fisheries target mainly orange roughy, alfonsinos, and dories, which form feeding and spawning aggregations around seamounts, rises and ridges. These topographic highs are a prominent feature of the SW Pacific, thus providing a large surface area for deep-sea fishing with

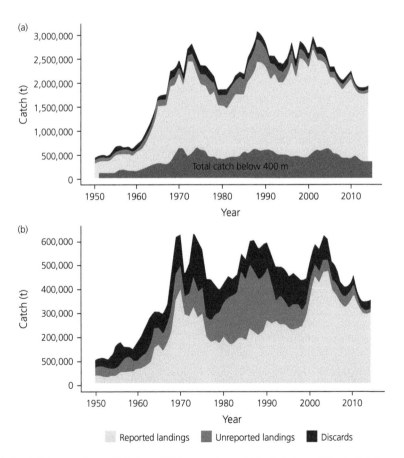

Figure 2 Catch data for a) all deep-sea demersal fisheries and b) deep-sea demersal fisheries below ~ 400m depth between years 1950–2015. Reported landings correspond to data from the Food and Agriculture Organization of the United Nations. Discards and unreported landings are based on reconstructed fisheries data from the Sea Around Us research initiative. Note difference in scales.

Table 2 Estimated total catches (tonnes) of deep-sea demersal fishes from deep (>400 m) and shallower (200–400 m) parts of Food and Agriculture Organization of the United Nations reporting areas Data based on the Food and Agriculture Organization of the United Nations reported landings and the reconstructed fisheries data from Sea Around Us research initiative.

FAO Fishing Region	Estimated Total Catch >400 m	Estimated Total Catch 200–400 m
Arctic Sea	12,617	
Atlantic, Antarctic	95,223	73
Atlantic, Eastern Central	134,600	638,185
Atlantic, Northeast	8,862,807	18,585,779
Atlantic, Northwest	3,749,962	14,061,792
Atlantic, Southeast	140,928	32,999,954
Atlantic, Southwest	586,350	15,773,859
Atlantic, Western Central	2368	6848
Indian Ocean, Antarctic	359,503	3083
Indian Ocean, Eastern	99,149	499,176
Indian Ocean, Western	36,109	4688
Mediterranean and Black Sea	256,111	1,479,343
Pacific, Antarctic	42,857	12
Pacific, Eastern Central	2175	372,398
Pacific, Northeast	308,787	10,420,560
Pacific, Northwest	5,639,395	2,487,274
Pacific, Southeast	287,715	25,655,253
Pacific, Southwest	4,281,410	18,970,010
Pacific, Western Central	7817	694

Abbreviation: FAO, Food and Agriculture Organization of the United Nations.

seamounts yielding up to 60 per cent of the catch in New Zealand (Clark and O'Driscoll 2003). However, there is a pattern of serial depletion of stocks from hills, pinnacles, and seamounts (Clark 1999; Clark et al. 2000). New Zealand's orange roughy fishery was first established on the Chatham Rise and adjacent seamounts. In the early years, catches from this area reached more than 50,000 tonnes, but as catches decreased dramatically, the fishery expanded to the Louisville Ridge seamount chain. Subsequently, it became clear that the roughy matures at the advanced age of 20–30 years and lives approximately 125 years (Andrews et al. 2009), making

them extremely vulnerable to overexploitation. Another prominent trawl fishery in the SW Pacific is that of the blue grenadier, often referred to as hoki, a species that was originally caught on the continental slope of New Zealand by Japanese trawlers who landed 99,000 tonnes in 1977. Once the New Zealand EEZ was established by the late 1980s, the fishery became nationalised with catches reaching 325,000 tonnes in 1998. Shortly after, an impoverished stock status was identified by management, which led to reduced Total Allowable Catches (TACs) in 2003 and 2007, ensuring the longevity of the fishery.

In the temperate regions, deep-sea fisheries are sustained by high primary productivity. In the North Atlantic, long-lining and then trawling for the Greenland halibut has culminated in the longest lasting high biomass deep-sea fishery (Victorero et al. 2018). The Greenland halibut fishery is very dynamic, with countries entering and leaving the fishery over the years, and producing management clashes, such as the 'turbot wars'—an international fishing dispute in which Canada reported illegal overfishing from the Spanish trawling fleet (Haedrich et al. 2001) and ended up seizing a Spanish fishing trawler. The Greenland halibut is also fished within the Barents Sea, which is a hotspot for shallow-water fisheries. Historically deepsea fish have been caught as bycatch within the prominent cod and haddock fisheries (Little et al. 2015).

Finally, it is worth considering the economical footprint of deep-sea fisheries. The economic importance of deep-sea fisheries is typically local, concerning a few nations at a certain place in time. For example, in 2010 Spain and France accounted, mainly through trawling, for approximately 75 per cent of the total European Union deep-sea catches from the NE Atlantic, equivalent to $90 million (Pew Environment Group 2012). In 2018 New Zealand's orange roughy fishery had an estimated asset value of $295 million (New Zealand Government 2019). In a global context, however, the catches over the period from 1950–2015 for deepsea fisheries only amounted to approximately 3 per cent of the total world fisheries catch. Therefore, it is apparent that their importance is quite small, and, thus, deep-sea fisheries lack economic significance on a global scale.

4.2 Environment and life histories/energetics of deep-sea demersal fishes

Understanding the rate of energetic processes and the life histories of deep-sea fishes is critical to understanding their potential for exploitation, sustainability, and recovery in the context of industrial fishing. The variables of interest include growth, metabolism, and reproduction, and, from a life-history perspective, the ages of maturity and longevity, which relate to natural mortality. Earlier examinations (e.g. Koslow 1996) showed diversity in these parameters even amongst deep-sea fishes, but a perception that all deep-sea demersal fishes are slow growing, late maturing and very long-lived was not uncommon. However, there are exceptions to this pattern and a simple dichotomy between deep- and shallow-living species is overly simplistic. Fishes live across a continuum of depth and associated environmental parameters such as food supply and temperature (Drazen and Haedrich 2012). Indeed, the energetics and life hisories of these species must be considered with the following questions in mind: Are all deep-living demersal fishes slow growing, long-lived, and slow to reproduce? How do energetic and life-history parameters vary with depth and environmental conditions? And ultimately, what controls these processes in deep-sea fishes and other animals?

Longevity, or maximum age, is a life-history parameter that, in unexploited stocks, is a predictor of natural morality rate, a key stock-assessment variable (Hoenig 1983). Great longevity has been found in other deep-sea organisms such as octocorals (~800–4000 years; octopods over a decade; Schwarz et al. 2018), and fishes such as orange roughy (>120 years; Andrews et al. 2009). However, there are many shallow-living species that are equally long-lived, including 400-year-old clams (Munro and Blier 2012) and 100-year-old estuarine sturgeon (Musick 1999). On average, though, deeper-living fishes are longer-lived than shallower-living ones. An increase in longevity with depth has been found in rockfishes (Cailliet et al. 2001), and a weak increase with depth across all

bony fishes (Drazen and Haedrich 2012). Furthermore, elasmobranchs living deeper than 200 m have greater longevity than pelagic or continental shelf species (Rigby and Simpfendorfer 2015). Thus, it follows that the deeper-living species have lower natural mortality rates, and they will be less likely to support high rates of fishing mortality for long periods of time. The reasons for progressively increasing longevity with depth is unclear. It may result from the environmental stability of the system or as a result of low predation rates from sharks and marine mammals—the likeliest predators for larger commercially exploited fishes. Sharks decline in abundance and diversity with depth (Priede et al. 2006), and marine mammal predation may also decline due to the increasing requirements for deep diving.

Growth rates are related to longevity, but this parameter is really an energetic one that describes how rapidly an animal gains body mass and reaches its maximum size. A number of deep-sea species grow slowly; deep-sea teleost growth rates decline exponentially with habitat depth (Figure 3), a robust pattern across phylogenetic groups. Demersal elasmobranchs inhabiting depths greater than 200 m grow more slowly than shelf or pelagic species (Rigby and Simpfendorfer 2015). Most of the data for these patterns are derived from temperate or subtropical systems where temperature declines with depth. Temperature has a large effect on intraspecific growth rates and should play a strong role in the patterns of growth across depths as well (e.g. Neuheimer and Groenkjaer 2012). However, other potential drivers could include decreases in food supply or links to metabolic rates. A full exploration of the potential drivers of slow growth is needed.

The reproductive parameters of deep demersal and benthic fishes also indicate that they have very low production potential. Associated with greater longevities, the age at maturity increases with habitat depth (Drazen and Haedrich 2012; Rigby and Simpfendorfer 2015) such that demersal teleosts with a minimum habitat depth of 0–200 m, have average age at maturity of 5–6 years, increasing to nearly 20 years for species living principally below 700 m (Figure 3). Since Marshall (1953), it has been noted that deep-sea fishes produce relatively few

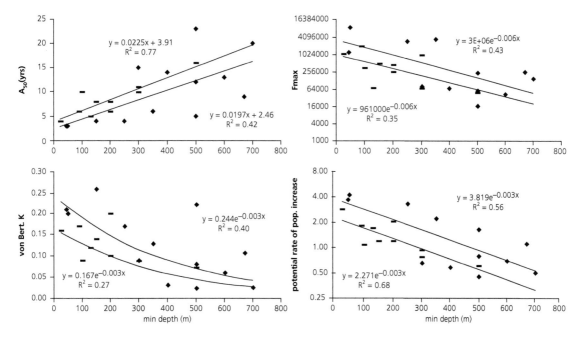

Figure 3 Life history attributes as a function of minimum depth of occurrence in demersal commercially exploited fishes. A50 is the age at 50% maturity, vonBert K is the von Bertalanffy growth coefficient or rate constant, Fmax is the maximum fecundity and potential rate of population increase (r1) is defined as ln(F50)/A50 where F50 is the fecundity at the size of 50% maturity (Jennings et al. 1998). Figure reproduced with permission from Drazen and Haedrich (2012).

but large eggs once they do begin to reproduce. Depth-related declines in maximum fecundity have been noted (Drazen and Haedrich 2012; Figure 3), and the relative annual fecundity of an assemblage of Mediterranean fishes has also been shown to decline with depth (Fernandez-Arcaya et al. 2016). However, these patterns are subtle, and phylogeny matters. Fernandez-Arcaya et al. (2016) found that within a single order, the Gadiformes, relative fecundity did not decline with depth. Within phylogenetic groups, deeper-living species have larger eggs even if fecundity does not change, suggesting greater energetic investment into reproduction to produce eggs and larvae with a higher success rate, despite lower food availability at greater depths. Deeper species may require larger eggs to provision offspring for greater vertical displacement of larvae and longer pelagic residence times. Alternatively, deeper-living species may have less energy for reproduction, despite a similar number of larger eggs, because they do not reproduce every year. Energetics modelling has suggested that the energy required for annual reproduction in grenadiers

would exceed the annual expenditure for routine metabolism, thus they are likely to spawn less frequently (Drazen 2002). There is a growing number of studies that find large numbers of females not in spawning condition during spawning periods (Bell et al. 1992; Rotllant et al. 2002; Fernandez-Arcaya et al. 2016), suggesting that substantial fractions of the adult population do not spawn every year. These observations match those from better-studied shallow-water species, such as cod, where females experiencing low-food conditions skip spawning (McBride et al. 2015). The implication is that reproductive output, and thus the potential for population growth, declines with habitat depth.

Metabolic rate is an energetic parameter that is not intuitively linked to fish productivity but is clearly related to the overall pace of life of these animals. Metabolism can be seen as the maintenance or operating costs of an animal, including locomotion, which is the third major energetic expenditure along with growth and reproduction. Temperature and body mass are prime determinants of an animal's metabolic rate. In fishes, metabolic rates decline tenfold with habitat depth from the surface to about 1000 m, even after taking into account declining water temperatures and variations in body size (Childress 1995; Drazen and Seibel 2007). Such declines occur in several taxa (not just fishes) in both eutrophic and oligotrophic settings, suggesting food supply is not a major determinant either. Depth-related declines, which are most pronounced in the shallowest 400–500 m, are evident from direct estimates of fish metabolism from *in situ* respirometry (Smith 1978; Drazen and Yeh 2012) and indirect measures of metabolic enzyme activities in teleosts and elasmobranchs (Sullivan and Somero 1980; Condon et al. 2012; Drazen et al. 2015). Unlike fishes, some benthic invertebrates and pelagic blind taxa (e.g. Cnidaria, chaetognaths), do not show depth-related declines in metabolic rates. The explanation for these patterns has been termed the 'visual interactions hypothesis' (Childress 1995; Seibel and Drazen 2007). It suggests that swimming animals that use vision to interact with predators and prey experience declining light levels with increasing depth and hence have shorter and shorter reactive distances. This situation then relaxes the selection for strong swimming capabilities because deeper species have less need for strong locomotory attack and escape behaviours. For instance, fishes at depth need burst only a short distance to hide in the dark effectively. Long prey pursuits and predator evasions are needed less and less. Deeper-living species with less locomotory machinery have lower metabolic overhead and reduced metabolic rates. Thus, while temperature is a driver of metabolism, so are light levels. Whether such metabolic reductions with depth correlate with reduced growth rates or whether lower metabolic costs provide opportunities for greater investment in growth or reproduction has barely been explored (Childress et al. 1980).

As might be expected, as metabolic and growth rates of demersal fishes (energy expenditures) decline with habitat depth, so do their feeding rates (energy acquisition). Early studies noted that amongst pelagic species, migrators consumed more than bathypelagic nonmigrators (Childress et al. 1980). Koslow (1996) also compared the feeding rates of slope-dwelling species and found their rates about a tenth of those of active seamount asso-

ciated fishes, such as orange roughy. A small number of feeding-rate estimates have since been made, mostly on species in the Mediterranean (e.g. Madurell and Cartes 2006). Despite the small number of measurements on demersal species (there are more on mesopelagic species; Drazen and Sutton 2017), it is clear that the feeding rates of species inhabiting the bathyal and abyssal environments are much lower compared with shallow-living fishes, even in the Mediterranean where temperatures remain relatively high below approximately 200 m depth.

In summary, the pace of life and productivity of fishes declines with habitat depth. There is no easy way to classify a fish as either 'deep' or 'shallow'; rather, a species will live in a depth range with depth as a convenient covariate of important habitat or environmental variables that drive its life history and energetic processes. However, much of the change in environmental conditions and thus energetic parameters occurs in the shallowest 400 to 500 m. Declining temperature, food supply, and light levels all play an important role alongside phylogenetic influences on life history and reproductive adaptations. This understanding can provide some expectation for the productivity of fish stocks. Deeper-living fishes are more likely to have slower growth, greater longevity, slower metabolism, and lower reproductive output. Consequently, sustainable fisheries are only likely at very low harvest rates, and, if overfished, stock recovery times are going to be very long.

4.3 Impacts of deep-sea fisheries and potential for recovery

4.3.1 Impacts on fish populations

The removal of biomass by deep-sea fishing is known to have profound effects on fish populations, including sharks, rays, and skates. The decline of certain fish species can be extremely rapid. Devine et al. (2006) showed that the abundances of five species from the North Atlantic had declined 87–98 per cent over a 17-year period. This makes such species critically endangered based on the International Union for Conservation of Nature (IUCN) criteria (Devine et al. 2006). Knowledge about recruitment of many of these species is rather limited, but long-term studies on one of the endangered species, the roundnose grenadier, revealed that there was a single large-scale recruitment event in the space of 30 years (Bergstad 2013). Baseline studies quantifying the impact of deep-sea fisheries through time are rare, but a study comparing the effects of a trawl fishery in the NE Atlantic over the span of 10 years showed a 53 per cent reduction in the abundance of demersal fish species (Bailey et al. 2009). The fishery affected many nontarget species, and, while the trawling was constrained to less than 1500 m depth and 52,000 km^2, its impacts are estimated to have extended down to 3086 m depth and an area of 142,000 km^2 as fish are highly mobile and move downslope as they mature (Bailey et al. 2009; Priede 2017). Historically, almost all deep-sea fisheries suffer from the same problem: the population does not stay at commercially exploitable levels for very long, creating what seems like a 'boom-and-bust' aspect to the fishery (Victorero et al. 2018). Seamount fisheries, in particular, usually last for only a decade or so (Clark et al. 2007). When a seamount fishery is discovered, catches are very high over the first few years, but then plateau or drop suddenly. The social aggregative behaviour of seamount fish over these fixed features makes them more catchable, and since the fishing usually removes spawning individuals, the population is prevented from replenishing itself, making such species extremely vulnerable to rapid depletion (Morato et al. 2006). A classic example of a seamount fishery is that of the pelagic armorhead on the southern Emperor seamounts, where the Soviet Union started a fishery in 1968 with 46,000 tonnes caught, but catches rose to almost 145,000 tonnes the next year. That year, Japan also started to fish armorhead on the Emperor seamounts, resulting in more than 500,000 tonnes removed from the seamounts in just 6 years. By 1976, the catch had dwindled to less than 30,000 tonnes, and the next year it was only 4000 tonnes. In the Southern Pacific, Australian vessels began fishing orange roughy on the Tasmanian seamounts in 1985, landing 64 tonnes. The catch peaked in 1990 at 58,600 tonnes and then dropped to 6600 tonnes in 1994, and less than 3000 tonnes from 1996 onwards. Clark et al. (2007) document similar patterns for the Madeira-

Canary seamounts, South Azores seamounts, Ob and Lena seamounts, Corner Rise seamounts, and the Namibian seamounts. Many of these fisheries were targeting orange roughy, alfonsinos, and oreos, which are species with peculiar life histories involving mating aggregations, late maturity, and/or slow growth.

Some fisheries have managed to persist for longer periods and may be sustained at levels much below virgin (unfished) stock biomass if management measures are able to prevent further depletion. Orange roughy, in particular, has suffered severe population losses over most of its range due to fishing. It is a very long-lived species, living to perhaps 125 years, becoming reproductively mature at age 20–30 years, and has moderate fecundity (70–100,000 eggs, compared to several million eggs for shallow-water species). It is also very widespread, being found in the North Atlantic, along the Western African margin, Southern Indian Ocean, and South Pacific Ocean. In most of these areas, orange roughy has been fished to very low population levels, resulting in lowered TACs being invoked by regional or national management agencies in some areas. The fishery has persisted longest in waters around and adjacent to New Zealand, possibly due to the large area involved and low TACs, which have kept the fishery from disappearing, although the TACs have been set higher than the actual catches for the past 10 years.

4.3.2 Impacts on habitat

Deep-sea trawlers target the continental slope, canyon margins, and seamounts, where they impact both the soft- and hard-bottom communities. In the NE Atlantic, the spatial extent of the bottom affected by trawling is one to three orders of magnitude higher than the combined impact of oil and gas, waste disposal, and telecommunication cables (Benn et al. 2010). There are a host of studies that have documented the impacts of trawling on habitats, beginning first with assessing bycatch (Connolly and Kelly 1996) and then documenting the physical alterations on the seafloor caused by the trawl gear either via side scan sonar (Friedlander et al. 1999; Puig et al. 2012) or bottom photographs (Roberts et al. 2000).

Dimech et al. (2012) found diversity indices to be higher at nontrawled versus trawled sites being fished for deep-water shrimp in the Mediterranean. Lower species richness, lower diversity, and lower evenness were observed at a canyon site that had a higher level of disturbance due to fishing (Ramirez-Llodra et al. 2010). Sediment remobilisation due to resuspension is one of the more important consequences of repeated trawling on soft sediment habitats. Puig et al. (2012) estimated that 2.4×10^{-4} km^3 of sediment had been removed annually from a trawling ground of about 4.2 km^2 along one flank of the La Fornera Canyon in the NW Mediterranean, drastically transforming the seafloor characteristics. This sediment was transported down the canyon walls to the deep-sea floor. The loss of surface sediments due to trawling leaves behind older sediment with lowered organic matter content and turnover, and concomitant reduced meiofauna biodiversity by 50 per cent and abundance by 80 per cent (Pusceddu et al. 2014). Relatively unstudied, the smaller organisms in the sediment might experience the greatest habitat loss due to trawling (Watling 2014). Off New Zealand, Baco et al. (2010) observed that the fauna of seep sites that had been subjected to more than 200 trawl hauls over the previous 18 years was concentrated in depressions and crevices, suggesting that trawling had removed everything not in a physical refuge.

Hard-bottom communities, such as cold-water corals and sponges, have received much more attention with regard to impacts of bottom trawling. Most likely, this increased attention is due to the fact that most of the fauna is large, often colourful and charismatic, but also fragile and, as a result of slow growth rates, have been living in one place for centuries or millennia (Fosså et al. 2002). Recovery of these long-lived species has been estimated to take many decades or centuries. Reed (2002) documented trawling damage to deep-water *Oculina* reefs off the Florida East Coast with resulting loss of habitat for important fish species. In the trawled areas of seamounts, the benthos has been dramatically reduced, for example, with the loss of coral habitat and a subsequent threefold decline in species richness (Niklitschek et al. 2010). Althaus et al. (2009) documented two orders of magnitude reduction of cover of the cold-water coral

Solenosmilia variabilis with subsequent threefold loss of megabenthic species richness on seamounts that had been trawled.

Probert et al. (1997) noted that trawling on the Chatham Rise off New Zealand would reduce overall biodiversity because of the removal of large sessile epifauna, and estimated that the coral patches would take about 4100 years to recover. Similarly, Koslow et al. (2001); Clark and O'Driscoll (2003); and Clark and Tittensor (2010) noted that routinely trawled seamounts generally had lost most or all of their coral and other suspension feeder community on the summits and upper flanks. Waller et al. (2007) and Watling et al. (2007) documented the lack of megafauna from areas that showed scars and other indicators of past trawling activity on the Corner Rise seamounts.

4.3.3 Potential for recovery of fish populations

As noted by Haedrich et al. (2001), most deep-sea fish stocks are fully exploited or overexploited before the requisite ecological data are gathered that could be used to manage effectively the stock for sustainable use. The fishery of concern for Haedrich et al. (2001) was that of the roundnose grenadier. While the fishery in the NE Atlantic started in 1965, there was virtually no information on the life history of this species until 1975; fecundity data were available only in 1994, and other biological data such as age, growth, and maturity in 1996 and 1997 (Haedrich et al. 2001). The lack of important data is the rule for deep-sea fisheries species, as one can see by quickly surveying the species in Fishbase.org. For example, there are few data for many species, such as the west coast sole, deep-water cape hake, and Benguela hake, all from the SW Africa continental slope; the capro dory from the SW Pacific; the white warehou from the SW and SE Pacific; and the red and black cusk eels, respectively, from the Peru–Chile slope. The lack of appropriate biological data has made modelling of stocks difficult, resulting in inaccurate catch forecasts and in the overexploitation of fish stocks. There is also a lack of data for mortality generated by the practice of discarding unwanted fish, which form an important part of the catch in trawl fisheries. These catches generally have not been recorded officially and thus not taken into account by fisheries management, making it harder to understand the true recovery of populations (Pawlowski and Lorance 2009). However, based on general patterns of the life histories of fish species across depths, it is clear that most species will have slower growth, slower reproductive capacity, and low potential for population growth compared with shallow-water stocks.

Few studies have been able to monitor deep-water fish stocks over sufficient time intervals and after fishery closures or reductions to evaluate recovery. Pelagic armorhead, for example, have shown little sign of recovery after several decades of reduced fishing pressure (Victorero et al. 2018). In the NE Atlantic, depleted spawning aggregations of the blue ling are yet to recover despite being closed to fishing since 1993 (ICES 2018), whereas reductions in fishing pressure may have stabilised the populations of several grenadier species on the slope (Neat and Burns 2010). Further study of populations in the Rockall Trough showed that over a 16-year period after fishing pressure had decreased nearly tenfold, some recovery was observed through increases in fish-size metrics (Mindel et al. 2017). This was evident in the upper portions of the slope (< 750 m) where the species present have the highest production capacity. Other studies have used life-history parameters, if known, in population models to estimate recovery times. For two grenadiers in the North Atlantic recovery times ranged from 14 to 80 years after all fishing had ceased (Baker et al. 2009) depending on model parameters. Though limited, both empirical and modelling approaches agree that decades to a century may be required for overexploited deep-water fish stocks to recover.

4.3.4 Recovery of impacted habitat

It is well known that most of the large coral specimens on seamounts, such as the bamboo corals, live for as many as 800 years (Watling et al. 2011), and some black corals up to 4200 years (Roark et al. 2009). Other corals may live for several decades, but the ages of seamount sponges are completely unknown. For these reasons, there has been conjecture that recovery of seamount communities from trawling impacts might take decades or longer

(Williams et al. 2010), but long-term studies are yet to be conducted. The available studies with deep-diving remotely operated vehicles, however, show that no significant recovery has happened over decadal time-scales. For example, Waller et al. (2007) dived on the summits of two seamounts in the Corner Rise group about 10 years after trawling ceased, and 30 years after the peak of trawling activity. They found some small specimens of sponges and plexaurid corals, most less than 15 cm in height. Similarly, seamounts off Tasmania showed no signs of recovery after 10 years (Althaus et al. 2009). Williams et al. (2010) found no indication of recovery in the megafaunal assemblages on seamounts off New Zealand and Australia after 5–10 years, although some small and flexible individual species seemed to increase in abundance, suggesting they might have survived the trawling impacts. More recently, Clark et al. (2019) studied trawled versus nontrawled seamounts along the northern edge of Chatham Rise, New Zealand, and found no evidence of recovery of the megafaunal assemblages 15 years after closure to trawling of one of the seamounts in 2001. Furthermore, the megabenthic community of the closed seamount was essentially similar to that of Graveyard seamount, which continues to have high levels of trawling pressure. Though less well studied, continental slope benthic habitats are also slow to recover. No recovery of *Lophelia* corals on the Darwin Mounds area were noted over an 8-year period (Huvenne et al. 2016).

4.4 Management and stakeholder processes

4.4.1 International debate and negotiations over deep-sea fisheries

In response to scientific concerns over the impacts of deep-sea fisheries on the fish populations, benthic ecosystems, and biodiversity, conservation and nonprofit organisations began engaging with policy-makers in the early 2000s. Together they called for action to address, in particular, bottom fishing on the high seas (WWF/IUCN 2001; UNICP 2003; Shotton 2005). These concerns culminated in a campaign in 2004 for a UN General Assembly (UNGA) moratorium on bottom-trawl fishing on the high seas until such fisheries were managed consistent with international legal obligations for sustainability and the protection of deep-sea ecosystems (Gianni 2004). In addition, several letters or statements of concern from scientists were transmitted to the UN, signed by over 1400 scientists worldwide (TerraNature 2004).

The rationale behind this campaign was that bottom-trawl fisheries on the high seas were largely unregulated and in contravention of most key provisions of the 1995 UN Fish Stocks Agreement (UNFSA 1995). These provisions include obligations to assess impacts of fisheries on the marine environment; prevent overfishing; minimise impacts of fisheries on nontarget, associated, and dependent species; protect biodiversity in the marine environment; protect habitats of special concern; and apply a precautionary approach, particularly when scientific information is uncertain, unreliable, or inadequate. The argument concluded that bottom trawling was the most damaging method of bottom fishing on the high seas and that the catch and value of high-seas bottom-trawl fisheries were miniscule when assessed against marine-capture fisheries worldwide, thus making a negligible contribution to global food security (Gianni 2004). It also highlighted that the majority of these vessels were flagged to only eleven, mostly developed, countries, which were already—or soon to be—legally bound by the provisions of the UNFSA (Gianni 2004). As a result of such efforts, the UNGA (2004) adopted the first of its substantive resolutions on deep-sea fisheries and called on states to:

take action urgently, and consider...the interim prohibition of destructive fishing practices, including bottom trawling that has adverse impacts on vulnerable marine ecosystems, including seamounts, hydrothermal vents and cold-water corals located beyond national jurisdiction, until such time as appropriate conservation and management measures have been adopted in accordance with international law.

Debate over the management of deep-sea fisheries on the high seas continued throughout 2005 and 2006 (FAO COFI 2005; UNFSA 2006). The dynamics involved countries that were not engaged in deep-

sea fishing but who supported the moratorium against a minority of politically influential nations whose vessels were fishing on the high seas, and countries with vested interests in bottom fishing who were interested in finding a compromise. In 2006, the UN Secretary General's report concluded that little had been done by relevant high-seas fishing nations to implement the urgent call for action in the 2004 UNGA resolution with respect to the management of deep-sea fisheries on the high seas. In December 2006, the UNGA formally adopted Resolution 61/105 (UNGA 2006), which (para 83) committed high-seas fishing nations individually and through the creation of regional fisheries management organisations (RFMOs) (Figure 4) to:

1. Conduct impact assessments of individual bottom fishing activities to determine whether they would cause significant adverse impacts (SAIs) on Vulnerable Marine Ecosystems (VMEs) and ensure that, if the assessment showed that fishing activities would have SAIs, they are managed to prevent such impacts, or else prohibited;
2. Identify and close areas of the high seas to bottom fishing where VMEs such as cold-water corals and seamount ecosystems are known or *likely* to occur, unless fishing in these areas can be managed to prevent SAIs on such ecosystems;
3. Establish and implement protocols to require vessels to cease fishing in areas where an encounter with VMEs occurs during fishing activities (i.e. corals or sponges are caught in the trawl gear);
4. Sustainably manage the exploitation of deep-sea fish stocks; and
5. Adopt and implement these measures, in accordance with the precautionary approach, ecosystem approach and international law, by no later than 31 December 2008.

Following the adoption of Resolution 61/105, the countries involved in the negotiations established a process under the auspices of the FAO to negotiate international guidelines for the implementation of the resolution. A primary focus was to establish agreed criteria for the three key operational commitments in the UNGA resolution: conducting impact assessments of deep-sea fisheries, identifying VMEs, and determining SAIs (FAO 2008). The

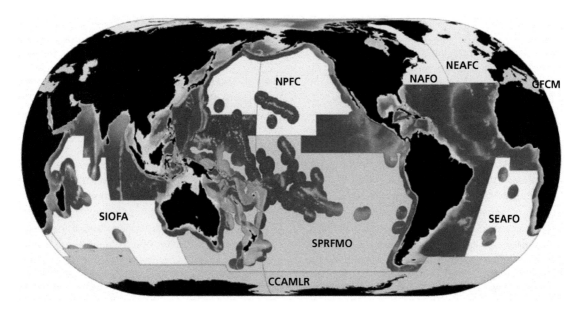

Figure 4 Map of the RFMOs of the world. Abbreviations are as follows: CCAMLR, Commission for the Conservation of Antarctic Marine Living Resources, GFCM, General Fisheries Commission for the Mediterranean, NAFO, Northwest Atlantic Fisheries Organization, NEAFC, North East Atlantic Fisheries Commission, NPFC, North Pacific Fisheries Commission, SEAFO, South East Atlantic Fisheries Organisation, SIOFA, South Indian Ocean Fisheries Agreement, SPRFMO, South Pacific Regional Fisheries Management Organisation.

FAO Guidelines were subsequently endorsed by the UNGA in 2009, in paragraphs 119 and 120 (UNGA 2009).

In 2009 and again in 2011 and 2016, the UNGA conducted further reviews of the implementation of previous resolutions on deep-sea fisheries and, on each occasion, reaffirmed, strengthened, and enhanced the call for action to protect deep-sea ecosystems on the high seas from the adverse impacts of deep-sea fisheries. Examples of the latter include the call to ensure vessels are not allowed to engage in bottom fishing unless impact assessments consistent with the Guidelines are first conducted, and to ensure the long-term sustainability of deep-sea fish stocks and nontarget species, as well as the rebuilding of depleted stocks (UNGA 2009); conduct cumulative impact assessments of bottom fisheries and review; update and revise impact assessments on a regular basis (UNGA 2011, para 129); use methods such as benthic ecosystem modelling, comparative benthic studies, and predictive modelling along with fisheries dependent information and scientific surveys to identify areas where VMEs are known or likely to occur; and to take into

account the potential impacts of climate change and ocean acidification in taking measures to manage deep-sea fisheries and protect VMEs (UNGA 2016, paras 181 and 185).

4.4.2 Implementation of the resolutions: protection of deep-sea ecosystems and sustainable deep-sea fisheries on the high seas

In response to the debate at the UN and the resolutions adopted by the UNGA, states and existing RFMOs began to establish enhanced conservation and management measures for deep-sea fisheries on the high seas. In areas of the high seas such as the North and South Pacific Oceans, where bottom fisheries were taking place but no RFMO had been established to manage these fisheries, states began negotiating treaties to establish such RFMOs. With the adoption of Resolution 61/105, the process accelerated, and by the end of 2008, most states and RFMOs had adopted regulations or so-called interim measures (pending the establishment of RFMOs) that incorporated the measures called for in the resolution. By the 2016 UNGA review, three new RFMOs were in operation bringing the total to eight RFMOs with the legal competence to manage bottom fisheries on the high seas (Figure 4), and all but one of the RFMOs had incorporated most of the objectives and actions called for in the UNGA resolutions, as well as key provisions of the Guidelines into their regulations for the management of deep-sea fisheries (Gianni et al. 2016). Representatives of the Deep Sea Conservation Coalition and member organisations (e.g. WWF, Greenpeace, Seas At Risk, New Zealand Ecology Action Centre, Oceana, Ecoceanos, and Pew Charitable Trusts) along with deep-sea scientists and fishing industry representatives have regularly participated in meetings, including annual meetings and scientific and technical committee meetings, of the majority of the RFMOs, as well as FAO workshops on deep-sea fisheries and the two UNGA stakeholder consultations, since the adoption of UNGA resolution 61/105.

However, while adoption of the measures, called for in the resolutions and the criteria established in the Guidelines, has largely occurred, with some notable exceptions, the implementation of the resolutions and criteria has been mixed. On the one hand, a number of RFMOs have confined bottom fishing to a historical 'footprint'—areas now designated as 'existing' fishing areas with significant restrictions imposed on any 'exploratory' fishing in areas outside the existing, authorised fishing areas. Several dozen known, likely and/or representative areas of VMEs have been closed by all eight RFMOs that regulate high-seas bottom fisheries. In addition, The Commission for the Conservation of Antarctic Marine Living Resources (CCAMLR) has banned all bottom trawling on the high seas in the Southern Ocean; the General Fisheries Commission for the Mediterranean (or GFCM) has banned bottom trawling below 1000 m depth in the Mediterranean; the Northwest Atlantic Fisheries Organization (or NAFO) has closed most seamounts to bottom fishing in the NW Atlantic; and the South Pacific Regional Management Organisation (SPRFMO), North East Atlantic Fisheries Commission (or NEAFC), South East Atlantic Fisheries Organisation (or SEAFO), and CCAMLR have prohibited bottom gillnet fishing. For areas of the high seas where there is no RFMO

to regulate bottom fisheries, the EU has adopted legislation incorporating the UNGA resolutions (EU 2008)
. In response, Spain has conducted a comprehensive impact assessment of its bottom fisheries on the high seas in the SW Atlantic and closed most areas below 300–400 m depth to bottom fishing to protect VMEs.

On the other hand, there are serious shortcomings in the implementation of the UNGA resolutions as of 2018. These include inadequate or partial impact assessments characterised by failure to follow the Guidelines, unresolved scientific uncertainties, unverified assumptions concerning risk, restricted interpretation of VMEs, and/or little to no mapping of VMEs in the authorised fishing areas. Few efforts have been made to conduct cumulative impact assessments to assess, for example, historical degradation of VMEs preceding the adoption of regulations and other potential stressors such as ocean acidification. Many areas where cold-water coral, sponge, and other VME–related species and habitats are likely to occur on the high seas, in particular those associated with seamounts and other underwater features in the SW Pacific and Southern Indian Ocean, remain open to, and vulnerable to, damage by deep-sea trawling (Clark et al. 2015; Wright et al. 2015; Gianni et al. 2016; Bell et al. 2019).

The reasons for these shortcomings are many and varied. Often, the cause is a combination of limited scientific information about deep-sea species and ecosystems and the impact of fishing coupled with an unwillingness of states and RFMOs to apply a precautionary approach in the face of scientific uncertainty, as is called for in the UNGA resolutions and required under international law (UNFSA, article 6). In addition, the primary business of most, if not all, RFMOs continues to be to agree on an acceptable level of catch or bycatch for commercial fish species and allocating quotas amongst the member countries of the RFMOs. Environmental issues and decision-making are often relegated to secondary or subordinate status.

Nonetheless, most of the states and RFMOs concerned continue to engage in the implementation of the UNGA resolutions at scientific, political, and regulatory levels, in part because the UNGA itself has continued to review their implementation (the next UNGA review is scheduled in 2020). The importance of effective implementation cannot be overstated. The approach adopted by the UNGA to the regulation of deep-sea fisheries, negotiated over many years, has set precedents for the management of deep-sea fisheries within EEZs. Equally, and indeed more important, is that the UNGA resolutions and FAO Guidelines be effectively implemented in practice, not only to ensure protection of deep-sea ecosystems and biodiversity from the harmful impacts of deep-sea fishing, but to serve as effective precedents for the management of other deep-sea activities.

4.5 The future of deep-sea fisheries

The future industry of deep-sea fishing is uncertain, though shallower deep-sea fisheries could be sustainable under proper management regimes. The latter species have the best potential to have stable stocks over the long term. Habitat issues will remain, however, for the trawl fisheries, since the physical damage imposed by trawls cannot be mitigated. Therefore, trawl fisheries should not be considered or labelled as sustainable since there is continual habitat impact. Nevertheless, fisheries for certain stocks, such as New Zealand's orange roughy and blue grenadier, and Greenland halibut off West Greenland, have obtained sustainability certifications, such as those generated by the Marine Stewardship Council (www.msc.org).

In deeper water, issues of, for example, fecundity and growth of deep-sea species are difficult to overcome. Not one of the species being fished deeper than 400 m has been able to last for more than a few decades without being overfished. Greenland halibut and orange roughy are widespread enough that their total demise may take a long time to occur, even though at many locations the populations are no longer economically viable. Other groups, such as deep-sea sharks, particularly the sleeper sharks and cat sharks, have very low fecundity levels. As with the Greenland halibut and orange roughy, however, it is their widespread distribution that has likely kept them from suffering catastrophic population declines.

On seamounts, in particular, the habitat consequences of deep-sea fishing can be severe, and as pointed out, there is no evidence that these deeper communities are showing any signs of recovery after as many as 15 years (Clark et al. 2019). It may be that shallower communities, such as those between 300 and 400 m on the Emperor seamounts where the pelagic armorhead fishery occurred, might be able to recover faster (Baco et al. 2019). In any case, given the relatively small amount of catches of deep-sea fish, especially by trawl gear below 400 m depth, careful consideration should be given to whether these catches warrant the severe amount of associated habitat damage. The small amount of fish caught are not important to global food security as has sometimes been claimed, and while the amount of profit can be considerable in certain cases, most deep-sea fishing survives only because of government subsidies, provided primarily to keep high-seas fishing companies profitable and people in the fishing business employed (Sumaila et al. 2010; Sala et al. 2018).

Regional fisheries management bodies are responsible for requiring that deep-sea fish are caught with minimal habitat disturbance and that VMEs are fully protected. However, there is much more habitat focused (rather than fish stock focused) work to do for many of the RFMOs (see proceedings of Scientific Committee meetings on the RFMO web pages). Watling and Auster (2017) argued that seamounts, which are now known to have a high coverage of VME indicator species, should be managed in their entirety as VMEs, a view that has not won wide acceptance. Instead, in the SPRFMO there is a move towards zoning individual seamounts, effectively putting only a small part of the seamount off limits to bottom trawling. A similar approach may be developing in the management considerations of the North Pacific Fisheries Commission as well. Hence, the outlook for deep-sea fish stocks and seamount habitats over the next few decades, as it stands, is uncertain.

Acknowledgements

The authors would like to thank Michelle Kay, Claire Nouvian, NOAA Office of Ocean Exploration, and Schmidt Ocean Institute for use of photos in

Figure 1; Deng Palomares of Sea Around Us for help with unreported deep-sea fish catches; and Tony Koslow, Tracey Sutton, and the Editors for keeping us accurate and honest, but of course, all errors remain ours. Lissette Victorero was supported by a grant from Laboratoires d'Excellences (LABEX) BCDiv (ANR-10-LABX-03).

References

Althaus, F., Williams, A, Schlacher, T.A., et al. (2009). Impacts of Bottom Trawling on Deep-coral Ecosystems of Seamounts Are Long-lasting. *Marine Ecology Progress Series*, 397, 279–94.

Andrews, A. H., Tracey, D. M., and Dunn, M. R. (2009). Lead–radium Dating of Orange Roughy (*Hoplostethus atlanticus*): Validation of a Centenarian Life Span. *Canadian Journal of Fisheries and Aquatic Science*, 66, 1130–40.

Baco, A. R., Roark, E. B., and Morgan, N. B. (2019). Amid Fields of Rubble, Scars, and Lost Gear, Signs of Recovery Observed on Seamounts on 30- to 40-Year Time Scales. *Science Advances*, 5, eaaw4513.

Baco, A. R., Rowden A. A., Levin L. A., Smith, C. R., and Bowden, D. A. (2010). Initial Characterization of Cold Seep Faunal Communities on the New Zealand Hikurangi Margin. *Marine Geology*, 272, 251–9.

Bailey, D. M., Collins, M. A., Gordon, J. D. M., Zurr, A. F., and Priede, I. G. (2009). Long-term Changes in Deep-water Fish Populations in the Northeast Atlantic: A Deeper Reaching Effect of Fisheries? *Proceedings of the Royal Society of London B: Biological Sciences*, 276, 1965–9.

Baker, K. D., Devine, J. A., and Haedrich, R. L., (2009). Deep-sea Fishes in Canada's Atlantic: Population Declines and Predicted Recovery Times. *Environmental Biology of Fishes*, 85, 79–88.

Bell, J. B., Guijarro-Garcia, E., and Kenny, A. (2019). Demersal Fishing in Areas Beyond National Jurisdiction: A Comparative Analysis of Regional Fisheries Management Organisations. *Frontiers in Marine Science*, 6, 596. doi: 10.3389/fmars.2019.00596.

Bell, J. D., Lyle, J. M., Bulman, C. M., et al. (1992). Spatial Variation in Reproduction, and Occurrence of Non-reproductive Adults, in Orange Roughy, *Hoplostethus atlanticus* Collett (Trachichthyidae), from South-eastern Australia. *Journal of Fish Biology*, 40, 107–22.

Benn, A. R., Weaver, P. P., Billett, D. S. M., et al. (2010). Human Activities on the Deep Seafloor in the North East Atlantic: An Assessment of Spatial Extent. *PLoS One*, 5, 1–15.

Bergstad, O. A. (2013). North Atlantic Demersal Deep-water Fish Distribution and Biology: Present Knowledge and Challenges for the Future. *Journal of Fish Biology*, 83, 1489–507.

Cailliet, G. M., Andrews A. H., Burton, E. J., et al. (2001). Age Determination and Validation Studies of Marine Fishes: Do deep-dwellers Live Longer? *Experimental Gerontology*, 36, 739–64.

Childress, J. J. (1995).Are There Physiological and Biochemical Adaptations of Metabolism in Deep-sea Animals? *Trends in Ecology and Evolution*, 10, 30–6.

Childress, J. J., Taylor, S. M., Cailliet, G. M., and Price, M. H. (1980). Patterns of Growth, Energy Utilization and Reproduction in Some Meso- and Bathypelagic Fishes off Southern California. *Marine Biology*, 61, 27–40.

Clark, M. (1999). Fisheries for Orange Roughy (*Hoplostethus atlanticus*) on Seamounts in New Zealand. *Oceanologica Acta*, 22, 593–602.

Clark, M. and O'Driscoll, R. (2003). Deepwater Fisheries and Aspects of their Impact on Seamount Habitat in New Zealand. *Journal of Northwest Atlantic Fisheries Science*, 31, 441–58.

Clark, M. R., Anderson, O. F., Chris Francis, R. I. C., and Tracy, D. M. (2000). The Effects of Commercial Exploitation on Orange Roughy (*Hoplostethus atlanticus*) from the Continental Slope of the Chatham Rise, New Zealand, from 1979 to 1997. *Fisheries Research*, 45, 217–38.

Clark, M. R., Bowden, D. A., Rowden, A. A., and Stewart, R. (2019). Little Evidence of Benthic Community Resilience to Bottom Trawling on Seamounts After 15 Years. *Frontiers in Marine Science*, 6, 1–16.

Clark, M. R. and Koslow, J. A. (2007). Impacts of Fisheries on Seamounts, in T. J Pitcher, T. Morato, P. J. B. Hart, et al. (eds.) *Seamounts: Ecology, Fisheries and Conservation*. Oxford: Blackwell Publishing, pp. 413–41.

Clark, M. R. and Tittensor, D. P. (2010). An Index to Assess the Risk to Stony Corals from Bottom Trawling on Seamounts. *Marine Ecology*, 31, 200–11.

Clark, M. R., Vinnichenko, V. I., Gordon, J. D. M., et al. (2007). Large-scale Distant-water Trawl Fisheries on Seamounts, in. T. J Pitcher, T. Morato, P. J. B. Hart, et al. (eds.) *Seamounts: Ecology, Fisheries and Conservation*. Oxford: Blackwell Publishing, pp. 361–99.

Clark, N. A., Ardron, J. A., and Pendleton, L. H. (2015). Evaluating the Basic Elements of Transparency of Regional Fisheries Management Organizations. *Marine Policy*, 57, 158–66.

Condon, N. E., Friedman, J. R., and Drazen, J. C. (2012). Metabolic Enzyme Activities in Shallow- and Deep Water Chondrichthyans: Implications for Metabolic and Locomotory Capacity. *Marine Biology*, 159, 1713–31.

Connolly, P. L. and Kelly, C. J. (1996). Catch and Discards from Experimental Trawl and Longline Fishing in the Deep Water of the Rockall Trough. *Journal of Fish Biology*, 49, 132–44.

Devine, J. A., Baker, K. D., and Haedrich, R. L. (2006). Deep-sea fishes qualify as endangered. *Nature*, 4398, 29.

Dimech, M., Kaiser, M. J., Ragonese, S., and Schembri, P. J. (2012). Ecosystem Effects of Fishing on the Continental Slope in the Central Mediterranean Sea. *Marine Ecology Progress Series*, 449, 41–54.

Drazen, J. C. (2002). Energy Budgets and Feeding Rates of *Coryphaenoides acrolepis* and *C . armatus*. *Marine Biology*, 140, 677–86.

Drazen, J. C., Friedman, J. R., Condon, N., et al. (2015). Enzyme Activities of Demersal Fishes from the Shelf to the Abyssal Plain. *Deep-Sea Research I*, 100, 117–26.

Drazen, J. C. and Haedrich, R. L. (2012). A Continuum of Life Histories in Deep-sea Demersal Fishes. *Deep-Research I*, 61, 34–42.

Drazen, J. C. and Seibel, B. A. (2007). Depth-related Trends in Metabolism of Benthic and Benthopelagic Deep-sea Fishes. *Limnology and Oceanography*, 52, 2306–16.

Drazen, J. C. and Sutton, T. T. (2017). Dining in The Deep: The Feeding Ecology of Deep-sea Fishes. *Annual Review of Marine Science*, 9, 337–66.

Drazen, J. C. and Yeh, J. (2012). Respiration of Four Species of Deep-sea Demersal Fishes Measured in Situ in the Eastern North Pacific. *Deep-Sea Research I*, 60, 1–6.

EU (European Union) (2008). *On the Protection of Vulnerable Marine Ecosystems in the High Seas from the Adverse Impacts of Bottom Fishing Gears. Council Regulation (EC) No. 734/2008*. https://eur-lex.europa.eu/legal-content/EN/TXT/PDF/?uri=CELEX:32008R0734&from=EN.

FAO (2008). *Deep-Sea Fisheries in the High Seas: A Trawl Industry Perspective on the International Guidelines for the Management of Deep-Sea Fisheries in the High Seas FAO Fisheries and Aquaculture Circular. No. 1036*. Rome: FAO.

FAO (2018). *The State of World Fisheries and Agriculture 2018—Meeting the Sustainable Development Goals*. Rome: FAO.

FAO COFI (2005). *Report of the Twenty-Sixth Session of the Committee on Fisheries*. 7–11 March. Rome : FAO COFI.

Fernandez-Arcaya, U., Drazen, J. C., Murua, H., et al. (2016). Bathymetric Gradients of Fecundity and Egg Size in Fishes: A Mediterranean Case Study. *Deep-Sea Research I*, 116, 106–17.

Fossa, J. H., Mortensen, P. B., and Furevik, D. M. (2002). The Deep-water Coral *Lophelia pertusa* in Norwegian Waters: Distribution and Fishery Impacts. *Hydrobiologia*, 471, 1–12.

Friedlander, A. M., Boehlert, G. W., Field, M. E., et al. (1999). Sidescan-sonar Mapping of Benthic Trawl Marks on the Shelf and Slope off Eureka, California. *Fishery Bulletin*, 97, 786–801.

Gianni, M. (2004). *High Seas Bottom Trawl Fisheries and Their Impacts on the Biodiversity of Vulnerable Deep-Sea Ecosystems: Options for International Action*. Gland, Switzerland: IUCN.

Gianni, M., Fuller, S. D., Currie, D. E. J., et al. (2016). How Much Longer Will It Take? A Ten-year Review of the Implementation of United Nations General Assembly Resolutions 61/105, 64/72 and 66/68 on the Management of Bottom Fisheries In Areas Beyond National Jurisdiction. *Deep Sea Conservation Coalition*, 1–80. http://www.savethehighseas.org/wp-content/uploads/2016/08/DSCC-Review-2016_Launch-29-July.pdf.

Haedrich, R. L., Merrett, N. R., and O'Dea, N. R. (2001). Can Ecological Knowledge Catch Up with Deep-water fishing? A North Atlantic Perspective. *Fisheries Research*, 51, 113–22.

Hoenig. J. M. (1983). Empirical Use of Longevity Data to Estimate Mortality Rates. *Fishery Bulletin*, 81, 898–902.

Huvenne, V. A. I., Bett, B. J., Masson, D. G., Le Bas, T. P., and Wheeler, A. J. (2016). Effectiveness of a Deep-sea Cold-water Coral Marine Protected Area, Following Eight Years of Fisheries Closure. *Biological Conservation*, 200, 60–9.

ICES (2016). *ICES Advice on Fishing Opportunities, Catch, and Effort, Greater North Sea Ecoregion, 9.3.28 Roundnose Grenadier* (Coryphaenoides rupestris) *in Division 3.a (Skagerrak and Kattegat)*. Copenhagen, Denmark: ICES.

ICES (2018). Blue Ling (*Molva dypterygia*) in Subarea XIV and Division Va (East Greenland, Iceland Grounds). ICES Advice on Fishing Opportunities, Catch, and Effort: Arctic Ocean, Greenland Sea, Icelandic Waters, Norwegian Sea, and Oceanic Northeast Atlantic Ecoregions. Published 13 June. Copenhagen, Denmark: ICES. https://doi.org/10.17895/ices.pub.4407.

Jennings, S., Reynolds, J. D., and Mills, S. C. (1998). Life History Correlates of Responses to Fisheries Exploitation Life History Correlates of Responses to Fisheries Exploitation. *Proceedings of the Royal Society of London B: Biological Sciences*, 265, 333–9.

Kiyota, M., Nishida, K., Murakami, C., and Yonezaki, S. (2015). History, Biology, and Conservation of Pacific Endemics 2. The North Pacific Armorhead, *Pentaceros wheeleri* (Hardy, 1983) (Perciformes, Pentacerotidae). *Pacific Science*, 70, 1–20.

Koslow, J. A. (1996). Energetic and Life-history Patterns of Deep-sea Benthic, Benthopelagic and Seamount-associated Fish. *Journal of Fish Biology*, 49, 54–74.

Koslow, J., Boehlert, G. W., and Gordon, J. D. M., et al. (2000). Continental Slope and Deep-sea Fisheries: Implications for a Fragile Ecosystem. *ICES Journal of Marine Science*, 57, 548–57.

Koslow, J., Gowlett-Holmes, K., Lowry, J., et al. (2001). Seamount Benthic Macrofauna off Southern Tasmania: Community Structure and Impacts of Trawling. *Marine Ecology Progress Series*, 213, 111–25.

Little, A. S., Needle, C. L., Hilborn, R., Holland, D. S., and Marshall, C. T. (2015). Real-time Spatial Management Approaches to Reduce Bycatch and Discards: Experiences from Europe and the United States. *Fish and Fisheries*, 16, 576–602.

Madurell, T. and Cartes, J. E. (2006). Trophic Relationships and Food Consumption of Slope Dwelling Macrourids from the Bathyal Ionian Sea (Eastern Mediterranean). *Marine Biology*, 148, 1325–38.

Marshall, N. B. (1953). Egg Size in Arctic, Antarctic and Deep-sea Fishes. *Evolution*, 7, 328–41.

McBride, R. S., Somarakis, S., Fitzhugh, G. R., et al. (2015). Energy Acquisition and Allocation to Egg Production in Relation to Fish Reproductive Strategies. *Fish and Fisheries*, 16, 23–57.

Merrett, N. R. and Haedrich, R. L. (1997). *Deep-Sea Demersal Fish and Fisheries*. London: Chapman & Hall.

Mindel, B. L., Neat, F. C., Webb, T. J., and Blanchard, J. L. (2017). Size-based Indicators Show Depth-dependent Change over Time in the Deep Sea. *ICES Journal of Marine Science*, 75, 113–21.

Morato, T., Cheung, W. W. L., and Pitcher, T. J. (2006). Vulnerability of Seamount Fish to Fishing: Fuzzy Analysis of Life-history Attributes. *Journal of Fish Biology*, 68, 209–21.

Munro, D. and Blier, P. U. (2012). The Extreme Longevity of *Arctica islandica* Is Associated with Increased Peroxidation Resistance in Mitochondrial Membranes. *Aging Cell*, 11, 845–55.

Musick, J. A. (1999). Ecology and Conservation of Long-lived Marine Animals, in J. A. Musick (ed.) *Life in the Slow Lane: Ecology and Conservation of Long-Lived Marine Animals*. Bethesda, MA: American Fisheries Society, pp. 1–10.

Neat, F. and Burns, F. (2010). Stable Abundance, but Changing Size Structure in Grenadier Fishes (Macrouridae) over a Decade (1998–2008) in Which Deepwater Fisheries Became Regulated. *Deep-Sea Research I*, 57, 434–40.

Neuheimer, A. B. and Groenkjaer, P. (2012). Climate Effects on Size-at-age: Growth in Warming Waters Compensates for Earlier Maturity in an Exploited Marine Fish. *Global Change Biology*, 18, 1812–22.

New Zealand Government (2019). https://www.stats.govt.nz/information-releases/environmental-economic-accounts-2019-tables. Accessed May 2019.

Niklitschek, E. J., Cornejo-Donoso, J., Oyarzún, C., Hernández, E., and Toledo, P. (2010). Developing Seamount Fishery Produces Localized Reductions in Abundance and Changes in Species Composition of Bycatch. *Marine Ecology*, 31, 168–82.

Pawlowski, L. and Lorance, P. (2009). Effect of Discards On Roundnose Grenadier Stock Assessment in the Northeast Atlantic. *Aquatic Living Resources*, 22, 573–82.

Pew Environment Group (2012). *Deep-sea Fisheries and Vulnerable Ecosystems in the Northeast Atlantic—How the EU Can Reform Its Deep-sea Management Regime*. Philadelphia: The Pew Environment Group.

Priede, I. G. (2017). *Deep-Sea Fishes: Biology, Diversity, Ecology and Fisheries*. Cambridge: Cambridge University Press.

Priede, I. G., Froese, R., Bailey, D. M., et al. (2006). The Absence of Sharks from Abyssal Regions of the World's Oceans. *Proceedings of the Royal Society of London B: Biological Sciences*, 273, 1435–41.

Priede, I. G., Godbold, J. A., King, N. J., et al. (2010). Deep-sea Demersal Fish Species Richness in the Porcupine Seabight, NE Atlantic Ocean: Global and Regional Patterns. *Marine Ecology*, 31, 247–60.

Probert, P. K., McKnight, D. G., and Grove, S. L. (1997). Benthic Invertebrate Bycatch from a Deep-water Trawl Fishery, Chatham Rise, New Zealand. *Aquatic Conservation of Marine and Freshwater Ecosystems*, 7, 27–40.

Puig, P., Canals, M., Company, J. B., et al. (2012). Ploughing the Deep Sea Floor. *Nature*, 489, 286–9.

Pusceddu, A., Bianchelli, S., Martín, J., et al. (2014). Chronic and Intensive Bottom Trawling Impairs Deep-sea Biodiversity and Ecosystem Functioning. *Proceedings of the National Academy of Science USA*, 111, 8861–6.

Ramirez-Llodra, E., Company, J. B., Sarda, F., and Rotllant, G. (2010). Megabenthic Diversity Patterns and Community Structure of the Blanes Submarine Canyon and Adjacent Slope in the Northwestern Mediterranean: A Human Overprint? *Marine Ecology*, 31, 167–82.

Reed, J. K. (2002). Deep-water *Oculina* Coral Reefs of Florida: Biology, Impacts, and Management. *Hydrobiologia*, 471, 43–55.

Rigby, C. and Simpfendorfer, C. A. (2015). Patterns in Life History Traits of Deep-water Chondrichthyans. *Deep-Sea Research II*, 115, 30–40.

Roark, E. B., Guilderson, T. P., Dunbar, R. B., Fallon, S. J., and Mucciarone, D. A. (2009). Extreme Longevity in Proteinaceous Deep-sea Corals. *Proceedings of the National Academy of Science USA*, 106, 5204–8.

Roberts, J. M., Harvey, S. M., Lamont, P. A., Gage, J. D., and Humphery, J. D. (2000). Seabed Photography, Environmental Assessment and Evidence for Deep-water Trawling on the Continental Margin West of the Hebrides. *Hydrobiologia*, 441, 173–83.

Rotllant, G., Moranta, J., Massuti, E. Sarda, F., and Morales-Nin, B. (2002). Reproductive Biology of Three Gadiform

Fish Species Through the Mediterranean Deep-sea Range (147–1850m). *Scientia Marina*, 66, 157–66.

Sala, E., Mayorga, J., Costello, C., et al. (2018). The Economics of Fishing the High Seas—Supplementary Materials. *Science Advances*, 4, 1–14.

Schwarz, R., Piatkowski, U., and Hoving, H. J. T. (2018). Impact of Environmental Temperature on the Lifespan of Octopods. *Marine Ecology Progress Series*, 605, 151–64.

Seibel, B. A. and Drazen, J. C. (2007).The Rate of Metabolism in Marine Animals: Environmental Constraints, Ecological Demands and Energetic Opportunities. *Philosophical Transactions of the Royal Society London, B*, 362, 2061–78.

Shotton, R. (ed.) (2005). Deep Sea 2003: Conference on the Governance and Management of Deep-sea Fisheries. Part 1: Conference reports. Queenstown, New Zealand, 1–5 December 2003. *FAO Fisheries Proceedings. No. 3/1*. Rome: FAO, 718.

Smith, K. L. (1978). Metabolism of the Abyssopelagic Rattail, *Coryphenoides armatus*, Measured in Situ. *Nature*, 274, 362–4.

Sullivan, K. M. and Somero, G. N. (1980). Enzyme Activities of Fish Skeletal Muscle and Brain as Influenced by Depth of Occurrence and Habits of Feeding and Locomotion. *Marine Biology*, 60, 91–9.

Sumaila, U. R., Khan, A., Teh, L., et al. (2010). Subsidies to High Seas Bottom Trawl Fleets and the Sustainability of Deep-sea Demersal Fish Stocks. *Marine Policy*, 34, 495–7.

TerraNature (2004). Scientists' Statement on Protecting the World's Deep-Sea Coral and Sponge Ecosystems. Presented to the 2004 Annual Meeting of the American Association for the Advancement of Science (AAAS), and the United Nations Convention on Biological Diversity (CBD). terranature.org/trawlingScientists_ban.htm. Accessed 6 December 2019.

UNFSA (1995). The United Nations Agreement for the Implementation of the Provisions of the United Nations Convention on the Law of the Sea of 10 December 1982 Relating to the Conservation and Management of Straddling Fish Stocks and Highly Migratory Fish Stocks: Articles 5 & 6.

UNFSA (2006). Review Conference. Report of the Review Conference on the Agreement for the Implementation of the Provisions of the United Nations Convention on the Law of the Sea of 10 December 1982 relating to the Conservation and Management of Straddling Fish Stocks and Highly Migratory. *A/CONF210/2006/15* paras 56, 57, 59.

UNGA (2004). Resolution 59/25: Sustainable Fisheries, Including Through the 1995 Agreement for the Implementation of the Provisions of the United Nations Convention on the Law of the Sea of 10 December 1982 Relating to the Conservation and Management of Straddling Fish, para 66.

UNGA (2006). Resolution 61/105: Sustainable Fisheries, Including Through the 1995 Agreement for the Implementation of the Provisions of the United Nations Convention on the Law of the Sea of 10 December 1982 Relating to the Conservation and Management of Straddling Fish. *A/RES/61/105*.

UNGA (2009). Resolution 64/72: Sustainable Fisheries, Including Through the 1995 Agreement for the Implementation of the Provisions of the United Nations Convention on the Law of the Sea of 10 December 1982 Relating to the Conservation and Management of Straddling Fish. *A/RES/64/72*.

UNGA (2011). Resolution 66/68: Sustainable Fisheries, Including Through the 1995 Agreement for the Implementation of the Provisions of the United Nations Convention on the Law of the Sea of 10 December 1982 Relating to the Conservation and Management of Straddling Fish. *A/RES/66/68*.

UNGA (2016). Resolution 64/72: Sustainable Fisheries, Including Through the 1995 Agreement for the Implementation of the Provisions of the United Nations Convention on the Law of the Sea of 10 December 1982 Relating to the Conservation and Management of Straddling Fish. *A/RES/71/123*.

UNICP (2003). United Nations Open-ended Informal Consultative Process on Oceans and the Law of the Sea: Fourth Meeting of the Consultative Process, 2–6 June. https://undocs.org/pdf?symbol=en/A/58/95.

Victorero, L., Watling, L., Palomares, M. L. D., and Nouvian, C. (2018). Out of Sight, but Within Reach: A Global History of Bottom-trawled Deep-sea Fisheries from >400 m Depth. *Frontiers in Marine Science*, 5, doi: 10.3389/fmars.2018.00098.

Waller, R., Watling, L., Auster, P., and Shank, T. (2007). Anthropogenic Impacts on the Corner Rise Seamounts, North-west Atlantic Ocean. *Journal of the Marine Biological Association of the United Kingdom*, 87, 1075–6.

Watling, L., Waller, R., and Auster, P. J. (2007). Corner Rise Seamounts: The Impact of Deep-sea Fisheries. *ICES Insights*, 2007, 10–4.

Watling, L. (2014). Trawling Exerts Big Impacts on Small Beasts. *Proceedings of the National Academy of Science USA*, 111, doi: 10.1073/pnas.1407305111.

Watling, L. and Auster, P. J. (2017). Seamounts on the High Seas Should Be Managed as Vulnerable Marine Ecosystems. *Frontiers in Marine Science*, 4, 1–4.

Watling, L., France, S. C., Pante, E., et al. (2011). *Biology of Deep-Water Octocorals*. Amsterdam, Netherlands: Elsevier Ltd.

Williams, A., Schlacher, T. A., Rowden, A. A., et al. (2010). Seamount Megabenthic Assemblages Fail to Recover from Trawling Impacts. *Marine Ecology*, 31, 183–99.

Wright, G., Ardron, J., Gjerde, K., Currie, D., and Rochette, J. (2015). Advancing Marine Biodiversity Protection Through Regional Fisheries Management: A Review of Bottom Fisheries Closures in Areas Beyond National Jurisdiction. *Marine Policy*, 61, 134–48.

WWF/IUCN (2001). *The Status of Natural Resources on the High-seas*. WWF/IUCN, Gland, Switzerland: WWF/IUCN.

Yeh, J. and Drazen, J. C. (2011). Baited-camera Observations of Deep-sea Megafaunal Scavenger Ecology on the California Slope. *Marine Ecology Progress Series*, 424, 145–56.

Zeller, D., Cashion, T., Palomares, M., and Pauly, D. (2017). Global Marine Fisheries Discards: A Synthesis of Reconstructed Data. *Fish and Fisheries*, 2017, 1–10. doi: 10.1111/faf.12233.

CHAPTER 5

Deep-sea mining: processes and impacts

Daniel O. B. Jones, Diva J. Amon, and Abbie S. A. Chapman

5.1 Deep-sea mining

Mineral resources are vital for modern society. A wide variety of metals, often in great quantities, are required to support modern human lifestyles for much of the global population and, at the moment at least, are necessary to make a successful transition to a green economy. Yet, high-quality terrestrial mineral resources are becoming depleted, and many mines today target much lower grade ores than were common in the past. The presence of high-quality mineral resources in the deep ocean have been known since the HMS *Challenger* expedition in the 1870s (Chapter 1), but the first quantifications of these extensive seabed mineral resources in the 1960s (Mero 1965) ignited a much wider interest. This attention coincided with the technological developments necessary to contemplate the harvesting of solid mineral resources from the deep seabed far from land. These developments and the fear of a deep-sea 'gold rush' (UNGA 1967) stimulated the initiation of international policy that aimed to ensure that deep-sea minerals could be harvested in a fair, environmentally sensitive, and globally equitable way (UNCLOS 1982) (Chapter 3).

Deep-sea mining prospecting currently focuses on three main mineral types: polymetallic nodules, seafloor massive sulphides (SMSs), and polymetallic (cobalt-rich) crusts (Figure 1).

Other resources in deep water have also been identified, including metal-rich sediments and rare earth elements (Amann 1985; Kato et al. 2011), phosphorites (Kudraß 1984), iron sands (Ellis et al. 2017), and even diamonds (Garnett 2002) currently mined in deep waters off Namibia and South Africa. Deep-sea mineral resources occur in a range of environments including abyssal plains, hydrothermal vents, seamounts, continental slopes, and shelves. To date, most of these resources have not been commercially exploited, but there is currently considerable industrial interest. Not surprisingly, there are concerns about environmental impacts and calls to reduce metal demand via the circular economy (Ghisellini et al. 2016).

5.2 Seafloor minerals

5.2.1 Abyssal Plains and polymetallic nodules

The deep-sea abyssal plains, with abundant polymetallic nodules, cover a vast area of the ocean's seafloor (Figure 1) and are one of the most pristine environments on the planet. The polymetallic nodules are potato-sized lumps of accreted metallic ore that lie mostly on the surface of the sediment but can be found to depths of 30cm. Nodules contain cobalt, copper, nickel and manganese, among other metals, and are usually found at depths between 4000 and 6000 m (International Seabed Authority 2010). These nodules take millions of years to form by precipitation of minerals around a core and can reach very high densities (up to 50 kg per square metre), carpeting the seafloor in some places (Figure 2). The most valuable nodules cover an area, roughly the size of the continental USA, known as the Clarion-Clipperton Zone (CCZ), in the northern equatorial

Daniel O. B. Jones, Diva J. Amon, and Abbie S. A. Chapman, *Deep-sea mining: processes and impacts* In: *Natural Capital and Exploitation of the Deep Ocean.*
Edited by: Maria Baker, Eva Ramirez-Llodra, and Paul Tyler, Oxford University Press (2020). © Oxford University Press.
DOI: 10.1093/oso/9780198841654.003.0005

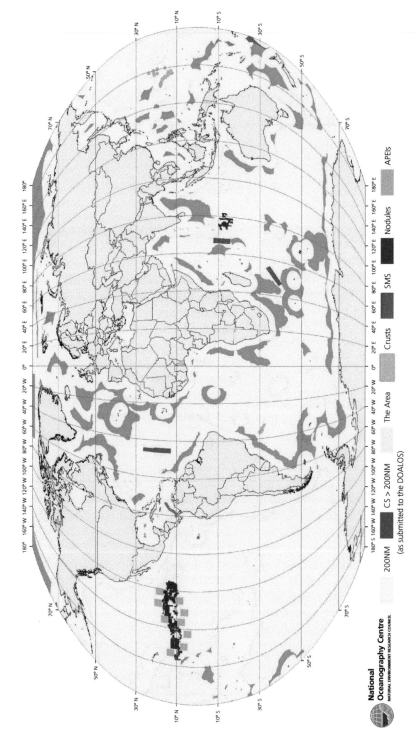

Figure 1 The locations of ISA exploration contract areas for the three main metal-rich mineral resource types in the "the Area" beyond national jurisdiction for seafloor massive sulphides (SMS), polymetallic nodules and crusts. The Areas of Particular Environmental Interest (APEIs) in the Clarion Clipperton Zone are indicated. Also shown are seabed areas within national jurisdiction (extending to 200 nautical miles and to the continental shelf beyond 200 nautical miles) and the Area. Image credit: Alan Evans, National Oceanography Centre, Southampton.

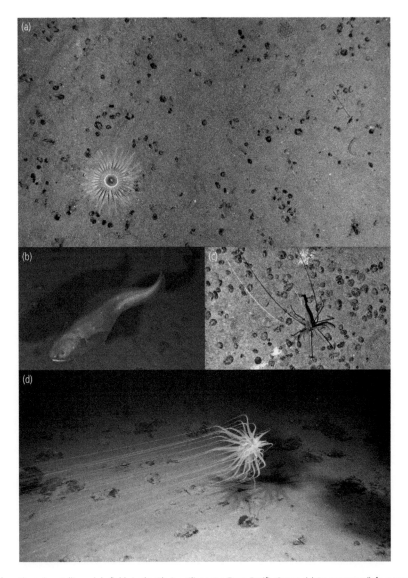

Figure 2 Fauna from the polymetallic nodule fields in the Clarion-Clipperton Zone, Pacific Ocean. (a) An anemone (left; approximately 20cm diameter) and small coral (right); (b) Abyssal fish *Bassozetus* sp.; (c) Decapod crustacean *Bathystylodactylus* sp. (approximately 10cm carapace length); (d) Cnidarian *Relicanthus* sp. (approximately 30cm tall) with very long tentacles streaming out into the seabed current. Image credits and copyright ©: (a and c) National Environment Research Council, RRS *James Cook* Cruise JC120; (b and d) Diva Amon and Craig Smith, University of Hawaii at Manoa.

Pacific Ocean (Figures 1 and 2). This area is not homogeneous and varies in topography, environmental conditions and biology (Simon-Lledó 2019a). The sediments around the nodules are typically very fine, although exposed bedrock locally outcrops.

5.2.2 Seamounts, ridges, and polymetallic crusts

Polymetallic crusts, rich in cobalt, accumulate on the flanks of seamounts and ridges, giving rise to hard, stable habitats at a range of water depths in the open ocean. Crusts form by precipitation in a slow process similar to nodule formation and can be up to 25 cm thick and cover tens to thousands of square kilometres, depending on the size of the seamount (Figures 1 and 3). They are found at water depths between 400 and 4000 m, but the thickest and most valuable, located in the Prime Crust Zone in the central-western Pacific Ocean, occur between 800 and 2500 m (Hein et al. 1992). Seamounts can have rugged topography, often with exposed rocks, although some have areas of sediment—particularly those with summit plateaus (guyots). Seamounts are also associated with complex hydrographic patterns, affecting the habitat, structure, and associated fauna (Xu and Lavelle 2017).

5.2.3 Hydrothermal vents and seafloor massive sulphides

SMSs are the sulphur-rich mineral precipitates that build up around areas where superheated fluids exit the seafloor and mix with cool seawater to form hydrothermal vents (Francheteau et al. 1979; Herzig and Hannington 1995; Humphris et al. 1995; Hoagland et al. 2010; Collins et al. 2013; Figure 4). Hydrothermal fluids are typically rich in sulphide, low in pH and have very variable temperatures (see e.g. Fisher et al. 2007), with the highest recorded temperature of 407 °C in the Equatorial Atlantic vents. SMSs are found in a variety of hydrothermal settings (with the majority occurring on mid-ocean ridges, back-arc spreading centres, and volcanic

Figure 3 Faunal communities from polymetallic-encrusted seamounts in the Pacific Ocean. (a) An abundant community of large corals with anemones, crinoids and ophiuroids; (b) A rattail fish (*Coryphaenoides* sp.); (c) A diverse community of corals with associated crinoids and ophiuroids; (d) An ophiuroid living commensally on a coral that is overgrown in some places by zoanthids; (e) A diverse and abundant coral and sponge community; (f) A community dominated by sponges. Image credits and copyright ©: NOAA Office of Ocean Exploration and Research.

Figure 4 Examples of hydrothermal vent communities. (a) Seafloor massive sulphides with associated communities of shrimp, crabs and snails discovered in 2016 at 3,863 m in the Mariana back-arc axis, West Pacific Ocean. Image credit: NOAA's Office of Ocean Exploration and Research. (b) A black coral observed at 2,227 m in the Endeavour rift valley, Northeast Pacific Ocean. Image credit: Ocean Networks Canada. (c) Squat lobsters and stalked barnacles dominate this chimney, attaining high biomass, in the E9 vent field of the East Scotia Ridge. Image credit: NERC ChEsSo Consortium. (d) Corals living on an extinct chimney at 2,203 m in Mothra vent field, Northeast Pacific Ocean. Image credit: Ocean Networks Canada. (e) *Ridgeia piscesae* tubeworm communities, likely hosting paralvinellid worms, scaleworms, limpets, and many other fauna in their bush-like structures found near a black smoker at 2,133 m at the Endeavour segment of the Juan de Fuca Ridge, Northeast Pacific Ocean. Image credit and copyright ©: Ocean Networks Canada.

arcs), at depths ranging from around 400 to 4960 m (Hannington et al. 2011; Connelly et al. 2012; Boschen et al. 2013; Figure 1). SMSs occur at different water depths and in varying stages of development: very active, high temperature (typically 260 to 400 °C) vent sites, and those more conducive to life around 60 °C; lower-temperature (ambient temperature to around 40 °C) sites characterised by 'shimmering' diffuse flow; and inactive vents at ambient temperatures (Rona 1985, 2003; Fisher et al. 2007; Boschen et al. 2013; Dunn et al. 2018). SMSs are therefore hosted in a spectrum of environments, with different temperature regimes, chemical fluxes, and stability. SMSs are sought for their high concentrations of iron, zinc, copper, and lead, but also gold and silver (Herzig et al. 1999; Baker and German 2009; Collins et al. 2013; Hein et al. 2013).

The environmental setting of SMS deposits contrasts strongly with that of polymetallic nodules. While nodules tend to be located in low-energy abyssal plains, covering large areas of seafloor (10s–1000s km²), individual accumulations of SMSs are present in relatively dynamic environments (affected by active volcanism, plume fallout, and slumping), on a relatively small area of the seabed (mounds may have diameters of ~100–200 m; Hannington et al. 2011; Collins et al. 2013). Some buried sulphides may be considerably larger, and structures associated with SMSs tend to be three-dimensionally extensive (Petersen et al. 2018; Figure 4). By contrast, hydrothermal vents can be stable for decades or more (e.g. Copley et al. 2007; Cuvelier et al. 2011; Du Preez and Fisher 2018).

5.3 Fauna living in association with mineral accumulations

5.3.1 Polymetallic nodules

The fauna of the abyssal plains with nodules shows extremely high biodiversity for many groups across a wide range of animal sizes (Paterson et al. 1998; Glover et al. 2002; Lambshead et al. 2003; Miljutin et al. 2015; Lindh et al. 2017; Simon-Lledó et al. 2019a). Regional diversity, however, is poorly characterised, and the connectivity among areas is only known for a handful of common species (Tabadoa et al. 2018). The visible life (megafauna) is composed primarily

of xenophyophores (giant single-celled foraminifera), cnidarians (e.g. corals and anemones), and sponges, but includes large crustaceans, echinoderms (e.g. sea cucumbers), and fishes (Amon et al. 2016; Vanreusel et al. 2016; Amon et al. 2017a; Amon et al. 2017b; Simon-Lledó et al. 2019a, b). Many organisms, large and small, live on the nodules themselves (e.g. Mullineaux 1987; Veillette et al. 2007; Lim et al 2017). Sediment-dwelling fauna are primarily nematodes, foraminiferans, polychaete worms, and crustaceans (Janssen et al. 2015; Pape et al. 2017; Goineau and Gooday 2019). The density of all faunal groups is generally low relative to other ecosystems (Simon-Lledó et al. 2019b).

5.3.2 Polymetallic crusts

Polymetallic crusts often have suspension feeders attached, such as sponges and cnidarians (e.g. corals) (Figure 3). Dense 'forests' of these fauna can occur, providing additional three-dimensional structure to the habitat that support a range of associated fauna, including crustaceans, echinoderms, and molluscs. Some coral and sponge individuals are known to grow very large (e.g. the octocoral *Iridogorgia magnispiralis* up to 5.7 m tall) and live for more than 4000 years (Roark et al. 2009; Jochum et al. 2012; Watling et al. 2013; Wagner and Kelley 2017). Communities inhabiting ferromanganese-encrusted seamounts and ridges are sensitive to mechanical disturbance and many have already been impacted by bottom trawling and other types of fishing (Schlacher et al. 2014). In addition, most studies have focused on megafaunal organisms (Gollner et al. 2017), leaving the macrofauna, meiofauna, and microbiota less well characterised (George 2013; Zeppilli et al. 2014).

5.3.3 Hydrothermal vents

SMSs have received particular attention from ecologists, as well as geologists, given their occurrence in areas of deep-sea hydrothermal venting. Active vents are well documented and typically host a variety of endemic and densely packed fauna with specific physiological adaptations. The biomass and productivity of deep-sea vent communities is higher than that of the wider deep ocean

(Grassle 1985; Zierenberg et al. 2000), but most of the species within these communities are rare, with low numbers of individuals, many of which are endemic to a handful of sites (Chapman et al. 2018; Van Dover et al. 2018).

Hydrothermal vent communities are also unique on larger spatial scales, with species, and even families, differing among regions across the globe (Bachraty et al. 2009; Moalic et al. 2011). For instance, there are well-studied tubeworm-dominated assemblages in the East Pacific, while snails and barnacles replace these worms in the West Pacific and Indian Oceans (Tunnicliffe et al. 1998; Ramirez-Llodra et al. 2007). The Atlantic Ocean vents are associated with high densities of shrimp (Tunnicliffe et al. 1998; Ramirez-Llodra et al. 2007), and Antarctic vent fields host high densities of squat lobsters (Rogers et al. 2012; Van Dover et al. 2018).

Where hydrothermal flow is waning, or venting has ceased altogether and chemical energy is supplied by sulphide weathering, replacing hydrothermal fluid, different communities thrive (Sylvan et al. 2012). Waning and inactive vents host less dense communities that have higher biodiversity than active vent sites (Tsurumi and Tunnicliffe 2003; Sylvan et al. 2012; Levin et al. 2016). Waning, or senescent, vents continue to host vent-associated fauna, but cannot support the large symbiont-hosting species typical of active vents (Van Dover et al. 2002; Tsurumi and Tunnicliffe 2003). Meanwhile, offering a new long-lasting substratum in ambient conditions, inactive vent sites enable non-endemic fauna, such as sponges, corals, and echinoderm assemblages to establish, with different sensitivities to mining processes (Levin et al. 2016; Van Dover 2019).

The species density, biodiversity, and biomass found at active and inactive vent sites requires improved understanding of these ecosystems and the determination of the risks of anthropogenic disruption (Washburn et al. 2019), although some of the impacts will likely differ as a result of the variable natural ecology described above (Van Dover 2014, 2018). For waning vents, for example, there are limited ecological studies to date (e.g. Van Dover et al. 2002; Tsurumi and Tunnicliffe 2003; Gollner et al. 2015b), and there are very few quanti-

tative studies of the ecology of inactive vent sites (see Van Dover 2019).

5.4 Regulations and jurisdictions

Many deep-ocean mineral resources are found in 'the Area', which encompasses all seafloor and subsoil thereof outside the jurisdiction of individual states (see Chapter 3). Activities related to the exploration and exploitation of mineral resources in the Area (i.e. the seafloor beyond national jurisdiction) are managed by the International Seabed Authority (ISA). The ISA was established in 1994, with the ratification of the United Nations Convention on the Law of the Sea (UNCLOS) and the Part XI Agreement, to issue contracts for seabed mining in the area beyond national jurisdiction, to receive royalties from mining, and to distribute those royalties for the benefit of developing countries that lack the technology and capital to carry out mining for themselves. UNCLOS declared the Area and its mineral resources as the 'common heritage of mankind', to be administered by the ISA for the benefit of humankind as a whole. UNCLOS requires (article 145) that necessary measures be taken to ensure effective protection for the marine environment from harmful effects that may arise from mining-related activities.

The ISA has developed regulations that govern the exploration of all three types of deep-sea mineral and is also in the process of developing regulations for exploitation (Bräger et al. 2018; Chapter 3). These will include details of the technical, financial, and environmental provisions for deep-sea mining in the Area. So far, the ISA has entered into twenty-nine 15-year exploration contracts (some of which have already been granted a 5-year extension). Eighteen of these contracts are for exploration for polymetallic nodules (sixteen in the Clarion-Clipperton Fracture Zone; one in the Western Pacific Ocean; and one in the Central Indian Ocean Basin). There are seven contracts for exploration for polymetallic sulphides in the South West Indian Ridge, Central Indian Ridge, and the Mid-Atlantic Ridge, and five contracts for exploration for cobalt-rich crusts in the Western Pacific Ocean and Western Atlantic. Contractors include state-run enterprises, as well as commercial organisations, with states sponsoring commercial applications to the ISA.

Deep-sea minerals are also found within national jurisdictions and thus can also fall under national regulations. Some nations have already developed plans and regulations for deep-sea mining, with these needing to be at least as stringent as the ISA Mining Code (Chapter 3). As yet, no commercial seafloor mining has occurred, but there are several projects, in shallow and deep waters, in development worldwide. For example, Papua New Guinea, New Zealand, the Kingdom of Tonga, Japan, and Vanuatu have all issued exploration permits to assess the value of SMSs found at deep-sea hydrothermal vents within their territorial waters (Boschen et al. 2013). In addition, a mining lease and environmental permit were granted to Nautilus Minerals for SMS mining at Solwara 1 in Papua New Guinea for exploitation of SMSs (Coffey Natural Systems 2008; Hoagland et al. 2010). Japan has also recently tested deep-sea mining equipment at 1600 m depth on an SMS deposit within their national jurisdiction (Narita et al. 2015; Okamoto et al. 2018), and the Norwegian Petroleum Directorate started exploration of the Arctic Mid-Ocean Ridge for the assessment of SMSs in 2018.

In addition to mining potentially occurring in the deep ocean within national jurisdictions, minerals are already being extracted in shelf waters, as there are fewer technical and regulatory challenges. The De Beers Group is mining for diamonds off Namibia (up to 150 m depth). Namibia also has extensive deep-water phosphate deposits (up to 225 m deep), which have attracted commercial interest by Namibia Marine Phosphate, but approval for this work has not yet been granted. There are also proposals to mine phosphate deposits off the coast of Mexico. These are also currently under review. In New Zealand, Trans-Tasman Resources was recently given consent by the Environmental Protection Authority to mine iron sands at depths less than 45 m, potentially clearing the way for mining of phosphates in deeper water.

5.5 Practicalities of deep-sea mining

Despite the ecological and geological differences between the ecosystems targeted, the approaches to mining these resources are anticipated to have some common stages (Figure 5). Some types of deep-ocean mining, such as the extraction of SMS, may be comparable to that currently conducted on land, and use similar equipment (Jones et al. 2018a). In the early stages of the industry development, it is likely that equipment design will have some similarities with existing land-based mining techniques, subsea trenching and dredging equipment, and remote system technology. In all cases, a seafloor collector will gather the mineral deposit from the seafloor. The minerals will be transferred via a vertical transport system (a riser pipe or other mechanism) to a surface vessel, where they will be dewatered and transferred to transport barges. The processed water, containing suspended sediment and mineral particulates, will either be discharged from the vessel at the sea surface or, more likely, carried via another vertical transport system to be discharged at depth (Weaver et al. 2018).

Whilst the mining of different deposit types shares some methods, the equipment used for extraction will differ. For example, the equipment produced for the Solwara 1 SMS project (off Papua New Guinea; Nautilus Minerals 2008) provides a good indication of the type of seafloor production tools that could be used for this type of mining. Three robotic tools on caterpillar tracks have been designed and built for Solwara 1 to extract the deposits. It is expected that the ground would be prepared for subsequent mining by using a cutting machine to flatten rough topography and create benches for the other machines to operate on. A second cutter would then mine along the benches. Both cutters would excavate rock by a continuous cutting process, where a rotating drum covered in picks cuts swathes approximately 1.8 m wide and approximately 650 mm deep (Nautilus Minerals 2008). A collecting machine would then hydraulically collect the disaggregated rock generated by the cutters off the seafloor as a slurry and pump it into the riser system. Ferromanganese-crust extraction is likely to employ similar cutting and collection machines as those proposed for use at SMS deposits.

Mining polymetallic nodules will require different seabed-mining technology adapted to collecting discrete nodules from on or in the surface of soft

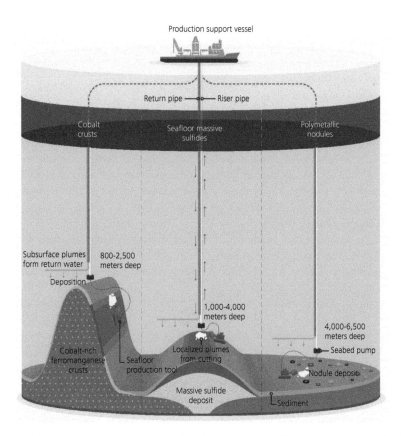

Production support vessel

Return pipe — Riser pipe

Cobalt crusts

Seafloor massive sulfides

Polymetallic nodules

Subsurface plumes form return water | 800-2,500 meters deep

Deposition

1,000-4,000 meters deep

4,000-6,500 meters deep

Seabed pump

Cobalt-rich ferromanganese crusts

Localized plumes from cutting

Nodule deposit

Seafloor production tool

Massive sulfide deposit

Sediment

Figure 5 Potential types of deep-sea mining operation. Image credit and copyright ©: 2017 The Pew Charitable Trusts. All Rights Reserved. Reproduced with permission. Any use without the express written consent of The Pew Charitable Trusts is prohibited.

sediment. Such a mining machine would consist of a vehicle carrying a collector, possibly on sled runners, which may be self-propelled at a speed of about 0.5 m per second, using tank-like tracks or screw-shaped tracks (Oebius et al. 2001; Jones et al. 2017). A mining operation may employ one or multiple collectors that are each likely to be over 10 m wide. The collector would recover nodules in surface sediments (<50 cm deep) by mechanical means or by separating them from the sediment using water jets. The seabed collecting devices would be connected to riser systems that pump the nodules from the seabed to the surface. During nodule mining operations, some of the flocculent surficial sediment would be resuspended by movement of the collector vehicle and hydraulic jets. Deeper sediment layers could be broken up into lumps that then might partly enter the collection system. Such residual sediment would be carried to the sea surface with the nodules and would likely be separated

from the nodules and discharged back near the seabed (Jones et al. 2017).

5.6 Environmental impacts of deep-sea mining

5.6.1 Wide-reaching impacts across depths and habitats

As with most industrial developments, mining in the deep sea will cause impacts to the environment. Mining of deep-ocean minerals will affect the composition, structure, and functioning of the biological communities that live on the minerals themselves, as well as the wider marine environment and nearby habitats. The impacts on fauna associated with different minerals will vary according to habitat and community. Major impacts will affect all areas, irrespective of the mineral being explored or exploited (Gollner et al. 2017). For example, in sedimented areas, the wide tracks of a mining vehicle would compact and move sediments, disturbing or crushing and killing organisms living in and on the sediment itself. In all areas, the mining machinery will introduce noise and light pollution to the dark, quiet deep sea, impacting biological communities at the seafloor and in the water column (Christiansen et al. 2019).

Sediment plumes as a function of mining operations may also have wide-reaching impact. They might be advected widely before settling on the seafloor, smothering fauna and physically damaging small organisms in the process—particularly fauna depending on particles from the water column for food, as these sediments clog and damage filtering, digestive, and respiratory systems (Aleynik et al. 2017; Boschen et al. 2013; Van Dover 2014; Gjerde et al. 2016; Gollner et al. 2017). Sediment plumes will originate from equipment disturbing the seabed, as well as surface dewatering processes (Nautilus Minerals 2008; Miller et al. 2018), where particles may be released in shallower ocean layers. Shallow–released particles may be more harmful to species than particles released at depth, as they can interact with shallow-water processes and organisms (e.g. plankton, birds, fish, marine mammals, and turtles), and humans (via contamination of, or impact on, commercial fishing stocks) (Niner et al. 2018).

Releasing sediment-laden water at depth could also have far-reaching impacts. Seabed communities may be smothered (Van Dover 2011; Miller et al. 2018); nutrients could be introduced to otherwise nutrient-poor systems; thermohaline circulation could be altered; toxic metals could be mobilised (Van Dover 2011; Miller et al. 2018); and deep-sea fisheries may be contaminated in a similar way to those at shallower depths. Models suggest that large sediment plumes will be created that spread over extensive areas, particularly in the case of polymetallic nodule mining (Aleynik et al. 2017), because the sediment particle grain size of the abyssal seafloor is typically <10 μm in diameter (Simon-Lledó et al. 2019a). Sediment plumes are likely to lead to high suspended sediment concentrations in the water column, with potential ramifications for midwater organisms, including larvae and plankton (Christiansen et al. 2019; Drazen et al. 2019).

Suspended sediment will eventually settle over at least twice the area, or more, of the operation (Gjerde et al. 2016; Aleynik et al. 2017).

5.6.2 Impacts of mining seafloor massive sulphides

SMSs are more three-dimensional than crusts or nodules and generally have a smaller footprint on the seafloor so mining of SMSs should leave a relatively small area of directly disturbed total seafloor but disturb a potential large proportion of the hydrothermal environment. The impacts of indirect disturbance, primarily from plumes, are expected to extend much further than the area directly disturbed and come with impacts unique to the setting (Van Dover 2014). SMSs comprise a range of trace metals that differ in quantity and type (Petersen et al. 2018). Given the relatively small yields of minerals that can be obtained from these SMSs, it has been argued that mining of active vent sites should be prohibited as the sulphides are home to high-density, endemic, and unique fauna, in many cases found nowhere else on Earth (Levin et al. 2016; Van Dover et al. 2018; Figure 4). Inactive hydrothermal vent sites might, therefore, seem a lower-impact mining option. Contrarily, the biodiversity of these sites is less well studied and is thought to be higher than at active vents (Van Dover 2011, 2019). Our understanding of the ecology of inactive vents is limited, and data on the composition and functioning of their microbial and faunal communities are scant. Studies of waning, or senescent, vents suggest that around half of the species inhabiting such ecosystems are vent endemic, and the other half are found in the wider environment, away from vent fluid (known as 'background fauna'; Van Dover et al. 2002; Tsurumi and Tunnicliffe 2003; Gollner et al. 2015b). Nevertheless, we lack information on the relationships between most inactive, waning, and active vent communities (Levin et al. 2016), making it difficult to assess the impacts of mining on habitats associated with inactive vents. Some fauna, particularly meiofauna (Gollner et al. 2015a; Gollner et al. 2015b; Plum et al. 2017), living at active vents are also found in peripheral areas surrounding vents, suggesting that connections between active, waning, and inactive environments and communities should be further explored to understand the reach of mining impacts.

Mining of SMSs will have similar impacts on fauna to extraction of other minerals. Animals will be destroyed directly, via the removal of their substratum or crushing, or indirectly, by sediment plumes. Although these impacts are common across different types of minerals, SMS mining is also expected to result in more chemical pollution owing to exposure by sulphides and the release of toxic heavy metals—harmful to peripheral fauna—when sulphur oxidises (Simpson and Spadaro 2016). Whilst these impacts may seem limited to vent fauna and animals living near these habitats, there is increasing evidence to show that nonvent organisms also use deep-sea hydrothermal vents. For example, some skates incubate their egg cases at active hydrothermal vent sites (Salinas-de-León et al. 2018). Nevertheless, the effects of mining on nonvent organisms remain difficult to quantify and monitor, as we are only now starting to learn about interactions between nonvent fauna and these extreme venting habitats. This is compounded by a divide in the extent of deep-sea biological research among different nations and the targets of hydrothermal vent mining prospects (Thaler and Amon 2019).

5.6.3 Environmental impacts of mining polymetallic nodules

Polymetallic nodule mining is expected to have a number of specific impacts on the highly diverse seafloor and water-column communities associated with this environment. Most obviously, polymetallic nodules provide a hard surface that is home to a wide variety of life, including sponges, corals, anemones, worms, foraminifera, nematodes, and microbes (Mullineaux 1987; Glover et al. 2002; Dahlgren et al. 2016; Lindh et al. 2017; Veillette et al. 2007). In turn, many larger organisms provide a substratum, or foundation, for other animals (e.g. sea stars and small crustacea on corals; Mullineaux 1987; Gooday et al. 2015; Amon et al. 2016). Polymetallic nodules take millions of years to form (Kuhn et al. 2017). Removing the polymetallic nodules will thus have major impacts on the associated fauna (Simon-Lledó et al. 2019c),

particularly as half of megafaunal species in the CCZ may directly depend on the polymetallic nodules (Amon et al. 2016; Vanreusel et al. 2016). A recently discovered example of this is the white 'Casper' octopus that lays its eggs on sponge stalks growing on polymetallic nodules and crusts (Purser et al. 2016). Deep-diving whales may also interact with the seafloor in the CCZ, possibly as part of their feeding behaviour (Marsh et al. 2018).

Polymetallic nodules occur on soft sediments in very stable environments with strong vertical stratification and low concentrations of organic matter (Mewes et al. 2014). Disturbance of sedimentary environments will lead to the disruption of the surface sediment (5–20 cm deep) and cause exposure of deeper sediment layers and compaction. These changes will impact the sediment geochemistry, which will kill the fauna living within the sediments and impair ecosystem recovery processes. In addition, the scale of polymetallic nodule mining will be particularly large, with the potential for areas of several hundred square kilometres to be disturbed each year by a single operation (Smith et al. 2008). Impacts on this scale are rare in deep-ocean environments, with the exception of benthic storms (Hollister and McCave 1984), and may lead to effects that can be seen at regional scales, such as population reductions or even species extinctions.

5.6.4 The effects of mining polymetallic crusts

The mining of polymetallic crusts will also have a variety of environmental impacts (Schlacher et al. 2014). The extraction process will remove mineral-rich surfaces of the seamounts, which support a benthic fauna that include corals, sponges, echinoderms, and other invertebrates, sometimes in very dense populations. Many of these animals are not yet known to science and may be long-lived and/or fragile, and larger individuals may be responsible for much of the reproductive output to maintain the population (Roark et al. 2009; Jochum et al. 2012). Fishes and other pelagic organisms are often present in large numbers and may be associated with benthic organisms (Schlacher et al. 2014). There is potentially a relatively high risk of species extinctions on seamounts as a result of isolation, endemicity, and uniqueness (Richer de Forges et al. 2000). Polymetallic-crust extraction will occur in areas affected by other human activities, particularly deep-sea fishing and ocean acidification (see Chapter 9), and that could result in cumulative negative impacts (Morato et al. 2010). The sediment plumes generated by mining operations may directly impact the fish and other pelagic organisms that congregate on and above seamounts (Barbier et al. 2014; Christiansen et al. 2019). Additionally, many commercially exploited fish species depend on seamount-associated rich benthic invertebrate assemblages as nursery grounds and as refuges to avoid predators. Thus, mining may also have secondary impacts on fish communities and the ecosystem services they provide.

5.7 Cross-ecosystem impacts: degradation and recovery

Polymetallic crusts and nodules host ecosystems that are not typically exposed to disturbances similar to those expected from mining. The low-energy availability limits biological rate processes (McClain et al. 2012), which will reduce the resistance and resilience of communities to mining. Although SMSs are established in the venting regions associated with volcanic activity, they can be found in areas of long-term stability (Copley et al. 2007; Du Preez and Fisher 2018). Across ecosystems and habitats, deep-sea mining will result in impacts, which may lead to species losses—particularly for rare and/or endemic taxa. Given the general slow pace of life, recovery from deep-sea-mining disturbance is expected to be slow (Gollner et al. 2017; Jones et al. 2017). With species and habitat losses may come changes to trophic interactions, population dynamics, and the composition of deep-sea communities (Boschen et al. 2013; Van Dover et al. 2018). Ecosystem functioning (Sweetman et al. 2019) will be impacted by mining activities, as community structure and biogeochemical cycling are altered (Paul et al. 2018; Stratmann et al. 2018). The potential for regional-scale impact to ecosystem functions is unknown. There will be associated loss of the potential ecosystem services offered by deep-sea communities, for example, those associated with novel genetic or biochemical components (see Chapter 7).

Despite expectations regarding ecosystem degradation and recovery following deep-sea mining, great uncertainty remains regarding the natural environment in and around the deep-ocean mineral deposits currently being considered for exploitation, as well as concerning the full impact of mining and the resilience of associated ecosystems (Gollner et al. 2017; Miller et al. 2018). Existing information on the ecological effects of mining and potential recovery times is limited, despite deep-ocean-mining-related research having been conducted since the 1970s (e.g. Japan Deep-Sea Impact Experiment, Benthic Impact Experiments, and Indian Deep-sea Environment Experiment; Jones et al. 2017). The most intensive assessment has been the disturbance and recolonisation experiment (DISCOL) that was carried out in an area of polymetallic nodules off Peru at a water depth of 4150 m in 1989 (Thiel and Schriever 1990). This experiment disturbed the seafloor across several kilometres with nearly eighty plough tracks. The experimental site and other similar seafloor areas were reinvestigated in 2015 through the EU–based intergovernmental Joint Programming Initiative Healthy and Productive Seas and Oceans (JPI-Oceans) Programme. Even after 26 years, there was little change to the disturbed tracks (Miljutin et al. 2011): they looked much the same as when they were first made (Boetius 2015). Detailed biological studies showed that while some mobile species moved back into the tracks, there was very little recolonisation of disturbed areas (Simon-Lledó et al. 2019c). Biogeochemical changes persisted (Paul at al. 2018), affecting important ecosystem functions (Stratmann et al. 2018), and even microbial communities show little sign of recovery (Gjerde et al. 2016). Recovery from commercial-scale mining is likely to be even slower, as both the temporal and spatial scales of disturbance will be much larger than those of the experiments, and there will be cumulative impacts affecting the deep sea, further complicating recovery (e.g. climate change—see Chapter 9). Regional-scale impacts could result in local extinctions and population declines, reducing biological connectivity and reproductive success, as larval supply decreases with distance from unaffected populations.

5.8 Knowledge gaps: a need to deepen understanding

A fundamental problem for predicting the impacts of deep-ocean mining is our limited knowledge about deep-sea ecosystems in general (Gollner et al. 2017). Many of the animals living in and on nodules, crusts, and sulphides (particularly on inactive vents) are poorly understood or entirely new to science; community and population dynamics also need to be further modelled and explored. As such, basic ecological information is missing (e.g. species identities, population sizes, behaviour, life history, and distributions). For the vast majority of organisms, we do not know how populations are connected or what is needed for the maintenance of viable communities (Mullineaux et al. 2018). Some species that have been studied show wide distributions and connectivity between populations on scales of hundreds of kilometres (Young et al. 2008; Breusing et al. 2015; Mitarai et al. 2016; Taboada et al. 2018), but assessments of polymetallic nodule systems show that there are also a large number of rare species, which tend to occupy a smaller geographic range (Glover et al. 2002; Janssen et al. 2015; Wilson 2017). In some cases, these patterns may be an artefact of limited sampling, but many species are known from only a few individuals that have poorly understood ecological roles, particularly for the smaller animals. In terrestrial ecosystems, conservation measures tend to focus on rare species, given the higher risk of extinction assumed for species with low abundances, small population sizes, and/or restricted geographic ranges; other species are then prioritised for the ecosystem functions they support (Pimm et al. 1988; Gaston 1994, 2003; Margules and Pressey 2000; Gaston and Fuller 2008; Leitão et al. 2016). Species that support ecosystem functioning, or contribute to the overall biodiversity of an area, can be used as indicators of ecosystem health. Accordingly, it seems that deep-sea 'indicator' species could prove helpful for developing conservation and management plans. However, identifying 'indicator' species in the deep sea is difficult, complicated by evidence such as the unique contributions of common and rare species in some systems (e.g. Chapman et al. 2018). This, in addition to our limited ecological knowledge of deep-sea species, relative to terrestrial taxa (Gooday and Goineau 2019), inhibits specific species-based conservation actions and our efforts to improve management. Nevertheless, recent progress in identifying potential indicator species in areas with massive sulphides (Sarrazin et al. 2015) and polymetallic nodules (Taboada et al. 2018) offer some promise for this type of conservation action.

5.9 Environmental management: reducing the impact of deep-ocean mining

5.9.1 Environmental management processes

Deep-sea mining is generally regarded as an inherently unsustainable and destructive process, but impacts may be reduced with good management (Durden et al. 2017; Jones et al. 2019). Before management strategies can be designed, however, fundamental research is needed to ascertain baseline conditions. This research should include high-resolution mapping and assessments of the spatial and temporal patterns in physical and chemical conditions and the faunal communities inhabiting the areas. Ecosystem functioning should also be studied to prevent mining-related ecosystem collapse and to ensure that the ecosystem services that we rely on will not be compromised during and after mining. Overall, this information will result in a better understanding of the communities that are at risk, and can be incorporated into robust environmental management plans.

Following thorough ecological-baseline research, the next stage is to evaluate potential impacts of mining operations by undertaking Environmental Impact Assessments (EIAs). A typical EIA assesses the risks of the project and sensitivities of the environment. It also identifies alternative project plans that may reduce or mitigate the impacts of the industry, helping to preserve unique and vulnerable communities (Durden et al. 2018). The risks are typically reduced by applying a four-stage mitigation hierarchy, whereby, in order of preference, risks are: (1) avoided (e.g. by moving the project away from a vulnerable habitat), (2) minimised (e.g. by introducing new technology to model and reduce the sediment plume generated by a mining vehicle), (3) restored, or (4) offset. The last two options, restoration and offsetting, are complex (Cuvelier et al. 2018) and often considered impractical for deep-sea mining at present as a result of a range of biological, technical, financial, and legal issues (Van Dover et al. 2017). Once a project's risks have been reduced as much as is practical, a decision can be made as to whether the economic, social, and political benefits of the project outweigh the costs, environmental or otherwise. If the project is approved, then plans can be made for ongoing environmental monitoring to identify and measure the impacts of the project. If these negative effects become too severe, the project can be curtailed. These management strategies should be continued throughout the life of the project and after it has been decommissioned.

5.9.2 Environmental management responsibilities

The mining company primarily carries out the environmental management of individual mining projects. However, additional regional management is necessary for sustainable mining on broader scales to achieve wider conservation objectives. Decisions about mine-site placement, the number of active mines, and the designation of marine protected areas are best made by the agency responsible for the regulation of mining within a region. In the case of deep-sea mining in areas beyond national jurisdiction, this is the ISA (Bräger et al. 2018). To date, the spatial allocation of exploration areas has been driven by contractor applications to the ISA in areas of interest in the global ocean. However, a regional management plan has been proposed for the CCZ (Wedding et al. 2013), which currently includes nine areas, known as Areas of Particular Environmental Interest (APEIs), where mining cannot currently occur. These APEIs were designated after the exploration areas were implemented in areas with the highest polymetallic nodule densities, and so are peripheral to the central CCZ. Each APEI consists of a 200 × 200 km² protected zone, surrounded by a 100 km buffer. The APEIs are designed to be large enough to maintain minimum viable population sizes for species within the proposed mining areas via self-recruitment after mining has ceased. Further spatial management includes Preservation Reference

Zones (or PRZs), which are areas established to monitor the effects of individual mining projects (alongside the Impact Reference Zones, or IRZ), and, by being representative areas where mining cannot occur, may also act as protected areas (Jones et al. 2018b). All other areas of mining interest, excluding the CCZ, do not have regional environmental management plans yet in place. These plans need to be developed prior to mining and should take into account a range of factors including the mining type, potential impacts, specific ecosystems, connectivity, vulnerability, and the optimal approaches for management. Initial scientific work to support this has begun (Dunn et al. 2018).

The high uncertainty associated with the impacts of mining, as well as the environments and ecosystems affected and how they will respond to disturbance, makes management of deep-ocean mining highly complex. It is likely that management will need to be adaptive to take into account the amount of additional information that will become available once mining starts. Uncertainty should be clearly documented and reduced as far as possible, for example, through further research targeting the areas and regions of exploitation interest. The precautionary approach (or 'precautionary principle') stemming from the 1992 Rio Declaration on Environment and Development Principle 15 indicates that positive action to protect the environment may be required before scientific proof of harm has been provided. This will be important when managing deep-sea mining activities, and could include protecting large and/or connected areas and careful evaluation of small mining projects before approving larger mines. These approaches have all been discussed, and some are underlying legal principles in UNCLOS. However, their implementation remains one of the key challenges associated with improving the sustainability of deep-sea mining.

5.10 Conclusions

Current interest in deep-sea mining is focused on three habitats for which we are lacking fundamental baseline knowledge about species composition, ecology, and natural environmental conditions. It is, however, without doubt that deep-sea mining has the potential to have far-reaching impacts on our oceans. While some impacts will be resource specific, mineral deposit extraction will broadly affect local and regional marine communities by removing suitable habitats, creating far-reaching sediment plumes, and reducing population sizes (even potentially causing extinctions). Deep-sea mining will impact habitats, which will take an unknown amount of time to recover; estimates for active vents range from decades to evolutionary timescales (Van Dover et al. 2018), whereas crusts and nodules are expected to also be on evolutionary timescales. The need for baseline information about reproduction and dispersal, growth, population sizes, diversity, distributions, connectivity, trophic interactions, and more is essential for successful EIAs and the sustainable management of these habitats during mineral extraction. As exploitation on such a large scale has never occurred before in the deep ocean, its environmental management is a nascent endeavour. For the impacts of deep-sea mining to be minimised, there is a requirement for cooperation between all stakeholders on a national and international level: industry, policymakers, scientists, NGOs, and members of the public whose livelihoods depend on ocean resources. Coherent local-to-regional-scale and strategic planning and management are important elements of this (Wedding et al. 2013), which need to be conducted in all areas in which there is interest in mining. The ISA and nations developing their seabed resources need to use all available tools and stakeholder expertise to stand by their commitments to ensure the harmful effects from deep-ocean mining are minimised and the industry proceeds in an informed and careful manner in the future. If deep-sea minerals are extracted for the benefit of humankind, careful consideration should be given concerning the trade-offs in terms of biotic natural capital and ecosystem services (see Chapter 2).

Acknowledgements

The authors would like to thank Alan Evans at the National Oceanography Centre for creating the map of mining activities. DJ received funding from the European Union Seventh Framework Programme (FP7/2007-2013) under the MIDAS (Managing Impacts of Deep-see Resource Exploitation) project, grant agreement 603418. DJ also acknowledges support from the UK Natural Environment Research Council (NERC) through National Capability funding to NOC (grant number NE/R015953/1). DA has received funding from the European Union's Horizon 2020 research and innovation programme under the Marie Sklodowska-Curie grant agreement number 747946. AC received support from SPITFIRE Doctoral Training Partnership (supported by the Natural Environmental Research Council, grant number: NE/L002531/1) and the University of Southampton. The funders had no role in the study, decision to publish, or preparation of the manuscript.

References

Aleynik, D., Inall, M. E., Dale, A., and Vink. A. (2017). Impact of Remotely Generated Eddies on Plume Dispersion at Abyssal Mining Sites in the Pacific. *Scientific Reports*, 7, 16959.

Amann, H. (1985). Development of Ocean Mining in the Red Sea. *Marine Mining*, 5, 103–16.

Amon, D. J., Ziegler, A. F., Dahlgren, T. G., et al. (2016). Insights into the Abundance and Diversity of Abyssal Megafauna in a Polymetallic-nodule Region in the Eastern Clarion-Clipperton Zone. *Scientific Reports*, 6, 30492.

Amon, D., Ziegler, A., Kremenetskaia, A., et al. (2017a). Megafauna of the UKSRL Exploration Contract Area and Eastern Clarion-Clipperton Zone in the Pacific Ocean: Echinodermata. *Biodiversity Data Journal*, 5, e11794.

Amon, D. J., Ziegler, A., Drazen, J. C., et al. (2017b). Megafauna of the UKSRL Exploration Contract Area and Eastern Clarion-Clipperton Zone in the Pacific Ocean: Annelida, Arthropoda, Bryozoa, Chordata, Ctenophora, Mollusca. *Biodiversity Data Journal*, 5, e14598.

Bachraty, C., Legendre, P., and Desbruyères, D. (2009). Biogeographic Relationships Among Deep-sea Hydrothermal Vent Faunas at Global Scale. *Deep-Sea Research Part I: Oceanographic Research Papers*, 56, 1371–8.

Baker, M. C. and German, C. R. (2009). Going for Gold! Who Will Win in the Race to Exploit Ores from the Deep Sea. *Ocean Challenge*, 16, 10–17.

Barbier, E. B., Moreno-Mateos, D., Rogers, A. D., et al. (2014). Ecology: Protect the Deep Sea. *Nature*, 505, 475–7.

Boetius, A. (2015). RV *Sonne Fahrtbericht*/Cruise Report SO242-2: JPI OCEANS *Ecological Aspects of Deep-sea*

Mining, DISCOL Revisited, Guayaquil—Guayaquil (Equador), 28 August–1 October 2015. GEOMAR Report, NSer 027. Keil, Germany: Helmholtz-Zentrum für Ozeanforschung.

Boschen, R. E., Rowden, A. A., Clark, M. R., and Gardner, J. P. A. (2013). Mining of Deep-sea Seafloor Massive Sulfides: A Review of the Deposits, Their Benthic Communities, Impacts from Mining, Regulatory Frameworks and Management Strategies. *Ocean and Coastal Management*, 84, 54–67.

Bräger, S., Romero Rodriguez, G. Q., and Mulsow, S. (2018). The Current Status of Environmental Requirements for Deep Seabed Mining Issued by the International Seabed Authority. *Marine Policy*, in press. doi: 10.1016/j.marpol.2018. 09.003.

Breusing, C., Johnson, S., Tunnicliffe, V., and Vrijenhoek, R. (2015). Population Structure and Connectivity in Indo-Pacific Deep-sea Mussels of the *Bathymodiolus septemdierum* Complex. *Conservation Genetics*, 16, 1415–30.

Chapman, A. S. A., Tunnicliffe, V., and Bates, A. E. (2018). Both Rare and Common Species Make Unique Contributions to Functional Diversity in an Ecosystem Unaffected by Human Activities. *Diversity and Distributions*, 24, 568–78.

Christiansen, B., Denda, A., and Christiansen, S. (2019). Potential Effects of Deep Seabed Mining on Pelagic and Benthopelagic Biota. *Marine Policy*, in press.

Coffey Natural Systems. (2008). *Environmental Impact Statement: Nautilus Minerals Niugini Limited, Solwara 1 Project*. Queensland, Australia: Coffey Natural Systems.

Collins, P., Croot, P., Carlsson, J., et al. (2013). A Primer for the Environmental Impact Assessment of Mining at Seafloor Massive Sulfide Deposits. *Marine Policy*, 42, 198–209.

Copley, J. T., Jorgensen, P. B. K., and Sohn, R. A. (2007). Assessment of Decadal-scale Ecological Change at a Deep Mid-Atlantic Hydrothermal Vent and Reproductive Time-series in the Shrimp *Rimicaris exoculata*. *Journal of the Marine Biological Association of the United Kingdom*, 87, 859–67.

Connelly, D., Copley, J., Murton, B., et al. (2012). Hydrothermal Vent Fields and Chemosynthetic Biota on the World's Deepest Seafloor Spreading Centre. *Nature Communications*, 3, 620. doi:10.1038/ncomms1636.

Cuvelier, D., Sarrazin, J., Colaço, A., et al. (2011). Community Dynamics over 14 Years at the Eiffel Tower Hydrothermal Edifice on the Mid-Atlantic Ridge. *Limnology and Oceanography*, 56, 1624–40.

Cuvelier, D., Gollner, S., Jones, D. O. B., et al. (2018). Potential Mitigation and Restoration Actions in Ecosystems Impacted by Seabed Mining. *Frontiers in Marine Science*, 5, 467.

Dahlgren, T., Wiklund, H., Rabone, M., et al. (2016). Abyssal Fauna of the UK-1 Polymetallic Nodule Exploration Area, Clarion-Clipperton Zone, Central Pacific Ocean: Cnidaria. *Biodiversity Data Journal*, 4, e9277.

Drazen, J. C., Smith, C. R., Gjerde, K., et al. (2019). Report of the Workshop Evaluating the Nature of Midwater Mining Plumes and Their Potential Effects on Midwater Ecosystems. *Research Ideas and Outcomes*, 5, e33527.

Du Preez, C. and Fisher, C. (2018). Long-Term Stability of Back-Arc Basin Hydrothermal Vents. *Frontiers of Marine Science*, 5, 54.

Dunn, D., Van Dover, C., Etter, R., et al. (2018). A Strategy for the Conservation of Biodiversity on Mid-ocean Ridges from Deep-sea Mining. *Science Advances*, 4, eaar4313.

Durden, J. M., Murphy, K., Jaeckel, A., et al. (2017). A Procedural Framework for Robust Environmental Management of Deep-sea Mining Projects Using a Conceptual Model. *Marine Policy*, 84, 193–201.

Durden, J. M., Lallier, L. E., Murphy, K., et al. (2018). Environmental Impact Assessment Process for Deep-sea Mining in 'the Area'. *Marine Policy*, 87, 194–202.

Ellis, J. I., Clark, M. R., Rouse, H. L., and Lamarche, G. (2017). Environmental Management Frameworks for Offshore Mining: The New Zealand Approach. *Marine Policy*, 84, 178–92.

Fisher, C. R., Takai, K., and Le Bris, N. L. (2007). Hydrothermal Vent Ecosystems. *Oceanography*, 20, 14–23.

Francheteau, J., Needham, H., Choukroune, P., et al. (1979). Massive Deep-sea Sulphide Ore Deposits Discovered on the East Pacific Rise. *Nature*, 277, 523–8.

Garnett, R. H. T. (2002). Recent Developments in Marine Diamond Mining. *Marine Georesources and Geotechnology*, 20, 137–59.

Gaston, K. (1994). *Rarity*. Dordrecht, Netherlands: Springer Netherlands.

Gaston, K. (2003). *The Structure and Dynamics of Geographic Ranges*. Oxford: Oxford University Press.

Gaston, K. and Fuller, R. (2008). Commonness, Population Depletion and Conservation Biology. *Trends in Ecology and Evolution*, 23, 14–19.

George, K. H. (2013). Faunistic Research on Metazoan Meiofauna From Seamounts—A Review. *Meiofauna Marine*, 20, 1e32.

Ghisellini, P., Catia, C., and Ulgiati. S. (2016). A Review on Circular Economy: The Expected Transition to a Balanced Interplay of Environmental and Economic Systems. *Journal of Cleaner Production*, 114, 11–32. https://doi.org/10.1016/j.jclepro.2015.09.007.

Gjerde, K. M., Weaver, P., Billett, D., et al. (2016a). *Implications of MIDAS Results for Policy Makers:*

Recommendations for Future Regulations. Romsey: Seascape Consultants.

Gjerde, K., Reeve, L., Harden-Davies, H., et al. (2016b). Protecting Earth's Last Conservation Frontier: Scientific, Management and Legal Priorities for MPAs Beyond National Boundaries. *Aquatic Conservation: Marine and Freshwater Ecosystems*, 26, 45–60.

Glover, A. G., Smith, C. R., Paterson, G. L. J., et al. (2002). Polychaete Species Diversity in the Central Pacific Abyss: Local and Regional Patterns, and Relationships with Productivity. *Marine Ecology Progress Series*, 240, 157–69.

Goineau, A. and Gooday, A. J. (2019). Diversity and Spatial Patterns of Foraminiferal Assemblages in the Eastern Clarion–Clipperton Zone (Abyssal Eastern Equatorial Pacific). *Deep-Sea Research I*, 149, 103036.

Gollner, S., Govenar, B., Fisher, C. R., and Bright, M. (2015a). Size Matters at Deep-sea Hydrothermal Vents: Different Diversity and Habitat Fidelity Patterns of Meio- and Macrofauna. Marine Ecology Progress Series, 520, 57–66.

Gollner, S., Govenar, B., Martinez Arbizu, P., et al. (2015b). Differences in Recovery Between Deep-sea Hydrothermal Vent and Vent-proximate Communities after a Volcanic Eruption. Deep-Sea Research I, 106, 167–82.

Gollner, S., Kaiser, S., Menzel, L., et al. (2017). Resilience of Benthic Deep-sea Fauna to Mining Activities. *Marine Environmental Research*, 129, 76–101.

Gooday, A. J. and Goineau, A. (2019). The Contribution of Fine Sieve Fractions (63–150 μm) to Foraminiferal Abundance and Diversity in an Area of the Eastern Pacific Ocean Licensed for Polymetallic Nodule Exploration. *Frontiers in Marine Science*, 6, 114. doi: 10.3389/fmars.2019.00114.

Gooday, A., Goineau, A., and Voltski, I. (2015). Abyssal Foraminifera Attached to Polymetallic Nodules from the Eastern Clarion Clipperton Fracture Zone: A Preliminary Description and Comparison with North Atlantic Dropstone Assemblages. *Marine Biodiversity*, 45, 1–22.

Gooday, A. J., Holzmann, M. C. C., Goineau, A., et al. (2017). Giant Protists (Xenophyophores, Foraminifera) Are Exceptionally Diverse in Parts of the Abyssal Eastern Pacific Licensed for Polymetallic Nodule Exploration. *Biological Conservation*, 207, 106–16.

Grassle, J. (1985). Hydrothermal Vent Animals: Distribution and Biology. *Science*, 229, 713–17.

Hannington, M., Jamieson, J., Monecke, T., Petersen, S., and Beaulieu, S. (2011). The Abundance of Seafloor Massive Sulfide Deposits. *Geology*, 39, 1155–8.

Hein, J. R., Schulz, M. S., and Gein, L. M. (1992). Central Pacific Cobalt-rich Ferromanganese Crusts: Historical Perspective and Regional Variability, in B. H. Keating

and B. R. Bolton (eds.) *Geology and Offshore Mineral Resources of the Central Pacific Basin.* New York: Springer-Verlag, pp. 261–83.

Hein, J., Mizell, K., Koschinsky, A., and Conrad, T. (2013). Deep-ocean Mineral Deposits as a Source of Critical Metals for High- and Green-technology Applications: Comparison with Land-based Resources. *Ore Geology Reviews*, 51, 1–14.

Herzig, P. and Hannington, M. (1995). Polymetallic Massive Sulfides at the Modern Seafloor: A Review. *Ore Geology Reviews*, 10, 95–115.

Herzig, P. M., Petersen, S., and Hannington, M. D. (1999). Epithermal-type Gold Mineralization at Conical Seamount: A Shallow Submarine Volcano South of Lihir Island, Papua New Guinea. Mineral Deposits: Processes to Processing. Proceedings of the Fifths Biennial SGA Meeting and the 10th IAGOD. Leiden, The Netherlands: Balkema, pp. 527–30.

Hoagland, P., Beaulieu, S., Tivey, M., et al. (2010). Deep-sea Mining of Seafloor Massive Sulfides. *Marine Policy*. 34, 728–32.

Hollister, C. D. and McCave, I. N. (1984). Sedimentation Under Deep-sea Storms. *Nature*, 309, 220–5.

Humphris, S. E., Herzig, P. M., Miller, D. J., et al. (1995). The Internal Structure of an Active Sea-floor Massive Sulphide Deposit. *Nature*, 377, 713–16.

International Seabed Authority (2010). *A Geological Model of Polymetallic Nodule Deposits in the Clarion Clipperton Fracture Zone.* Kingston, Jamaica: International Seabed Authority.

Janssen, A., Kaiser, S., Meißner, K., et al. (2015). A Reverse Taxonomic Approach to Assess Macrofaunal Distribution Patterns in Abyssal Pacific Polymetallic Nodule Fields. *PLoS ONE*, 10, e0117790.

Jochum, K. P., Wang, X., Vennemann, T. W., Sinha, B., and Müller, W. E. G. (2012). Siliceous Deep-sea Sponge *Monorhaphis chuni*: A Potential Paleoclimate Archive in Ancient Animals. *Chemical Geology*, 300–1, 143–51.

Jones, D. O. B., Kaiser, S., Sweetman, A. K., et al. (2017). Biological Responses to Disturbance from Simulated Deep-sea Polymetallic Nodule Mining. *PLoS ONE*, 12, e0171750.

Jones, D. O. B., Amon, D. J., and Chapman, A. S. A. (2018a). Mining Deep-Ocean Mineral Deposits: What Are the Ecological Risks? *Elements*, 14, 325–30.

Jones, D. O. B., Ardron, J. A., Colaço, A., and Durden, J. M. (2018b). Environmental Considerations for Impact and Preservation Reference Zones for Deep-sea Polymetallic Nodule Mining. *Marine Policy*, in press.

Jones, D. O. B., Durden, J. M., Murphy, K., et al. (2019). Existing Environmental Management Approaches Relevant to Deep-sea Mining. *Marine Policy*, 103, 172–81.

Kato, Y., Fujinaga, K., Nakamura, K., et al. (2011). Deep-sea Mud in the Pacific Ocean as a Potential Resource for Rare-earth Elements. *Nature Geoscience*, 4(8), 535. doi: 10.1038/NGEO1185.

Kudraß, H. R. (1984). The Distribution and Reserves of Phosphorite on the Central Chatham Rise (SONNE-17 Cruise 1981). *Geologische Jahrbuch, Reihe D*, 65, 179–94.

Kuhn, T., Wegorzewski, A., Rühlemann, C., and Vink, A. (2017). Composition, Formation, and Occurrence of Polymetallic Nodules, in R. Sharma (ed.) *Deep-Sea Mining: Resource Potential, Technical and Environmental Considerations*. New York: Springer International Publishing, pp. 23–63.

Lambshead, P. J. D., Brown, C. J., Ferrero, T. J., et al. (2003). Biodiversity of Nematode Assemblages from the Region of the Clarion-Clipperton Fracture Zone, an Area of Commercial Mining Interest. *BMC Ecology*, 3, 1–12.

Levin, L. A., Baco, A. R., Bowden, D., et al. (2016). Hydrothermal Vents and Methane Seeps: Rethinking the Sphere of Influence. *Frontiers in Marine Science*, 3, 72.

Leitão, R., Zuanon, J., Villéger, S., et al. (2016). Rare Species Contribute Disproportionately to the Functional Structure of Species Assemblages. *Proceedings of the Royal Society B*, 283, 20160084. http://dx.doi.org/10.1098/rspb.2016.0084.

Lim, S-C., Wiklund, H., Glover, A. G., Dahlgren, T. G., and Tan, K-S. (2017). A New Genus and Species of Abyssal Sponge Commonly Encrusting Polymetallic Nodules in the Clarion-Clipperton Zone, East Pacific Ocean. *Systematics and Biodiversity*, 15, 507–19.

Lindh, M. V., Maillot, B. M., Shulse, C. N., et al. (2017). From the Surface to the Deep-Sea: Bacterial Distributions across Polymetallic Nodule Fields in the Clarion-Clipperton Zone of the Pacific Ocean. *Frontiers in Microbiology*, 8, 1696. doi: 10.3389/fmicb.2017.01696.

Margules, C. and Pressey, R. (2000). Systematic Conservation Planning. *Nature*, 405, 243–53.

Marsh, L., Huvenne, V. A. I., and Jones, D. O. B. (2018). Geomorphological Evidence of Large Vertebrates Interacting with the Seafloor at Abyssal Depths in a Region Designated for Deep-sea Mining. *Royal Society Open Science*, 5, 180286.

McClain, C. R., Allen, A. P., Tittensor, D. P., and Rex, M. A. (2012). Energetics of Life on the Deep Seafloor. *Proceedings of the National Academy of Sciences USA*, 109, 15366–71.

Mero, J.L. (1965). *The Mineral Resources of the Sea*. Amsterdam, Netherlands: Elsevier.

Mewes, K., Mogollón, J. M., Picard, A., et al. (2014). Impact of Depositional and Biogeochemical Processes on Small Scale Variations in Nodule Abundance in the Clarion-Clipperton Fracture Zone. *Deep-Sea Research I*, 91, 125–41.

Miljutin, D. M., Miljutina, M. A., Arbizu, P. M., and Galéron, J. (2011). Deep-sea Nematode Assemblage Has not Recovered 26 Years after Experimental Mining of Polymetallic Nodules (Clarion-Clipperton Fracture Zone, Tropical Eastern Pacific). *Deep-Sea Research I*, 58, 885–97.

Miljutin, D., Miljutina, M., and Messié, M. (2015). Changes in Abundance and Community Structure of Nematodes from the Abyssal Polymetallic Nodule Field, Tropical Northeast Pacific. *Deep-Sea Research I*, 106, 126–35.

Miller, K., Thompson, K., Johnston, P., and Santollo, D. (2018). An Overview of Seabed Mining Including the Current State of Development, Environmental Impacts, and Knowledge Gaps. *Frontiers in Marine Science*, 4, 418.

Mitarai, S., Watanabe, H., Nakajima, Y., Shchepetkin, A., and McWilliams, J. (2016). Quantifying Dispersal from Hydrothermal Vent Fields in the Western Pacific Ocean. *Proceedings of the National Academy of Sciences USA*, 113, 2796–981.

Moalic, Y., Desbruyères, D., Duarte, C. M., et al. (2011), Biogeography Revisited with Network Theory: Retracing the History of Hydrothermal Vent Communities. *Systematic Biology*, 61, 127–37.

Morato, T., Hoyle, S. D., Allain, V., and Nicol, S. J. (2010). Seamounts Are Hotspots of Pelagic Biodiversity in the Open Ocean. *Proceedings National Academy Sciences of the USA*, 107, 9707–11.

Mullineaux, L. S. (1987). Organisms Living on Manganese Nodules and Crusts: Distribution and Abundance at Three North Pacific Sites. *Deep-Sea Research*, 34, 165–84.

Mullineaux, L. A., Metaxas, A., Beaulieu, S., et al. (2018). Exploring the Ecology of Deep-sea Hydrothermal Vents in a Metacommunity Framework. *Frontiers in Marine Science*, 5, 49.

Nautilus Minerals (2008). *Environmental Impact Statement Solwara 1 Project*. Brisbane, Australia: Coffey Natural Systems Pty Ltd.

Narita, T., Oshika, J., Okamoto, N., Toyohara, T., and Miwa, T. (2015). Summary of Environmental Impact Assessment for Mining Seafloor Massive Sulfides in Japan. *Journal of Shipping and Ocean Engineering*, 5, 103–14.

Niner, H. J., Ardron, J. A., Escobar, E. G., et al. (2018). Deep-sea Mining With no Net Loss of Biodiversity—An Impossible Aim. *Frontiers in Marine Science*, 5, 53.

Oebius, H. U., Becker, H. J., Rolinski, S., and Jankowski, J. A. (2001). Parametrization and Evaluation of Marine Environmental Impacts Produced by Deep-sea Manganese Nodule Mining. *Deep-Sea Research II*, 48, 3453–67.

Okamoto, N., Shiokawa, S., Kawano, S., et al. (2018). Current Status of Japan's Activities for Deep-sea Commercial Mining Campaign, in *Proceedings of the*

2018 OCEANS—MTS/IEEE Kobe Techno-Oceans (OTO), 28–31 May. New York: Institute of Electrical and Electronics Engineers, pp. 1–7.

Pape, E., Bezerra, T. N., Hauquier, F., and Vanreusel, A. (2017). Limited Spatial and Temporal Variability in Meiofauna and Nematode Communities at Distant but Environmentally Similar Sites in an Area of Interest for Deep-Sea Mining. *Frontiers in Marine Science,* 4, 205.

Paul, S. A. L., Gaye, B., Haeckel, M., Kasten, S., and Koschinsky, A. (2018). Biogeochemical Regeneration of a Nodule Mining Disturbance Site: Trace Metals, DOC and Amino Acids in Deep-sea Sediments and Pore Waters. *Frontiers in Marine Science,* 5, 117.

Paterson, G. L. J., Wilson, G. D. F, Cosson, N., and Lamont, P. A. (1998). Hessler and Jumars (1974) Revisited: Abyssal Polychaete Assemblages from the Atlantic and Pacific. *Deep-Sea Research II,* 45, 225–51.

Petersen, S., Lehrmann, B., and Murton, B. J. (2018). Modern Seafloor Hydrothermal Systems: New Perspectives on Ancient Ore-Forming Processes. *Elements,* 14, 307–12.

Pimm, S. L., Jones, H. L., and Diamond, J. (1988). On the Risk of Extinction. *The American Naturalist,* 132(6), 757–85.

Plum, C., Pradillon, F., Fujiwara, Y., and Sarrazin, J. (2017). Copepod Colonization of Organic and Inorganic Substrata at a Deep-sea Hydrothermal Vent Site on the Mid-Atlantic Ridge. *Deep-Sea Research II,*137, 335–48.

Purser, A., Marcon, Y., Hoving, H-J. T., et al. (2016). Association of Deep-sea Incirrate Octopods with Manganese Crusts and Nodule Fields in the Pacific Ocean. *Current Biology,* 26, R1268–R1269.

Ramirez-Llodra, E., Shank, T., and German, C. (2007). Biodiversity and Biogeography of Hydrothermal Vent Species: Thirty Years of Discovery and Investigations. *Oceanography,* 20, 30–41.

Richer De Forges, B., Koslow, J. A., and Poore, G. C. B. (2000). Diversity and Endemism of the Benthic Seamount Fauna in the Southwest Pacific. *Nature,* 405, 944–7.

Roark, E. B., Guilderson, T. P., Dunbar, R. B., Fallon, S. J., and Mucciarone, D. A. (2009). Extreme Longevity in Proteinaceous Deep-sea Corals. *Proceedings of the National Academy of Sciences USA,* 106, 5204–8.

Rogers, A., Tyler, P., Connelly, D., et al. (2012). The Discovery of New Deep-sea Hydrothermal Vent Communities in the Southern Ocean and Implications for Biogeography. *PLoS Biology,* 10, e1001234.

Rona, P. A. (1985). Black Smokers on the Mid-Atlantic Ridge. *Eos, Transactions of the American Geophysical Union,* 66, 682.

Rona, P. A. (2003). Resources of the Sea Floor. *Science,* 299, 673–4.

Salinas-de-León, P., Phillips, B., Ebert, D., et al. (2018). Deep-sea Hydrothermal Vents as Natural Egg-case Incubators at the Galapagos Rift. *Scientific Reports,* 8, 1788.

Sarrazin, J., Legendre, P., de Busserolles, F., et al. (2015). Biodiversity Patterns, Environmental Drivers and Indicator Species on a High-temperature Hydrothermal Edifice, Mid-Atlantic Ridge. *Deep-Sea Research II,* 121, 177–92.

Schlacher, T. A., Baco, A. R., Rowden, A. A., et al. (2014). Seamount Benthos in a Cobalt-rich Crust Region of the Central Pacific: Conservation Challenges for Future Seabed Mining. *Diversity and Distributions,* 20, 491–502.

Simon-Lledó, E., Bett, B. J., Huvenne, V. A. I., et al. (2019a). Megafaunal Variation in the Abyssal Landscape of the Clarion Clipperton Zone. *Progress in Oceanography,* 170, 119–33.

Simon-Lledó, E., Bett, B. J., Huvenne, V. A. I., et al. (2019b). Ecology of a Polymetallic Nodule Occurrence Gradient: Implications for Deep-sea Mining. *Limnology and Oceanography,* 64, 1883–94.

Simon-Lledó, E., Bett, B. J., Huvenne, V. A. I., et al. (2019c). Biological Effects 26 Years After Simulated Deep-sea Mining. *Scientific Reports,* 9, 8040.

Simpson, S. and Spadaro, D. (2016). Bioavailability and Chronic Toxicity of Metal Sulfide Minerals to Benthic Marine Invertebrates: Implications for Deep Sea Exploration, Mining and Tailings Disposal. *Environmental Science & Technology,* 50, 4061–70.

Smith, C. R., Levin, L. A., Koslow, A., Tyler, P. A., and Glover, A. G. (2008). The Near Future of the Deep Seafloor Ecosystems, in N. Polunin (ed.) *Aquatic Ecosystems: Trends and Global Prospects.* Cambridge: Cambridge University Press, pp. 334–49.

Stratmann, T., Mevenkamp, L., Sweetman, A. K., Vanreusel, A., and van Oevelen, D. (2018). Has Phytodetritus Processing by an Abyssal Soft-Sediment Community Recovered 26 Years After an Experimental Disturbance? *Frontiers in Marine Science,* 5, 59.

Sweetman, A. K., Smith, C. R., Shulse, C. N., et al. (2019). Key Role of Bacteria in the Short-term Cycling of Carbon at the Abyssal Seafloor in a Low Particulate Organic Carbon Flux Region of the Eastern Pacific Ocean. *Limnology and Oceanography,* 64, 694–713.

Sylvan, J., Toner, B., and Edwards, K. (2012). Life and Death of Deep-sea Vents: Bacterial Diversity and Ecosystem Succession on Inactive Hydrothermal Sulfides. *mBio,* 3, e00279–11.

Taboada, S., Riesgo, A., Wiklund, H., et al. (2018). Implications of Population Connectivity Studies for the Design of Marine Protected Areas in the Deep Sea: An Example of a Demosponge from the Clarion-Clipperton Zone. *Molecular Ecology,* 27, 4657–79.

Thaler, A. D. and Amon, D. (2019). 262 Voyages Beneath the Sea: A Global Assessment of Macro- and Megafaunal Biodiversity and Research Effort at Deep-sea Hydrothermal Vents. *PeerJ*, 7, e7397.

Thiel, H. and Schriever, G. (1990). Deep-sea Mining, Environmental Impact and the DISCOL project. *Ambio*, 19, 245–50.

Tunnicliffe, V., McArthur, A., and McHugh, D. (1998). A Biogeographical Perspective of the Deep-sea Hydrothermal Vent Fauna. *Advances in Marine Biology*, 34, 353–442.

Tsurumi, M. and Tunnicliffe, V. (2003). Tubeworm-associated Communities at Hydrothermal Vents on the Juan de Fuca Ridge, Northeast Pacific. *Deep-Sea Research Part I: Oceanographic Research Papers*, 50, 611–29.

United Nations (1982). *United National Convention on the Law of the Sea. Part IX.*

UNGA (1967). United Nations General Assembly, First Committee, 1515th meeting, 1 November, Agenda Item 92. New York: UNGA.

Van Dover, C. L. (2019). Inactive Sulfide Ecosystems in the Deep Sea: A Review. *Frontiers in Marine Science*, 6, 461.

Van Dover, C. L. (2014). Impacts of Anthropogenic Disturbances at Deep-sea Hydrothermal Vent Ecosystems: A Review. *Marine Environmental Research*, 102, 59–72.

Van Dover, C. L. (2011). Mining Seafloor Massive Sulphides and Biodiversity: What Is at Risk? *ICES Journal of Marine Science*, 68, 341–8.

Van Dover, C. L., German, C. R, Speer, K. C., Parson, L. M., and Vrijenhoek, R. C. (2002). Evolution and Biogeography of Deep-sea Vent and Seep Invertebrates. *Science*, 295, 1253–7.

Van Dover, C. L., Ardron, J. A., Escobar, E., et al. (2017). Biodiversity Loss from Deep-sea Mining. *Nature Geoscience*, 10, 464–5.

Van Dover, C. L., Arnaud-Haond, S., Gianni, M., et al. (2018). Scientific Rationale and International Obligations for Protection of Active Hydrothermal Vent Ecosystems from Deep-sea Mining. *Marine Policy*, 90, 20–8.

Vanreusel, A., Hilario, A., Ribeiro, P. A., Menot, L., and Arbizu, P. M. (2016). Threatened by Mining, Polymetallic Nodules Are Required to Preserve Abyssal Epifauna. *Scientific Reports*, 6, 26808.

Veillette, J., Sarrazin, J., Gooday, A. J., et al. (2007). Ferromanganese Nodule Fauna in the Tropical North Pacific Ocean: Species Richness, Faunal Cover and Spatial Distribution. *Deep-Sea Research I*, 54, 1912–35.

Wagner, D. and Kelley, C. D. (2017). The Largest Sponge in the World? *Marine Biodiversity*, 47, 367–8.

Watling L., Rowley, S., and Guinotte, J. (2013) The World's Largest Known Gorgonian. *Zootaxa*, 3630(1), 198–9.

Washburn, T. W., Turner, P. J., Durden, J. M., et al. (2019) Ecological Risk Assessment for Deep-sea Mining. *Ocean & Coastal Management*, 176, 24–39.

Weaver, P. P. E., Billett, D. S. M., and Van Dover, C. L. (2018). Environmental Risks of Deep-sea Mining, in M. Salomon and T. Markus (eds.) *Handbook on Marine Environment Protection, Science, Impacts and Sustainable Management*. New York: Springer-Verlag, pp. 215–45.

Wedding, L. M., Friedlander, A. M., Kittinger, J. N., et al. (2013). From Principles to Practice: A Spatial Approach to Systematic Conservation Planning in the Deep Sea. *Proceedings of the Royal Society B: Biological Sciences*, 280, 20131684.

Wilson, G. D. F. (2017). Macrofauna Abundance, Species Diversity and Turnover at Three Sites in the Clipperton-Clarion Fracture Zone. *Marine Biodiversity*, 47, 323–47.

Xu, G. and Lavelle, J. W. (2017). Circulation, Hydrography, and Transport over the Summit of Axial Seamount, a Deep Volcano in the Northeast Pacific. *Journal of Geophysical Research: Oceans*, 122, 5404–22.

Young, C., Fujio, S., and Vrijenhoek, R. (2008). Directional Dispersal Between Mid-ocean Ridges: Deep-ocean Circulation and Gene Flow in *Ridgeia piscesae*. *Molecular Ecology*, 17, 1718–31.

Zeppilli, D., Bongiorni, L., Serrato Santos, R., and Van reusel, A. (2014). Changes in Nematode Communities in Different Physiographic Sites of the Condor Seamount (North-East Atlantic Ocean) and Adjacent Sediments. *PLoS ONE*, 9, (12), e115601.

Zierenberg, R. A., Adams, M. W. W., and Arp, A. J. (2000). Life in Extreme Environments: Hydrothermal Vents. *Proceedings of the National Academy of Sciences USA*, 97, 12961–2.

CHAPTER 6

The natural capital of offshore oil, gas, and methane hydrates in the World Ocean

Angelo F. Bernardino, Erik E. Cordes, and Thomas A. Schlacher

6.1 The natural capital of hydrocarbon reserves

Oil and gas (hydrocarbon) resources are an immensely valuable natural commodity for the global economy and an appreciable natural capital (as defined in Chapter 1) to humankind. The natural capital of hydrocarbon reserves can be roughly estimated in terms of their material (natural stocks) or financial contributions to industries, transportation, and overall national economies. So here we follow the concepts of Costanza et al. (1997) and discuss the natural capital value of hydrocarbon resources (i.e. oil and gas) as any financial or ecosystem service benefits generated by the physical stock of oil and gas that is used to support welfare to humankind. Costanza et al. (1997) subsequently discuss the importance of hydrocarbon energy sources and contrast the use of fossil fuels with their environmental costs globally. We also put the exploration of offshore and deep-sea (>150 m depth) hydrocarbon reserves into context of the typical industrial (i.e. operational and accidental) impacts to deep-sea ecosystems, mainly focusing on hydrocarbon seeps and vulnerable invertebrate taxa such as cold-water corals. Although deep-sea ecosystems associated with or under direct influence of most offshore oil and gas exploration hold value through provision, regulating, and cultural ecosystem services (e.g. Levin et al. 2016), we do not attempt to quantify the value of these services in

this chapter. We focus on evaluating the stock value of hydrocarbon offshore exploitation and production of oil and gas, attempt to compile recent published synthesis of threats and impacts that arise from this activity on deep oceans below 150 m, and provide a critical view of benefits and costs of hydrocarbon-based energy in light of current and future environmental costs of the global dependence on this type of natural capital exploration and exploitation from the depths of our oceans.

Hydrocarbons provide at least 57 per cent of the global energy consumption (BP 2018), and most economies are closely tied to the reliable and continued supply of oil and gas products, for at least the next couple of decades. Whilst the direct value of hydrocarbon reserves can be calculated using market principles, there is also a much broader indirect socioeconomic effect associated with the production, transport, and consumption of oil and gas (Polasky and Segerson 2009). A significant part of many countries' gross domestic product (GDP) is linked to hydrocarbons; for example, in 2015, hydrocarbon energy sources accounted for U$43.3 billion per million tonnes of oil equivalent. Notwithstanding a continuing rise in the global oil consumption (BP 2018), the financial value of hydrocarbon reserves has declined by at least 50 per cent in the last 40 years as renewable forms of energy become more widespread.

Whilst being a valuable financial commodity, the exploration of oil and gas may cause environmental

Angelo F. Bernardino, Erik E. Cordes, and Thomas A. Schlacher *The natural capital of offshore oil, gas, and methane hydrates in the World Ocean* In: *Natural Capital and Exploitation of the Deep Ocean.* Edited by: Maria Baker, Eva Ramirez-Llodra, and Paul Tyler, Oxford University Press (2020). © Oxford University Press.
DOI: 10.1093/oso/9780198841654.003.0006

harm and, therefore, may impact other ecosystem services that are important to humankind (Costanza et al. 1997). Impacts can range from local habitat modifications during drilling to massive spills and, ultimately, global changes to the climate (Millennium Ecosystem Assessment 2005; IPCC 2014). Further, Earth's overall natural capital including the stocks of nonrenewables such as minerals and hydrocarbons may continue to decrease, and with the costs of maintaining ecosystem services that may be impacted in the process of exploiting a specific natural capital such as hydrocarbon resources. Thus, meaningful assessments of the true value of hydrocarbon reserves must include their economic benefits versus their costs in terms of diminishing the value of other components of the biosphere.

6.1.1 Oil and gas reserves in offshore systems

Deep-water oil and gas production is closely linked to new enabling technology in exploration and extraction (see Chapter 1), and the industry can be very profitable in a number of regions. The industry has a global activity and is currently exploring the vast reaches of the deep seabed with support of technological advances (Macreadie et al. 2018). The expansion of the offshore oil and gas industry to deeper ocean basins may potentially increase the

risks, resulting in more severe environmental harm over potentially larger areas (Davies et al., 2007; Cordes et al. 2016b; Bernardino and Sumida 2017; Venegas-Li et al. 2019). As oil and gas drilling moves further offshore into deeper waters, the probability of an accident may increase (Muehlenbachs et al. 2013). On the other hand, technological advances, improved risk evaluation, and an efficient legal and regulatory framework may decrease accidents and their environmental impacts (Baram 2010).

Annual oil production (2017) is estimated at 25–27 million barrels globally, with deep-water sources (i.e. those below 150 m depth) comprising a significant part of revenues (Table 1). A sharp increase in offshore exploration in the last decade is particularly evident for Brazil, the United States (Gulf of Mexico), Angola, and Norway. For these nations, 60 to 80 per cent of their maritime oil production is deeper than 150 m (EIA 2016). Deep-water production in the Gulf of Mexico accounted for at least 66 per cent of the US oil and gas production in 2017, with over 85 per cent of the technically recoverable reserves (>62.9 BBOE) being located in deep waters (Kaiser and Narra 2018). Global oil reserves are estimated at 1696.6×10^{12} barrels, of which $92\,649 \times 10^3$ barrels are extracted daily (BPD; BP 2018); at current rates of production, global reserves are estimated to last ca. 50 years (BP 2018), and they represent ca. 2 per cent of the global annual

Table 1 Oil reserves, production, and estimated offshore and deep (>125 m) and ultra-deep (>1500 m) production globally and at selected countries with significant offshore production in the world

	Oil Reserves (Thousand Million Barrels)	Oil Production (tbd)	Estimated Offshore Production (tbd)	% of Production from Deep and Ultra-deep Offshore	Offshore Value (Billion U$ per y at U$50 per Barrel)	GDP (Billion U$ in 2017)
Global	1696.6	92,649	27,000	<1	492.7	84,835
Saudi Arabia	266.2	11,951	3510	n/a	64	683.8
USA	50.0	13,057	1800	>60 %	32.8	19,390
Brazil	12.8	2734	2200	>90 %	40.1	2055
Mexico	7.2	2224	2000	n/a	36.5	1149
Norway	7.9	1969	1890	>80 %	34.5	399
Angola	9.5	1674	1500	>80 %	27.4	124

Data compiled from EIA (2016); BP, (2018); and World Bank (2017) GDP reports.
U$50 per barrel.
Abbreviations: GDP, gross domestic product; tbd, thousand barrels per day.

GDP. Offshore operations make up nearly 30 per cent of oil production (Narula 2018), and the high degree of dependence on hydrocarbon sources by the global economy implies that offshore operations will become more important in the coming decades.

6.1.2 The potential of deep-sea gas hydrate reservoirs

Methane hydrates are another form of marine hydrocarbon that can be mined in solid form (associated with water ice) at depths below 250 m (MacDonald 1990; Levin 2005). Methane hydrates are mainly produced by anaerobic microbial methanogenesis under low temperatures and high pressure, limiting their potential recovery in solid form; this has to date prevented larger commercial exploration from offshore sources and its wider use as an energy source (Sloan 1990; Gornitz and Fung 1994; Narula 2018).

Methane hydrates are found on the margins of all continents, with most reservoirs located deeper than 1000 m and exhibiting a hydrate subseafloor layer of 300 to 700 m (Gornitz and Fung 1994). Extensive methane hydrate areas include the NE and NW margin of Canada (Labrador Sea), Bering Sea, Indian margin, W and E African margin, Argentinean margin, China Sea, and Norwegian Sea.

The amount of methane hydrate reservoirs available for extraction from the deep ocean is modelled based on the biological and thermogenic processes associated with its formation. Methane hydrates can be recovered in the deep ocean from the surface seafloor layer (2–10 cm) as small ice aggregates, or as deep as 400 m into the seafloor. The volume of methane gas trapped in marine sediments is estimated at ca. 19.5×10^{15} m^3, based on the distribution of deep-sea organic-rich sediments on continental margins that are in the range of optimal low temperature and depth (pressure) required for the formation of methane hydrates (MacDonald 1990). Other estimates of oceanic methane gas reservoirs range widely from 3.1 to 7600×10^{15} m^3 (Gornitz and Fung 1994). Gornitz and Fung (1994) estimated the global oceanic reservoirs of methane gas at 26.4×10^{15}m^3, with the largest reservoirs being potentially located in equatorial regions (10° S to 20° N) and at high latitudes (>40° N) in the Northern Hemisphere.

The modelled oceanic methane gas reservoirs ($26,400 \times 10^{12}$ m^3) are 163 times greater than the currently known natural gas reserves (193.5×10^{12} m^3; BP 2018), illustrating an enormous potential as a future energy source. Although these reserves offer a large potential commodity, the technological challenges and higher production costs if compared to other hydrocarbons, make the exploitation of these reservoirs currently unfeasible and uneconomical (Narula 2018). However, China, Japan, and the United States have begun exploration tests for methane hydrates, suggesting that current technological and market barriers may be overcome in the near future.

6.2 The ecology of offshore hydrocarbon-associated ecosystems: a brief sketch

Marine offshore ecosystems associated with hydrocarbon encompass a variety of assemblage types, often structured by differences in the extent to which the underlying hydrocarbons influence the biota, and the resistance of the biota to potential toxic effects. 'Hydrocarbon habitats' typically include sites where pore water fluxes are enriched in methane (Levin 2005), and occur in heterogeneous settings along the continental shelf-slopes in all of the explored regions of the World Ocean (Sibuet and Olu 1998). Hydrocarbon seeps are associated with the seafloor slopes at both passive and active continental margins and are commonly associated with subseafloor accumulations of gas hydrates in salt-dominated provinces with overlying hydrocarbon reservoirs, or with the high organic content in the fans of submarine canyons that is broken down anaerobically to form methane (LeBris et al. 2016). The assemblages at and near seeps mainly include macrofauna of the soft sediments (e.g. annelid polychaetes, bivalves, and peracarid crustaceans; Levin 2005; Bernardino et al. 2012), larger epibenthic communities dominated by symbiotic and habitat-forming species (e.g. vesicomyid clams, bathymodiolin mussels, and siboglinid tubeworms), and cold-water corals, often being located in peripheral areas where the chemical signal of the hydrocarbons wanes (Cordes et al. 2009; Cordes et al. 2010).

Effects on fauna from dissolved and gaseous hydrocarbons that seep into the sediment and the water column above (hyperbenthic layer) can represent a potential source of carbon, energy, toxins, or a mixture of all three (Levin et al. 2005). 'Fresh' hydrocarbons (i.e. those not having undergone microbial degradation) are a complex mixture of aromatic compounds that can be directly toxic to most species. In addition, most of the hydrocarbon oxidation typically occurs in subsurface layers where oxygen concentrations are low, and if the electron donor is sulphate (SO_4^{2-}), the by-products include hydrogen sulphide species (HS^- and H_2S), both being toxic molecules that interfere with the function of the cytochrome-C oxidase enzyme, which is essential for aerobic respiration (reviewed in Cordes et al. 2009).

If an organism can tolerate the exposure to potentially harmful chemicals, there is abundant energy and carbon to be found in the hydrocarbons habitats. Numerous microbes can metabolise hydrocarbons (e.g. anaerobic methanotrophic archaea (ANMEs), methanotrophs, and methylotrophs), some forming symbiotic relationships with eukaryotic hosts (Dubilier et al. 2008). Hosts can be tubeworms such as siboglinid polychaetes that lack a digestive tract and rely completely on sulphide-oxidising bacteria, contained in a specialised organ of the worms, for their nutrition (Fisher 1990). Mussels (bathymodiolin bivalves) and clams (several groups) at seeps can contain methane- and/or sulphide-oxidising bacterial symbionts in their gills, and exhibit drastically reduced digestive tracts and enlarged gills to house the symbionts (Duperron et al. 2009). Another notable example of seep-associated fauna is the 'ice-worm', which is one of the only animal species known to inhabit gas hydrates directly (Fisher et al. 2000). It uses its large parapodia (paddle-like appendages) to fan seawater over exposed gas hydrates, which leads to the formation of small furrows in the hydrate and elevation of the concentration of methane in the seawater. The microbes on the chaetae of the worm use the methane as an energy source and produce the food for the worm. As long as the hydrocarbon seep persists, the ener-

getic input of the hydrocarbons can be utilised by this array of symbionts to provide an abundant and reliable food source in a generally food-limiting deep ocean (Carney 1994).

Many of the symbiont-bearing organisms at hydrocarbon seeps can reach very large sizes and typically are long-lived. Tubeworms can stand 2 m above the seafloor, and can extend at least as far down into the sediments (Cordes et al. 2005a). They form large aggregations that can cover many square metres of the seafloor, and in rich hydrocarbon provinces, such as the Gulf of Mexico (MacDonald et al. 1990), the Pacific margin of Costa Rica (Sahling et al. 2008), offshore east Africa (Olu et al. 2007), and off New Zealand (Bowden et al. 2013), can form continuous fields that extend for hundreds of metres. Mussels and clams can also form extensive beds, locally increasing habitat heterogeneity in the benthos (Cordes et al. 2010).

In addition to the physical habitat structure formed, the high biomass of organisms can represent a potential food resource. Whilst there are no known lethal predators of tubeworms (Bergquist et al. 2007), there are some species that live as epizoites on the worms and consume part of them (as quasi-parasites). Several species of worms (*Galapagomystides* spp.) live on the anterior end of tubeworms, feeding on their blood and the large haemoglobin molecules contained within the tubeworms' vascular system (Becker et al. 2013). There are also species of polynoid polychaetes that feed on the gills of tubeworms (Bergquist et al. 2007). Mussels can release large amounts of mucous with particles to clear their gills and prevent them from clogging; these 'pseudo-faeces' may form a food source for other consumer species associated at or near mussel beds (Becker et al. 2013). Many of the bathymodiolin mussel species contain internal parasites that cause damage to their gills while they feed on particles trapped on the gills; other fauna will use this food source after it is released (Becker et al. 2013).

Extensive and persistent biogenic seep habitats are occupied by a wide variety of other species. During the early stages of seep development, mussels will settle in areas where hard substrata are available, while mobile clams will occupy soft-sediment areas (Fisher 1990; Bergquist et al. 2005). Species typically associated with the biogenic physical matrix formed by the tubeworms and mussels are primarily seep-endemic species with a number of specialised adaptations that allow them to persist in the relatively chemically harsh conditions at the seeps (Sibuet and Olu 1998; Bernardino et al. 2012). As the local source of hydrocarbons becomes depleted, the unique faunal assemblages associated with the seeps wane (Cordes et al. 2006).

The rate of seepage may decline over time because of a shifting in the underlying plumbing that is delivering the hydrocarbons to the surface (Aharon et al. 1997; Sahling et al. 2008), or due to the direct influence of the symbiotic fauna on the biogeochemistry (Cordes et al. 2005b). By-products of anaerobic methane oxidation are carbonate (CO_3^{2-}) and bicarbonate (HCO_3^-; Aharon and Fu 2000; Boetius et al. 2000). When these interact with the calcium dissolved in the overlying seawater, and the saturation state is high enough, they will precipitate out of solution as calcium carbonate. This biologically mediated process is referred to as authigenic carbonate precipitation. These carbonates provide the primary hard substrate at seeps, while the biogenic structures provide the habitat heterogeneity to support a higher diversity of organisms (Cordes et al. 2010). Hydrocarbons can exert an influence on deep-sea communities long after seepage has ceased, mainly through the formation of carbonates. Carbonates can persist at relic seep sites for millennia, with some age estimates of authigenic carbonates at seeps in the Gulf of Mexico at over 40,000 years (Aharon et al. 1997). They may become buried, or form in the subsurface, and then later exhumed by benthic storms or submarine landslides. Because most of the symbiotic species at seeps require a hard substratum for settlement, they typically occupy these carbonates.

As seepage wanes and sulphide and methane concentrations in the water column become undetectable, cold-water corals can begin to grow on the authigenic carbonate substrate (Cordes et al. 2008, 2009). They can form coral gardens made up of a mixture of octocorals and black corals, or coral mounds and reefs made up of stony corals (scleractinian anthozoans; Cordes et al. 2016a). Similar to tubeworms, mussels, and clams, corals increase habitat heterogeneity by creating large, complex and persistent structures (Cordes et al. 2010). These architecturally complex structures can create habitat for a cornucopia of species, rivalling biodiversity of shallow-water coral reefs (Mortensen et al. 1995). Additionally, the cover and abundance of megabenthos associated with cold-water corals can exceed that of the surrounding soft-sediment communities by an order of magnitude (Mortensen et al. 1995).

Deep-water corals have been reported from a variety of deep-sea settings, including frequent observations at hydrocarbon seeps. The relationship of cold-water corals with hydrocarbon seeps has repeatedly been recognised. For example, the occurrence of *Desmophyllum pertusum* (formerly *Lophelia pertusa*) reefs with pockmarks and gas hydrates in Norway led to the hypothesis that the species depends on methane as a carbon source (Hovland and Thomsen 1997). However, stable isotope signatures from the Gulf of Mexico hydrocarbon province showed that *D. pertusum* was highly unlikely to assimilate carbon from methane (Becker et al. 2009). Additional studies have shown that *D. pertusum* has some phylotypes in the microbiome that may be capable of methane oxidation, and that they are capable of chemosynthesis under controlled laboratory conditions (Middelburg et al. 2015). Among the octocorals, stable isotope data demonstrated that one species (out of many examined) in the Gulf of Mexico, *Callogorgia delta*, contained significant inputs of 'light' seep carbon, and habitat suitability modelling showed that it is more abundant near seeps (Quattrini et al. 2013). It appears likely that the primary mechanism that causes the observed association of corals with reefs is the provision of hard (carbonate) substratum rather than being a source of carbon. Hard substrata are uncommon in most of the deep sea; thus, seeps can be viewed to create a 'habitat subsidy' for large epibenthos. Deep-sea corals have also been observed associated with oil and gas structures in the deep sea (Gass and Roberts 2006; Larcom et al. 2014; Jones et al. 2019).

6.3 Operational impacts

6.3.1 Physical and chemical impacts on organisms and ecosystems

Specialised seep communities are co-located with subsurface reserves of hydrocarbons. Therefore, extracting these reserves may cause environmental harm. Direct impacts may arise from drilling activities, including (but not limited to) the release of drilling muds and the generation of sediment plumes from the drilling itself (Gray et al. 1990). Once in the production phase, the placement and later decommissioning of seafloor infrastructure and pipelines connecting the production platforms to the shore-side refineries can further impact benthic communities (Ulfsnes et al. 2013; Jones et al. 2019).

Direct physical and chemical impacts of typical drilling operations have been observed for deep-sea coral communities, but are rarely reported for seep-endemic communities such as mussels, clams, and tubeworms. The microbial communities at seeps consist of numerous members capable of limiting methane flux to the ocean and atmosphere (Boetius and Wenzhöfer 2013), and metabolising hydrocarbons and other compounds released by standard drilling operations (Orcutt et al. 2010). The large symbiotic fauna at seeps (tubeworms, mussels, and clams) have evolved under continual exposure to hydrocarbons and other chemical compounds considered toxic for other organisms. They appear to be able to tolerate elevated concentrations of reduced compounds as well as low dissolved oxygen concentrations (Tunnicliffe 1991; Fisher et al. 2000; Hourdez et al. 2002). Deep-sea corals also appear to tolerate by-products of typical drilling operations to variable degrees. For example, *D. pertusum*, one of the most widespread deep-sea scleractinian corals, showed declines in health only when it was completely buried by particulate material (Brooke et al. 2009), and little to no response to elevated concentrations of drilling muds *in situ* (Allers et al. 2013).

Accidental release of large volumes of hydrocarbons, and associated 'remedial' actions, can severely impact deep-water communities. The largest offshore marine oil spill, the *Deepwater Horizon* blowout, which occurred in April 2010, is a case study to illustrate the environmental harm caused to deep-sea ecosystems (Joye et al. 2016). In this spill, much of the released oil rose to the surface, where it interacted with plankton to form an oiled marine snow that subsequently rained down on the deep-sea benthos (Passow et al. 2012). In addition, a deep-water plume consisting of microdroplets of oil (<500 μm in diameter) was present in the deep water-column (Reddy et al. 2012). Moreover, chemical dispersants that were used in an attempt to break up the surface oil into smaller droplet sizes may have contributed to the impacts on deep-sea coral communities (DeLeo et al. 2016), and altered the pelagic microbial community and its ability to degrade the oil (Kleindienst et al. 2015).

Deep-sea corals were impacted during the spill between 1000 and 1900 m depth up to a distance of 22 km from the well head (Fisher et al. 2014a). Detrimental effects ranged from small areas covered in flocculent material to large portions of the colony covered in flocs and exhibiting lost tissue and mucous production (White et al. 2012). Mesophotic coral communities along the edge of the continental shelf between 100 and 200 m depth also exhibited signs of impact from the spill (Etnoyer et al. 2016). Time-series studies of benthic megafauna and corals impacted during the spill revealed a lower diversity of invertebrates and a limited capacity to recover from impact after 8 years of observations, with continued degradation of the coral colonies that suffered the most severe impacts (Girard et al. 2019; McClain et al. 2019; Figure 1). However, it is notable that at most of the deeper (>1000 m) sites where impacts to coral communities were observed, there were also seep fauna nearby that did not exhibit any visual signs of changes that could be attributed to oil (Cordes, pers. obs.).

Laboratory experiments have demonstrated the mechanism of the corals' response to oil and dispersant exposure. Octocoral sea fans, including those impacted during the *Deepwater Horizon* oil spill, have shown only minor or no effect from the exposure to hydrocarbons, either as suspended droplets or as the water-accommodated fraction (DeLeo et al. 2016). However, when dispersant was added to those mixtures, the corals exhibited distinct signs of stress and varying degrees of polyp

Figure 1 Impacts from the Deepwater Horizon oil spill on the deep-sea octocoral Paramuricea biscaya at the Mississippi Canyon 294 site. The initial photo was taken in November 2010, approximately 4 months after the initial explosion of the oil rig. Subsequent photos were taken at different times of the year over the next 7 years. Note that within 1 year, the associated ophiuroid was no longer present and hydroids had colonised large portions of the impacted colony. In subsequent years, limited recovery can be observed as portions of the colony began to re-grow. Photo series courtesy of Fanny Girard and supported by NSF, the NOAA NRDA programme, and the GoMRI funded ECOGIG consortium.

mortality (DeLeo et al. 2016). Studies of gene expression via RNA sequencing revealed that the coral colonies impacted in the *Deepwater Horizon* spill activated pathways related to tissue repair and melanin production (as a barrier to further damage), but also initiated apoptosis of compromised tissues and downregulated genes related to toxin processing (DeLeo et al. 2018).

The deep-sea soft-sediment communities also exhibited significant shifts in composition, diversity, and ecosystem function following the *Deepwater Horizon* spill. The macrofaunal and meiofaunal communities associated with deep-sea coral sites consisted of an elevated abundance of nematodes and opportunistic polychaete families, as well as declines in taxa such as amphipods that are known to be sensitive to disturbance (Fisher et al. 2014a). This was also the case in the broad sedimented background of the deep Gulf of Mexico, with ele-

vated polyaromatic hydrocarbon concentrations and significant alterations of the macrofauna and meiofauna community structure within 6 km in every direction of the *Deepwater Horizon* location, and extending to nearly 40 km to the southwest in the direction of the subsurface plume (Montagna et al. 2013).

6.3.2 Long-term and climate impacts

Energy generation involves a range of environmental impacts to ecosystem services, so the sustainability of the offshore hydrocarbon industry needs to comply with standards of environmental performance and human health whilst carrying out its regular operations. The hydrocarbon industry has a number of impacts on ecosystem services, including supporting (e.g. nutrient cycling and chemosynthesis), regulating (maintenance of physical,

Table 2 Global oil consumption, average market prices, total value, and social costs of carbon–based CO_2 emissions on yearly consumption

Year	Oil Consumed (Billion Barrels y $^{-1}$)	Average Market Price (US$ Barrel^{-1})	Market Value (Trillion US$)	SCC (Trillion US$)
2011	32.13	111.26	3.575	1.727
2017	35.84	54.19	1.942	1.927

Sources: BP (2012, 2018); EPA (2018); van den Bergh and Botzen (2014).

Emissions = 0.43 tCO_2/barrel; SCC = U$125 tCO_2.

Abbreviation: SCC, social costs of carbon emissions.

chemical, and biological conditions), provisioning (food production), and cultural (physical and spiritual interactions with biota) (Papathanasopoulou et al. 2015).

Climate impacts from the hydrocarbon industry are directly related to the burning of fossil fuels and release of greenhouse gases (GHG) such as carbon dioxide (CO_2), methane (CH_4), and nitrous oxide (N_2O). The (GHG) emissions to the atmosphere from use of hydrocarbon energy (fossil fuels and industrial processes) in 2011 reached 34.8 $GtCO_2$, which is 65 per cent of global GHG emissions (IPCC 2014). The global oil consumption in 2011 is estimated at 32,132.4 million barrels (BP 2012), which would result in 13.82 $GtCO_2$ emitted to the atmosphere from oil alone (EPA 2018). Considering an annual global offshore crude oil production of 25 million barrels per day (USEIA 2016), it is estimated that 3.92 $GtCO_2$ would be emitted from the direct burning of the offshore production. In the United States alone, 96 per cent of total CO_2 emissions originate from fossil fuel burning (EPA 2018), revealing the strong footprint of the hydrocarbon industry on climate change.

The costs of CO_2 emissions on a broad range of potential damages associated with climate-change effects on human society can be estimated by estimating social costs of carbon emissions (SCC). SSC is proxy of the cost of emitting more CO_2 to the atmosphere given, for example, the impacts on the biodiversity and ecosystems, economic implications for green energy development, and costs of extreme weather (van den Bergh and Botzen 2014). A lower bound (i.e. conservative) SCC is US$125 per tCO_2 (van den Bergh and Botzen 2014). This bound results on approximately US$1.73 trillion of costs associated with oil consumption in 2011, or 2.4

per cent of the global GDP in that year. The market prices of the 32.1 billion barrels produced in 2011 at the record rates of US$111.26 per barrel result in a market value of US$3.57 trillion. However, the SCC from CO_2 emissions from the oil produced in 2011 are near half (48.2 per cent) of the market value and can be much worse during years of lower oil market rates (Table 2). At prices near US$54 per barrel, the social costs associated with CO_2 emissions are similar to the crude oil industry revenues. It is noteworthy that the industrial production supported by fossil fuels leads to a significantly higher monetary value (~43 per cent of global GDP in 2015; Narula 2018), but the total GHG emitted from those activities and, therefore, the SCC arising from those emissions are also much higher if compared to other energy sources (IPCC 2014). Although the economic dependence on hydrocarbons is unlikely to fall in the next decades at current market prices, the expected rise in atmospheric GHG and predicted impacts of climate-change effects will likely dramatically increase the social and environmental costs of hydrocarbons, and a new regulatory framework will be needed to account for those costs globally.

6.4 Best practices for exploitation and management

Environmental management of hydrocarbon exploitation is principally concerned with controlling risks and mitigating environmental impacts (Cordes et al. 2016b). Strategies are designed that aim for exploration and decommissioning to meet relevant environmental standards. As most operations are within each nation's Economic Exclusive Zones (EEZs), environmental regulations for offshore hydrocarbon activities vary among countries. In some regions, such as the NE Atlantic, countries propose environmental legislation through the OSPAR (Convention for the Protection of the Marine Environment of the North-East Atlantic) commission (see Chapter 3). In the United States, for example, government agencies (e.g. the Bureau of Ocean and Energy Management (BOEM) and Ministry of Environmental Affairs) act as executing bodies for environmental laws pertaining to the industry.

The use of scientific data is a critical requirement to obtain baseline data that are robust and defensible in offshore lease areas before exploitation takes place. These data must adequately capture the range of temporal variability and the range of spatial heterogeneity in the region. In essence, any impacts (or lack thereof) can only be demonstrated by comparison with good baseline data. Thus, if these are inadequate, impact assessments will be weak or invalid. Several countries use a case-by-case project approach to characterise broadly biological and environmental variables at target sites. These may be useful for regional baselines at broad spatial scales (100s of km), but typical impacts from drilling or operational activities are restricted to within 5 km near installations (Cordes et al. 2016b). This implies that, at the regional scales of over 10s of km, it is imperative to map in detail and sample ecosystems at high intensity and resolution, especially those known to be particularly vulnerable, and including both epifaunal and infaunal organisms. Ideally, monitoring programmes should include more than one ecosystem type or assemblage type to cover the full range of environmental sensitivities and responses. Good candidates are deep-sea coral reefs that commonly occur near offshore platforms and have been proposed to serve as 'deep sentinels' of hydrocarbon contamination and can be monitored by nondestructive high-definition video images (Figure 1) and recently developed genetic profiling methods (Fisher et al. 2014a; DeLeo et al. 2018; Girard and Fisher 2018). A random, or stratified random, spatial design encompassing multiple coral reef colonies is usually needed (Larsson et al. 2013; Jarnegren et al. 2017; Girard and Fisher 2018).

Spatial management at the basin scale is probably the most key process in need of further development

and application in the offshore hydrocarbon industry. All environmental management (irrespective of system) requires data on the types and distribution of ecological assets. As a result, offshore areas of high biological or ecological interest need to be mapped, studied, and included in spatial management plans of continental margins before the exploitation blocks are leased. There are excellent examples of Strategic Environmental Assessment practices for the offshore industry in Norway (Fidler and Noble 2012; Olsen et al. 2016), and the use of restricted areas on the US margin that exclude marine sanctuaries, submarine canyons, and seamounts from any offshore activity (BOEM 2019). After exploitation, the decommissioning of submerged infrastructure and oil platforms is usually required by law. Drill cuttings near offshore platforms accumulate significant quantities of metals and hydrocarbons that need to be removed during decommissioning (Breuer et al. 2004). To reduce the risks of long-term impacts caused by abandoned offshore structures and pollutants, the decommissioning process also needs Environmental Impact Assessments based on scientific standards to ensure recovery of exploited offshore ecosystems. Scientific guidelines for the sustainability of those activities have been proposed (Fortune and Paterson 2018; Jones et al. 2019).

6.5 Spatial overlap between ecological assets and oil leases creates challenges

Given the high reliance on hydrocarbon energy, it is unlikely that offshore operations will cease to operate in the next few decades. Most exploratory offshore activities occur within marine EEZs, where licencing hydrocarbon exploration may cover significant expanses (up to 100 per cent) of the deep-sea seafloor within a particular EEZ (Venegas-Li et al. 2019). Some licenced offshore blocks can overlap (sometimes significantly so) with areas of appreciable marine biodiversity, assemblages of vulnerable fauna, or other habitats considered ecologically important or vulnerable (Dunn et al. 2014; Almada and Bernardino 2017; Harfoot et al. 2018; Bernardino et al. 2019; Venegas-Li et al. 2019).

The large area of leases inside EEZs (from 5 to 100 per cent; Venegas-Li et al. 2019) is a critical concern for the management of deep-sea ecosystems (Girard et al. 2019). Yet, a number of EEZs lack sufficiently detailed spatial information about the occurrence and status of deep-sea ecosystems to conduct reliable spatial risk mapping; this can result in risk analyses that are overly 'optimistic' if vulnerable assets are not accurately mapped (Venegas-Li et al. 2019). More broadly, it can be argued that systematic conservation planning using modern statistical and spatial principles of conservation investment is not common in the global offshore industry; this is somewhat surprising given the long duration, considerable depth range, and broad geographic reach of the activity. There are established practices for the systematic planning of the spatial arrangement of conservation actions and investments in many other marine sectors (Engelhard et al. 2017), and it would be sensible to apply these frameworks to the oil and gas industry to achieve better conservation outcomes and reduce potential social and economic dissatisfaction with these actions.

Despite holding the largest hydrocarbon reserves in the world, oceanic methane hydrate reservoirs are still largely ignored in projections for hydrocarbon exploration and conservation needs. Given the predicted technological exploration advances and market conditions, the global deep-sea methane reserves may be explored from massively large expanses of the deep seafloor along continental margins, targeting an average 400 m subseafloor layer where hydrates are stable (MacDonald 1990). There is a global estimated area with methane hydrate occurrence of 13×10^6 km^2 at depths between 800 and 2900 m (Gornitz and Fung 1994), which is near ten-times larger than the deep-sea mining areas under commercial applications (1.4×10^6 km^2; Miller et al. 2018). As a result, industrial exploitation of these deep-sea methane hydrate reservoirs would not only severely increase the risks and scale of direct impacts to deep-sea ecosystems, but also increase potential environmental damage from massive emissions of GHG to the atmosphere.

6.6 Ecosystem recovery from operational impacts

The recovery time and restoration options available after an impact attributed to the offshore hydrocar-

bon industry are very poorly known for most deep-sea species, assemblages, and ecosystems. The *Deepwater Horizon* blowout was the largest accident of this type, spilling an estimated 4.9 million barrels of oil and gas at a depth of 1500 m over 87 days. After dispersion, onshore and offshore transport, and evaporation, it is estimated that between 14 and 31 per cent of the 2 million barrels of oil was deposited on the seafloor, covering an area of 3200 km^2 (Valentine et al. 2014; Chanton et al. 2015). Several studies have shown stress, tissue loss, and oil and dispersant sedimentation over coral colonies as far as 20 km from the spill site at over 1500 m depth (White et al. 2012; Fisher et al. 2014b). Benthic (meiofauna and macrofauna) sediment communities were also impacted by the deposition of oil-related compounds in patchy seafloor areas that extended for over 170 km^2 (Montagna et al. 2013).

There is considerable evidence for the persistence of impacts associated with oil and surfactant (dispersant) on deep-sea corals for more than 7 years after the DWH spill (Hsing et al. 2013; Girard and Fisher 2018). Coral colonies (*Paramuricea biscaya*) were impacted by the deposition of oil and dispersants, and the effects on colonies were heterogeneous (from no effect to strong impacts), suggesting variable tolerance to the oil or a patchy contamination of individual colonies (Hsing et al. 2013). Overall, coral colonies that were impacted did not show evidence of recovery and exhibited lower growth or were colonised by hydroids, suggesting unhealthy conditions and lower individual fitness 7 years after the accident (Dark patterns in colonies from Figure 1; Hsing et al. 2013; Girard and Fisher 2018). Girard et al. (2019) documented significantly slower growth for impacted coral colonies, and a remarkably slow growth rate of octocorals (0.14 to 2.5 centimetres per colony per year), and DeLeo et al. (2018) found wide alterations of gene expression related to oxidative stress, toxin processing, and tissue regeneration along with hypermelanisation of tissues as a response to direct oil and dispersant exposure. These results indicate that the initial acute impacts that followed the accident caused rapid coral tissue damage to some colonies and altered physiology; recovery from these effects may take decades or longer. The slower growth of impacted colonies will likely impact coral reproduction and overall benthic biodiversity in the long term as larger colonies have higher reproductive output and harbour higher numbers of associated fauna (Page and Lasker 2012; Girard et al. 2019).

There is further evidence for the pervasive effects of polycyclic aromatic hydrocarbons and dispersants on deep-sea sponge beds and associated communities (Vad et al. 2018). The experience with the *Deepwater Horizon* in 2010 suggests that decades will be necessary for the recovery of assemblages impacted by large oil spills in the deep-sea (Montagna et al. 2013; Girard and Fisher 2018). Notwithstanding the *Deepwater Horizon* spill being the most closely studied in the deep sea, many questions regarding the long-term environmental harm remain unanswered, and it is uncertain how many coral colonies have been impacted, and these impacts remain undocumented.

6.7 Conclusions

Arguably, the offshore oil and gas industry will operate as long as there is a demand for oil and gas and as long as the reserves are economically and technically feasible to exploit. However, the natural capital offered by hydrocarbon reserves needs to be contrasted with its environmental harm and social costs; therefore, there is a need to reconsider the 'business-as-usual' model of the oil and gas industry. It stands to reason, and is a legal requirement, that the industry should operate within the legal framework and regulations set by the governments that grant the exploration leases. It follows that the risk of significant environmental damage will therefore be largely determined by law and how the law is executed and policed in any particular area.

There are broad international differences in policing how regulations are met, but in most cases, it requires some form of 'biological monitoring' and reporting. The objectives of monitoring programmes vary widely, and, hence, the designs also vary widely. This means that it can be difficult to attribute 'ecological impacts' unequivocally to the offshore oil and gas industry and to gauge how reliable and sensitive any monitoring is. Ideally, international standards for operating practices and environmental management strategies are highly desirable (Cordes et al. 2016b). Improving technical training of stakeholder staff and supporting an independent Environmental Impact Assessment review body are potential pathways to improve management practices in the offshore hydrocarbon industry (Barker and Jones 2013).

It must be recognised that the industry is largely determined to achieve better environmental performance and minimise ecological impacts. There are, in our opinion, however, several areas that can be improved to achieve more robust environmental protection. Whilst at the global scale, society shall be encouraged to switch to alternative, renewable energy sources to lower impacts related to the emissions of burnt oil and gas, it is possible to lower the environmental risks associated with exploration and production. Ultimately, better risk control will largely come from safer technology and better environmental laws that are consistently enforced (Chapter 3). Thus, 'environmental futures' for the offshore oil and gas industry rely on meeting challenges in the engineering domain and expectations of society at large.

Acknowledgements

The authors acknowledge funding agencies that have supported deep-sea research, and ship time that have allowed collection of data. We also acknowledge many deep-sea biologists that have contributed to the science that is synthesised in this work. AFB is funded by the Brazilian National Research Agencies (CNPq and CAPES) and Fundação do Amparo a Pesquisa do Espírito Santo (FAPES). TAS was supported by the Australian Research Council, Collaborative Research Networks, and Yaroomba Beach.

References

Aharon, P. and Fu, B. (2000). Microbial Sulfate Reduction Rates and Oxygen Isotope Fractionations at Oil and Gas Seeps in Deepwater Gulf of Mexico. *Geochimica et Cosmochimica Acta*, 62, 233–46.

Aharon, P., Schwarcz, H. P., and Roberts, H. H. (1997). Radiometric Dating of Submarine Hydrocarbon Seeps in the Gulf of Mexico. *Geological Society of America Bulletin*, 109, 568–79.

Allers, E., Abed, R. M., Wehrmann, L. M., et al. (2013). Resistance of *Lophelia pertusa* to Coverage by Sediment and Petroleum Drill Cuttings. *Marine Pollution Bulletin*, 74(1), 132–40.

Almada, G. V. M. B. and Bernardino, A. F. (2017). Conservation of Deep-sea Ecosystems Within Offshore Oil Fields on the Brazilian Margin, SW Atlantic. *Biological Conservation*, 206, 92–101. http://doi.org/10.1016/j.biocon.2016.12.026.

Baram, M. (2010). *Preventing Accidents in Offshore Oil and Gas Operations: The US Approach and Some Contrasting Features of the Norwegian Approach*. Deepwater Horizon Study Group—Working Paper. https://ccrm.berkeley.edu/pdfs_papers/DHSGWorkingPapersFeb16-2011/PreventingAccidents-in-OffshoreOil-and-GasOperations-MB_DHSG-Jan2011.pdf.

Barker, A. and Jones, C. (2013). A Critique of the Performance of EIA within the Offshore Oil and Gas Sector. *Environmental Impact Assessment Review*, 43, 31–9. doi: 10.1016/j.eiar.2013.05.001.

Becker, E. L., Cordes, E. E., Macko, S. A., and Fisher, C. R. (2009). Importance of Seep Primary Production to *Lophelia pertusa* and Associated Fauna in the Gulf of Mexico. *Deep-Sea Research I*, 56, 786–800. doi:10.1016/j.dsr.2008.12.006.

Becker, E. L., Cordes, E. E, Macko, S. A., Lee, R. W., and Fisher, C. R. (2013). Using Stable Isotope Contents in Animal Tissues to Infer Trophic Interactions and Feeding Biology in Seep Communities on the Gulf of Mexico Lower Slope. *PLoS ONE*, 8, e74459. doi:10.1371/journal.pone.0074459.

Bergquist, D. C., Eckner, J. T., Urcuyo, I. A., et al. (2007). A Local Hydrothermal Vent Food Web: Application of Stable Isotopes to a Complex Community. *Marine Ecology Progress Series*, 330, 49–65.

Bergquist, D. C., Fleckenstein, C., Knisel, J., et al. (2005). Variations in Seep Mussel Bed Communities Along Physical and Chemical Environmental Gradients. *Marine Ecology Progress Series*, 293, 99–108. doi: 10.3354/meps293099.

Bernardino, A. F., Levin, L. A., Thurber, A. R., and Smith, C. R. (2012). Comparative Composition, Diversity and Trophic Ecology of Sediment Macrofauna at Vents, Seeps and Organic Falls. *PLoS ONE*, 7(4), e33515. doi: 10.1371/journal. pone.0033515.

Bernardino, A. F. and Sumida, P. Y. G. (2017). Deep Risks from Offshore Development. *Science*, 358(6361), 312. doi: 10.1126/science.aaq0779.

Bernardino, A. F., Gama, R. N., Mazzuco, A. C. A., Omena, E. P., and Lavrado, H. P. (2019). Submarine Canyons Support Distinct Macrofaunal Assemblages on the Deep SE Brazil Margin. *Deep-Sea Research I*, 149, 103052. https://doi.org/10.1016/j.dsr.2019.05.012.

BOEM (2019). Areas Under Restriction for Offshore Exploration. Bureau of Ocean and Energy Management. https://www.boem.gov/Areas-Under-Moratoria/. Accessed 22 March 2019.

Boetius, A., Ravenschlag, K., Schubert, C. J., et al. (2000). A Marine Microbial Consortium Apparently Mediating Anaerobic Oxidation of Methane. *Nature*, 407, 623–6.

Boetius, A. and Wenzhöfer, F. (2013). Seafloor Oxygen Consumption Fuelled by Methane from Cold Seeps. *Nature Geoscience*, 6(9), 725.

BP (2012). *BP Statistical Review of World Energy*. June 2012. https://www.bp.com/en/global/corporate/energy-economics/statistical-review-of-world-energy.html.

BP (2018). *BP Statistical Review of World Energy*. 67th edn. June 2018. https://www.bp.com/content/dam/bp/business-sites/en/global/corporate/pdfs/energy-economics/statistical-review/bp-stats-review-2018-full-report.pdf.

Bowden, D. A., Rowden, A. A., Thurber, A. R., et al. (2013). Cold Seep Epifaunal Communities on the Hikurangi Margin, New Zealand: Composition, Succession, and Vulnerability to Human Activities. *PLoS ONE*, 8(10), e76869.

Breuer, E., Stevenson, A. G., Howe, J. A., Carroll, J., and Shimmield, G. B. (2004). Drill Cutting Accumulations in the Northern and Central North Sea: A Review of Environmental Interactions and Chemical Fate. *Marine Pollution Bulletin*, 48, 12–25.

Brooke, S. D., Holmes, M. W., and Young, C. M. (2009). Sediment Tolerance of Two Different Morphotypes of the Deep-sea Coral *Lophelia pertusa* from the Gulf of Mexico. *Marine Ecology Progress Series*, 390, 137–44.

Carney, R. S. (1994). Consideration of the Oasis Analogy for Chemosynthetic Communities at Gulf of Mexico Hydrocarbon Vents. *Geo-Marine Letters*, 14, 149–59.

Chanton, J., Zhao, T., Rosenheim, B. E., et al. (2015). Using Natural Abundance Radiocarbon to Trace the Flux of Petrocarbon to the Seafloor Following the Deepwater Horizon Oil Spill. *Environmental Science and Technology*, 49(2), 847–54. doi: 10.1021/es5046524.

Cordes, E. E, Arthur, M. A., Shea, K., and Fisher, C. R. (2005a). Modelling the Mutualistic Interactions Between Tubeworms and Microbial Consortia. *PLoS Biology*, 3, 497–506.

Cordes, E. E., Hourdez, S., Predmore, B. L., Redding, M. L., and Fisher, C. R. (2005b). Succession of Hydrocarbon Seep Communities Associated with the Long-lived Foundation Species *Lamellibrachia luymesi*. *Marine Ecology Progress Series*, 305, 17–29.

Cordes, E. E., Bergquist, D. C., Predmore, B. L., et al. (2006). Alternate Unstable States: Convergent Paths of Succession in Hydrocarbon-seep Tubeworm-associated Communities. *Journal of Experimental Marine Biology and Ecology*, 339, 159–76. doi: 10.1016/j.jembe.2006.07.017.

Cordes, E. E., McGinley, M., Podowski, E. L., et al. (2008). Coral Communities of the Deep Gulf of Mexico. *Deep-Sea Research I*, 55, 777–87. doi: 10.1016/j.dsr.2008.03.005.

Cordes, E. E., Bergquist, D. C., and Fisher, C. R. (2009). Macro-ecology of Gulf of Mexico Cold Seeps. *Annual Review in Marine Sciences*, 1, 143–68. doi: 10.1146/annurev.marine.010908.163912.

Cordes, E. E., Cunha, M.M., and Galeron, J., (2010). The Influence of Geological, Geochemical, and Biogenic Habitat Heterogeneity on Seep Biodiversity. *Marine Ecology*, 31, 51–65. doi: 10.1111/j.1439-0485.2009.00334.x.

Cordes, E. E., Arnaud-Haond, S., and Bergstad, O-A. (2016a). Cold-Water Corals, in L. Innis and A. Simcock, (eds.) *The First Global Integrated Marine Assessment: World Ocean Assessment I*. New York: United Nations General Assembly, pp. 1–28

Cordes, E. E., Jones, D. O. B., Schlacher, T. A., et al. (2016b). Environmental Impacts of the Deep-water Oil and Gas Industry: A Review to Guide Management Strategies. *Frontiers in Environmental Sciences*, 4, 58. doi: 10.3389.fenvs.2016.00058.

Costanza, R., d'Arge, R., de Groot, R., et al. (1997). The Value of the World's Ecosystem Services and Natural Capital. *Nature*, 387, 253–60.

Davies, A. J., Murray Roberts, J., and Hall-Spencer, J. (2007). Preserving Deep-sea Natural Heritage: Emerging Issues in Offshore Conservation and Management. *Biological Conservation*, 138, 299–312.

DeLeo, D. M., Ruiz-Ramos, D. V., Baums, I. B., and Cordes, E. E. (2016). Response of Deep-water Corals to Oil and Chemical Dispersant Exposure. *Deep-Sea Research II*, 129, 137–47. doi: 10.1016/j.dsr2.2015.02.028.

DeLeo, D. M., Herrera, S., Lengyel, S. D., et al. (2018). Gene Expression Profiling Reveals Deep-sea Coral Response to the Deepwater Horizon Oil Spill. *Molecular Ecology*, 27(20), 4066–77. doi: 10.1111/mec.14847.

Dubilier, N., Bergin, C., and Lott, C. (2008). Symbiotic Diversity in Marine Animals: The Art of Harnessing Chemosynthesis. *Nature Reviews Microbiology*, 6(10), 725.

Duperron, S., Lorion, J., Samadi, S., Gros, O., and Gaill, F. (2009). Symbioses Between Deep-Sea Mussels (Mytilidae: Bathymodiolinae) and Chemosynthetic Bacteria: Diversity, Function and Evolution. *Comptes Rendus Biologies*, 332(2–3), 298–310.

Dunn, D. C., Ardron, J., Bax, N., et al. (2014). The Convention on Biological Diversity's Ecologically or Biologically Significant Areas: Origins, Development, and Current Status. *Marine Policy*, 49, 137–45.

EIA (2016). Offshore Oil Production in Deepwater and Ultra-deepwater Is Increasing. *Energy Information Administration Report*. https://www.eia.gov/todayin-energy/detail.php?id=28552.

EPA (2018). Inventory of US Greenhouse Gas Emissions and Sinks: 1990–2016. *Annex 2 (Methodology for Estimating CO^2 Emissions from Fossil Fuel Combustion)*, Table A-41 and Table A-50. US EPA #430-R-18-003. Washington, DC: US Environmental Protection Agency, p. 101.

Engelhard, S. L., Huijbers, C. M., Stewart-Koster, B., et al. (2017). Prioritising Seascape Connectivity in Conservation Using Network Analysis. *Journal of Applied Ecology*, 54, 1130–41.

Etnoyer, P. J., Wickes, L. N., Silva M., et al. (2016). Decline in Condition of Gorgonian Octocorals on Mesophotic Reefs in the Northern Gulf of Mexico: Before and After the Deepwater Horizon Oil Spill. *Coral Reefs*, 35(1), 77–90.

Fidler, C. and Noble, B. (2012). Advancing Strategic Environmental Assessment in the Offshore Oil and Gas Sector: Lessons from Norway, Canada, and the United Kingdom. *Environmental Impact Assessment Review*, 34, 12–21. doi: 10.1016/j.eiar.2011.11.004.

Fisher, C. R. (1990). Chemoautotrophic and Methanotrophic Symbioses in Marine Invertebrates. *Reviews in Aquatic Science*, 2, 399–436.

Fisher, C. R., MacDonald, I. R., Sassen, R., et al. (2000). Methane Ice Worms: *Hesiocaeca methanicola* Colonizing Fossil Fuel Reserves. *Naturwissenschaften*, 87, 184–7.

Fisher, C. R., Demopoulos, A. W. J., Cordes, E. E., et al. (2014a). Coral Communities as Indicators of Ecosystem-Level Impacts of the Deepwater Horizon Spill. *BioScience*, 64, 796–807.

Fisher, C. R., Hsing, P., Kaiser, C. L., et al. (2014b). Footprint of Deepwater Horizon Blowout Impact to Deep-water Coral Communities. *Proceedings of the National Academy of Sciences USA*, 111(32), 11744–9.

Fortune, I. S. and Paterson, D. M. (2018). Ecological Best Practice in Decommissioning: A Review of Scientific Research. *ICES Journal of Marine Science*, doi: 10.1093/icesjms/fsy130.

Gass, S. E. and Roberts, J. M. (2006). The Occurrence of the Cold-water Coral *Lophelia pertusa* (Scleractinia) on Oil and Gas Platforms in the North Sea: Colony Growth, Recruitment and Environmental Controls on Distribution. *Marine Pollution Bulletin*, 52, 549–59.

Girard, F. and Fisher, C. R. (2018). Long-term Impact of the Deepwater Horizon Oil Spill on Deep-sea Corals Detected After Seven Years of Monitoring. *Biological Conservation*, 225, 117–27. doi: 10.1016/j.biocon.2018.06.028.

Girard, F., Cruz, R., Glickman, O., Harpster, T., and Fisher, C. R. (2019). In Situ Growth of Deep-sea Octocorals After the Deepwater Horizon Oil Spill. *Elementa: Science of the Anthropocene*, 7(1), 12. http://doi.org/10.1525/elementa.349.

Gornitz, V. and Fung, I. (1994). Potential Distribution of Methane Hydrates in the World's Oceans. *Global Biogeochemical Cycles*, 8(3), 335–47.

Gray, J. S., Clarke, A. J., Warwick, R. M., and Hobbs, G. (1990). Detection of Initial Effects of Pollution on Marine Benthos: An Example from the Ekofisk and Eldfisk Oilfields, North Sea. *Marine Ecology Progress Series*, 66, 285–99. doi: 10.3354/meps066285.

Harfoot, M. B. J., Tittensor, D. P., Knight, S., et al. (2018). Present and Future Biodiversity Risks from Fossil Fuel Exploitation. *Conservation Letters*, 11, e12448. doi: 10.1111/conl.12448.

Hsing, P-Y., Fu, B., Larcom, E. A., Berlet, S. P., et al. (2013). Evidence of Lasting Impact of the *Deepwater Horizon* Oil Spill on a Deep Gulf of Mexico Coral Community. *Elementa*, 1, 000012. https://www.elementascience.org/articles/10.12952/journal.elementa.000012/.

Hourdez, S., Weber, R. E., Green, B. N., Kenney, J. M., and Fisher, C. R. (2002). Respiratory Adaptations in a Deep-sea Orbiniid Polychaete from Gulf of Mexico Brine Pool NR1: Metabolic Rates and Hemoglobin Structure/Function Relationships. *Journal of Experimental Marine Biology and Ecology*, 205, 1669–81.

Hovland, M. and Thomsen, E. (1997). Cold-water Corals—Are They Hydrocarbon Seep Related? *Marine Geology*, 137(1–2), 159–64.

IPCC (2014). Climate Change 2014: Synthesis Report, in R. K. Pachauri and L. A. Meyer (eds.) *Contribution of Working Groups I, II and III to the Fifth Assessment Report of the Intergovernmental Panel on Climate Change*. Geneva, Switzerland: IPCC, p. 151.

Jarnegren, J. J., Brooke, S., and Jensen, H. (2017). Effects of Drill Cuttings on Larvae of the Cold Water Coral *Lophelia pertusa*. *Deep-Sea Research II*, 137, 454–62. doi: 10.1016/j.dsr2.2016.06.014.

Jones, D. O., Gates, A. R., Huvenne, V. A. I., Phillips, A. B., and Bett, B. J. (2019). Autonomous Marine Environmental Monitoring: Application in Decommissioned Oil Fields. *Science of the Total Environment*, 668, 835–53.

Joye, S. B., Bracco, A., Ozgokmen, T., et al. (2016). The Gulf of Mexico Ecosystem, Six Years After the Macondo Oil Well Blowout. *Deep-Sea Research II*, 129, 4–19. doi: 10.1016/j.dsr2.2016.04.018.

Kaiser, M. J. and Narra, S. (2018). A Retrospective of Oil and Gas Development in the US Outer Continental Shelf Gulf of Mexico, 1947–2017. *Natural Resources Research*. https://doi.org/10.1007/s11053-018-9414-3.

Kleindienst, S., Paul, J., and Joye, S. B. (2015). Using Dispersants Following Oil Spills: Impact on the Composition and Activity of Microbial Communities. *Nature Reviews Microbiology*, 13, 388–96.

Larcom, E. A., McKean, D. L., Brooks, J. M., and Fisher, C. R. (2014). Growth Rates, Densities, and Distribution of *Lophelia pertusa* on Artificial Structures in the Gulf of Mexico. *Deep-Sea Research I*, 85, 101–9.

Larsson, A. I., van Oevelen, D., Purser, A., and Thomsen, L. (2013). Tolerance to Long-term Exposure of Suspended Benthic Sediments and Drill Cuttings in the Cold-water Coral *Lophelia pertusa*. *Marine Pollution Bulletin*, 70, 176–88. doi: 10.1016/j.marpolbul.2013.02.033.

Le Bris, N, Arnaud-Hoand, S., Beaulieu, S., et al. (2016) Hydrothermal Vents and Cold Seeps, in L. Inniss and A. Simcock (eds.) *The First Global Integrated Marine Assessment: World Ocean Assessment I*. New York: United Nations General Assembly, Chapter 46.

Levin, L. A. (2005). Ecology of Cold Seep Sediments: Interactions of Fauna with Flow, Chemistry and Microbes. *Oceanography and Marine Biology: An Annual Review*, 43, 1–46.

Levin, L. A., Baco, A. R., Bowden, D. A., et al. (2016). Hydrothermal Vents and Methane Seeps: Rethinking the Sphere of Influence. *Frontiers in Marine Science*, 3, 72. https://doi.org/10.3389/fmars.2016.00072.

MacDonald, G. J. (1990). Role of Methane Clathrates in Past and Future Climates. *Climatic Change*, 16, 247–81.

MacDonald, I. R., Guinasso, N. L., Reilly, J. F., et al. (1990). Gulf of Mexico Hydrocarbon Seep Communities: VI. Patterns in Community Structure and Habitat. *Geo-Marine Letters*, 10(4), 244–52.

McClain, C., Nunnally, C., and Benfield, M. C. (2019). Persistent and Substantial Impacts of the Deepwater Horizon Oil Spill on Deep-sea Megafauna. *Royal Society Open Science*, 6, 191164. doi: 10.1098/rsos.191164.

Macreadie, P. I., McLean, D. L., Thomson, P. G., et al. (2018). Eyes in the Sea: Unlocking the Mysteries of the Ocean Using Industrial, Remotely Operated Vehicles (ROVs). *Science of the Total Environment*, 634, 1077–91.

Middelburg, J. J., Mueller, C. E., Veuger, B., et al. (2015). Discovery of Symbiotic Nitrogen Fixation and Chemoautotrophy in Cold-water Corals. *Scientific Reports*, 5, 17962.

Millennium Ecosystem Assessment (2005). *Ecosystems and Human Well-being*. Synthesis. Washington, DC: Island Press

Miller, K. A., Thompson, K. F., Johnston, P., and Santillo, D. (2018). An Overview of Seabed Mining Including the Current State of development, Environmental Impacts and Knowledge Gaps. *Frontiers in Marine Science*, 4, 418. doi: 10.3389/fmars.2017.00418.

Montagna, P. A, Baguley, J. G., Cooksey, C., et al. (2013). Deep-sea Benthic Footprint of the Deepwater Horizon Blowout. *PLoS ONE*, 8, e70540.

Mortensen, P. B., Hovland, M., Brattegard, T., and Farestveit, R. (1995). Deep Water Bioherms of the Scleractinian Coral *Lophelia pertusa* (L.) at 64°N on the

Norwegian Shelf: Structure and Associated Megafauna. *Sarsia*, 80, 145–58.

Muehlenbachs L., Cohen M. A., and Gerarden T. (2013). The Impact of Water Depth on Safety and Environmental Performance in Offshore Oil and Gas Production. *Energy Policy*, 55, 699–705. doi: 10.1016/j.enpol.2012.12.074.

Narula, K. (2018). The Maritime Dimension of Sustainable Energy Security. *Lecture Notes in Energy*, Vol. 68. New York: Springer. https://doi.org/10.1007/978-981-13-1589-3.

Olsen, E., Holen, S., Hoel, A. H., Buhl-Mortensen, L., and Røttingen, I. (2016). How Integrated Ocean Governance in the Barents Sea Was Created by a Drive for Increased Oil Production. *Marine Policy*, 71, 293–300. doi: 10.1016/j.marpol.2015.12.005.

Olu-Le Roy, K., Caprais, J-C., Fifis, A., et al. (2007). Cold-seep Assemblages on a Giant Pockmark off West Africa: Spatial Patterns and Environmental Control. *Marine Ecology*, 28(1), 115–30.

Orcutt, B. N., Joye, S. B., Kleindienst, S., et al. (2010). Impact of Natural Oil and Higher Hydrocarbons on Microbial Diversity, Distribution, and Activity in Gulf of Mexico Cold-seep Sediments. *Deep-Sea Research II*, 57(21–23), 2008–21.

Page, C. A. and Lasker, H. R. (2012). Effects of Tissue Loss, Age and Size on Fecundity in the Octocoral *Pseudopterogorgia elisabethae*. *Journal of Experimental Marine Biology and Ecology*, 434, 47–52. doi: 10.1016/j.jembe.2012.07.022.

Papathanasopoulou E., Beaumont N., Hooper T., Nunes J., and Queiros A. M. (2015). Energy Systems and Their Impacts on Marine Ecosystem Services. *Renewable and Sustainable Energy Reviews*, 52, 917–26.

Passow, U., Ziervogel, K., Asper, V., and Diercks, A. (2012). Marine Snow Formation in the Aftermath of the Deepwater Horizon Oil Spill in the Gulf of Mexico. *Environmental Research Letters*, 7, 035031. doi: 10.1088/1748-9326/7/3/035301

Polasky, S. and Segerson, K. (2009). Integrating Ecology and Economics in the Study of Ecosystem Services: Some Lessons Learned. *Annual Review of Resource Economics*, 1, 409–34 doi: 10.1146/annurev.resource.050708.144110.

Quattrini, A. M., Georgian, S. E., Byrnes, L., et al., (2013). Niche Divergence by Deep-sea Octocorals in the Genus *Callogorgia* Across the Continental Slope of the Gulf of Mexico. *Molecular Ecology*, 22, 4123–40. doi: 10.1111/mec.12370.

Reddy, C. M., Arey, J. S., Seewald, J. S., et al. (2012). Composition and Fate of Gas and Oil Released to the Water Column During the Deepwater Horizon Oil Spill.

Proceedings of the National Academy of Sciences of the USA, 109, 20229–34.

Sahling, H., Masson, D. G., Ranero, C. R., et al. (2008). Fluid Seepage at the Continental Margin Offshore Costa Rica and Southern Nicaragua. *Geochemistry, Geophysics, Geosystems*, 9, 5. https://doi.org/10.1029/2008GC001978.

Sibuet, M. and Olu, K. (1998). Biogeography, Biodiversity and Fluid Dependence of Deep-sea Cold-seep Communities at Active and Passive Margins. *Deep-Sea Research II*, 45(1–3), 517–67.

Sloan, E. D. D. (1990). *Clathrate Hydrates of Natural Gases*. New York: Marcel Dekker.

Tunnicliffe, V. (1991). The Biology of Hydrothermal Vents: Ecology and EVolution. *Oceanography and Marine Biology: An Annual Review*, 29, 319–407.

Ulfsnes, A., Haugland, J. K., and Weltzien, R. (2013). Monitoring of Drill Activities in Areas with Presence of Cold Water Corals. *Det Norske Veritas (DNV) Report*: 2012–1691. Stavanger, Norway: Det Norsk Veritas, Stavanger.

UNEP (2006). Accounting for Economic Activities in Large Marine Ecosystems and Regional Seas. Geneva: UNEP.

USEIA (2016). US Energy Information Administration—Offshore Production Nearly 30 per cent of Global Crude Oil Output in 2015. https://www.eia.gov/todayinenergy/detail.php?id=28492. Assessed 21 March 2019.

van den Bergh, J. C. J. M. and Botzen, W. J. W. (2014). A Lower Bound to the Social Cost of CO_2 Emissions. *Nature Climate Change*, 4, 253–8. doi: 10.1038/NCLIMATE2135.

Vad, J., Kazanidis, G., Henry, L., et al. (2018). Potential Impacts of Offshore Oil and Gas Activities on Deep-Sea Sponges and the Habitats they Form. *Advances in Marine Biology*, 79, 33–60.

Venegas-Li, R., Levin, N., Morales-Barquero, L., et al. (2019). Global Assessment of Marine Biodiversity Potentially Threatened by Offshore Hydrocarbon Activities. *Global Change Biology*, 25(6), 2009–20. doi: 10.1111/gcb.14616.

Valentine, D. L., Fisher, G. B., Bagby, S. C., et al. (2014). Fallout Plume of Submerged Oil from Deepwater Horizon. *Proceedings of the National Academy of Sciences USA*, 111(45), 15906–11. doi: 10.1073/pnas.1414873111.

White, H. K, Hsing, P., Cho, W., et al. (2012). Impact of the Deepwater Horizon Oil Spill on a Deep-water Coral Community in the Gulf of Mexico. *Proceedings of the National Academy of Sciences*, 109, 20303–8.

World Bank (2017) World Bank Annual Report 2017. Washington, DC: World Bank. doi: 10.1596/978- 1-4648-1119-7.

The exploitation of deep-sea biodiversity: components, capacity, and conservation

Harriet Harden-Davies

7.1 Introduction

Humans have been harnessing the natural properties of marine organisms for millennia—initially in their unprocessed form for sustenance, and more recently via extracted products as biomaterials, functional food ingredients, and medicines. As accelerating scientific and technological advances open up the deep ocean, potential avenues to exploit components and characteristics of marine biodiversity are revealed. For example, a compound originally derived from the deep-sea sponge *Halichondria okadai*, Halichondrin-B, is used as the model for the anticancer drug *Halaven* and is widely regarded as the most complex natural product ever synthesised (Newman 2016). To keep pace with such innovations and to promote equitable and sustainable activities, the international legal framework has evolved over recent decades to address the conservation and sustainable use of biodiversity together with the sharing of benefits arising from the utilisation of genetic resources. Gaps remain, however, particularly for the deep, remote, and technologically demanding ocean areas beyond national jurisdiction (ABNJ) that account for more than 60 per cent of the global ocean.

The question of how to share benefits from marine genetic resources is one of the most contentious issues in ongoing negotiations for the development of a new international legally binding instrument for the conservation and sustainable use of marine

biodiversity in ABNJ under the *1982 United Nations Convention on the Law of the Sea* (UNCLOS). The development of this agreement for biodiversity beyond national jurisdiction (BBNJ agreement), which began preparations in 2016 and is scheduled to conclude in 2020, is a historic opportunity to strengthen the international framework for marine biodiversity in this vast deep-ocean space (Gjerde et al. 2019). It has also illuminated a number of challenges, such as delineating components of biodiversity with disjointed definitions in legal and scientific terminologies, achieving a common understanding of the activities and potential outcomes of exploitation, and aligning opportunities to support the conservation and sustainable use of biodiversity. To consider the exploitation of deep-ocean biodiversity, it is necessary to consider what is meant by exploitation, and what, in terms of deep-sea biodiversity, can be exploited. Science and technology is, and should be, at the heart of this discussion.

In this chapter, first, the exploitable components of biodiversity are identified through a discussion of the definitions of relevant terms such as 'genetic resources'. Unique characteristics of deep-sea biodiversity, as a source of novel genetic and biochemical properties, are discussed before examining different perceptions of the value of such resources as well as illustrative examples of utilisation. Second, the scientific and technological activities involved with the exploration and potential exploitation of deep-sea biodiversity are examined, and

Harriet Harden-Davies, *The exploitation of deep-sea biodiversity: components, capacity, and conservation* In: *Natural Capital and Exploitation of the Deep Ocean.* Edited by: Maria Baker, Eva Ramirez-Llodra, and Paul Tyler, Oxford University Press (2020). © Oxford University Press.
DOI: 10.1093/oso/9780198841654.003.0007

the potential benefits to be captured from genetic resources are discussed. Third, issues associated with the conservation and sustainable use of deep-sea biodiversity are discussed. Given the close alignment between potential exploitation of biodiversity and the conservation and sustainable use of biodiversity, the development of the BBNJ agreement provides a central focus for the discussion in this chapter.

7.2 Exploitable components of deep-sea biodiversity

The deep ocean is the largest biosphere on the planet, but it remains largely unexplored. The following section considers the definition of 'biodiversity' and related components, particularly genetic resources, that could be subject to exploitation. The specific interest in deep-sea biodiversity as a source of inspiration for technological innovation is discussed. The value of genetic and biological resources is then examined (in instrumental, inherent and intrinsic terms), and illustrative examples of the utilisation of marine genetic resources are provided.

7.2.1 Deep-sea biodiversity as inspiration for innovation

Biodiversity, biological resources, and genetic resources are inextricably linked. This interconnection is illustrated by the definitions contained in article 2 of the *1992 Convention on Biological Diversity* (CBD) (Table 1). Biological diversity is, essentially, the variability among living organisms. Genetic variability is the fundamental basis of biodiversity, and in turn, biodiversity is a source of biological and genetic resources. Further, the CBD definition of 'biological resources' includes 'genetic resources'. Genetic resources, for example, are defined by the CBD as 'genetic material of actual or potential value'. This reference to 'value' reflects the potential utilisation of biodiversity for various scientific and technological applications. Consequently, the CBD and the *2010 Nagoya Protocol on Access to Genetic Resources and the Fair and Equitable Sharing of Benefits Arising from Their Utilization to the Convention on Biological Diversity* (Nagoya Protocol) established a

framework for access and benefit sharing of genetic resources. Although this framework is applicable only in areas within national jurisdiction, not in ABNJ, the definitions established by the CBD and Nagoya Protocol have formed the basis of the discussions of definitions in the BBNJ negotiations (UN 2017).

There are, however, ambiguities in the CBD definitions, which raise a number of questions relating to the scope of what is included in genetic resources and precisely what elements are included in the regimes for access and sharing of benefits arising from the utilisation of such resources. The CBD definition of 'genetic resources' is restricted to material containing functional units of heredity, i.e. DNA. The CBD did refer to 'derivatives', in the definition of 'biotechnology', but this term was not defined. Presumably for this reason, the Nagoya Protocol included a definition of 'derivatives' in order to capture the compounds produced by organisms, such as natural products (Table 1). This illustrates the challenges of accurately describing complex scientific processes and natural phenomena in legal terminology. Viewed collectively, it seems that the material scope of the access and benefit-sharing regime under the CBD and Nagoya Protocol incorporates genetic and other biochemical elements of marine organisms, including genes (i.e. DNA and RNA) and primary and secondary metabolites.

The negotiations for the BBNJ agreement have not been conclusive on this point. Rather, the scope of marine genetic resources has been a point of contention in the process: does the material scope include derivatives, as well as genes, and does the geographic scope include the high seas (water column beyond national jurisdiction) and the Area (seabed beyond national jurisdiction)? A range of options were put forward during the preparatory phase for the BBNJ negotiations, with several delegations calling for a distinction between organisms used as a commodity (particularly fish) and organisms used for research on their genetic properties (UN 2017). This is another illustration of the difficulty in developing scientifically satisfactory legal definitions. It also highlights the challenges in delineating the scope of genetic resources whereby subtle nuances relating to utilisation can have sig-

Table 1 Definitions of terms

Term	Definition	Source
Biological diversity	'Variability among living organisms from all sources including, *inter alia*, terrestrial, marine and other aquatic ecosystems and the ecological complexes of which they are part: this includes diversity within species, between species and of ecosystems'	CBD art. 2
Biological resources	'Includes genetic resources, organisms or parts thereof, populations, or any other biotic component of ecosystems with actual or potential use or value for humanity'	CBD art. 2
Genetic material	'Any material of plant, animal, microbial, or other origin containing functional units of heredity'	CBD art. 2
Genetic resources	'Genetic material of actual or potential value'	CBD art. 2
Derivative	'… a naturally occurring biochemical compound resulting from the genetic expression or metabolism of biological or genetic resources, even if it does not contain functional units of heredity…'	Nagoya Protocol art. 2(e)

nificant legal and regulatory implications. An instructive example of this is the difficulty in applying the UNCLOS definition of 'sedentary species' in practice, due to the multifaceted life-cycles of deep-sea organisms (for a discussion see Mossop 2010). As a result of the ambiguities and blurred distinctions between the CBD definitions, the term 'genetic resources of ABNJ' could be considered to include all marine life in ABNJ (Marciniak 2017).

Ocean life, particularly in the deep ocean, is thought to be far more diverse than in terrestrial habitats. The total number of described marine species is approximately 250,000 (Mora et al. 2011), with around 2000 new species described each year; however, it is estimated that 90 per cent of total marine species are yet to be described. Around 26,000 species have been described from the deep ocean, and estimates of the total number of deep-sea species vary from 500,000 to over 10 million species (see Chapter 1), with an expected one million new species discoveries in the deep pelagic 'twilight zone' alone (Robison 2009). Marine microbes are the most diverse and abundant form of marine life—one litre of seawater could contain up to one

billion bacterial cells and an order of magnitude more viruses (Bowler et al. 2009). Although the diversity of microbes remains poorly understood (Abida et al. 2013), they are thought to represent a significant source of 'genomic innovation' (Sogin et al. 2006) and biological inspiration. Much remains to be discovered before humans fully grasp the diversity, abundance, connectivity and functionality of deep-ocean life, especially in ABNJ (Snelgrove et al. 2018).

Despite the knowledge gaps, it is known that deep-ocean life has evolved to thrive in environmental conditions of darkness, high pressure, reduced oxygen levels, and temperature extremes ranging from −2 °C to 150 °C (Danovaro et al. 2014). Unique adaptations of deep-ocean organisms include physiological adaptations to withstand pressure and temperature extremes, adaptations to low-light through advanced chemosensory capabilities or the ability to produce light through bioluminescence (Robison 2009), and lower metabolic rates to conserve energy where necessary. While most organisms cope with food limitations by grazing on organic matter sinking from sunlit surface waters above, others, have established symbiotic or trophic relationships with chemoautotroph microorganisms that use the chemical energy sources of hydrothermal vents, cold seeps, and organic falls, such as methane and hydrogen sulphide, to synthesise organic matter and support productive communities. Such adaptations and the existence of so-called 'extremophiles' has led many to view the deep ocean as a source of interest and intrigue for deep-sea biodiscovery. As noted by Vierros et al. (2016), genetic resources from ABNJ could include 'material from deep-sea animals, microbes or other organisms, and parts thereof containing functional units of heredity of actual or potential value'.

As a result, the deep ocean represents a rich collection of genetic diversity and has been described as a 'unique reservoir for a broad range and diversity of molecules of interest to further scientific knowledge and develop new products that improve human well-being' (Broggiato et al. 2014). Marine species are thought to be more likely to yield previously undescribed chemicals (natural products) than terrestrial species (Arrieta et al. 2010), partly due to the comparatively recent origins of the marine biodiscovery research field. Marine microbes are already a large source of marine natural products and proteins (Arrieta et al. 2010; Abida et al. 2013) but are still considered generally unexplored as a source of novel bioactive compounds (Gerwick and Fenner 2013; Newman and Cragg 2016; Carroll et al. 2019).

Research into marine natural products has developed rapidly since it began in the late 1940s, with researchers using SCUBA diving equipment to access sponge species (Bergmann and Feeney 1951). According to Carroll et al. (2019), more than 1400 new marine natural products were described in 2017, a 17 per cent increase from the previous year. Approximately 30,000 of the 1 million natural products described are from marine origins (Martins et al. 2014) with approximately 600 of these deriving from the deep ocean (Skropeta 2011). Research into deep-sea natural products, though fairly nascent at less than 5 per cent of the total number of marine natural products, is increasing (Skropeta 2008; Skropeta and Wei 2014; Harden-Davies 2016). The number of deep-sea natural products described between 2009 to 2013 grew by 188, including compounds derived from depths greater than 5000 m (Skropeta and Wei 2014). In addition to natural products, enzymes are also a target for biodiscovery research, while physiological characteristics provide a source of inspiration for new biomaterials and biological tools, as discussed further in this Chapter. Research into the genetic and other biochemical properties of deep-sea organisms appears to be growing; however, access to deep-sea biodiversity is hampered by limited availability of specialised scientific equipment and other research-capacity constraints. Consequently, the full potential value of deep-sea biodiversity remains unknown.

7.2.2 'Actual or potential' value

The CBD definitions of both genetic and biological resources hinge on the phrases 'actual or potential value' in the case of genetic material, and 'actual or potential use or value for humanity' in the case of biological resources. The value of biological resources can be instrumental, inherent, or intrinsic

(Rayfuse 2007); the following subsections explore these different perspectives (Table 2).

Instrumental value

Instrumental value refers to direct use such as pharmaceuticals and chemicals, as well as indirect uses such as services provided by biological resources and ecosystems (Rayfuse 2007). The 'medicinal, agricultural or other economic value' of species and communities are also referred to in Annex I of the CBD in the context of identifying and monitoring components of biodiversity. The Nagoya Protocol recognises the importance of genetic resources to 'food security, public health, biodiversity conservation, and the mitigation of and adaptation to climate change', in the Preamble.

Marine genetic resources are one of the ecosystem services provided by the deep sea (Armstrong et al. 2012; Rogers et al. 2014) (see Chapter 2, Table 1). Genes are the building blocks of life (Marcot 2007), providing the functional units of diversity. This value is explicitly referenced in The Global Plan of Action for Conservation, Sustainable Use and Development of Forest Genetic Resources (FAO 2014), which describes genetic diversity as the 'mainstay of biological stability' and notes that genetic resources are crucial to the adaptation and protection of ecosystems. Genetic diversity is critical for preserving the adaptive potential to stressors such as climate change (Taylor et al. 2017). As noted in Marlow et al. (2019): 'marine genetic resources, derived from the immense biodiversity of the ocean, not only represent an alluring potential source of fundamental scientific discovery and commercial bioproducts – they provide an integral part of the global systems that make Earth habitable'. Functions of healthy ocean ecosystems range from the recycling of nutrients to sequestering carbon and underpinning the adaptive capacity and resilience of such ecosystems to change. The value of such services is arguably intrinsic and inherent, as well as instrumental.

The natural innovations of marine life are a source of inspiration for various applications that imitate or harness the genetic and other biochemical properties of organisms, including for biomaterials, food, industrial processes, bioremediation, cosmetics, pharmaceuticals, and scientific research. Examples include:

Biomaterials and biomimicry

The characteristics of marine organisms can be imitated or mimicked by science (Whitesides 2015). For example, the dermal denticles of shark skin is a source of inspiration for new materials that reduce biofouling and reduce drag through water (Eadie and Ghosh 2011); the shape of whale and fish fins inspires the design of some propulsion technologies; and the camouflaging capabilities of cephalopods (octopus, squid, and cuttlefish) is used to inspire materials that autonomously sense and adapt to the colour of surroundings (Yu et al. 2014). Yet more examples of bioinspiration for new biomaterial research and development include the silicate fibres from sponge spicules used for 'bioinspired' research into biofabrication for purposes of biosilica-mediated regeneration of tooth and bone defects (Müller et al. 2009); the structure of and biomineralisation process underpinning marine mollusc shells (Morris et al. 2016); jellyfish collagen used for cartilage repair, wound healing applications, and regenerative medicine research (Pugliano et al. 2017); and starfish and squid inspired early work in soft robotics (Shepherd et al 2011). Biogenic polymers produced by a deep-sea sponge have been used in research into bone substitutes and to heal bone defects (Wang et al. 2013). Another example from the deep sea is the iron-plated natural armour of the scaly-foot snail *Crysomallon squamiferum*, found at southern Indian Ocean hydrothermal vents, which provides the basis for biomimicry investigations and the development of new materials, mechanical design principles, and engineering (Yao et al. 2010).

Cosmetics

Several products marketed for antioxidant, antiageing, and the treatment of skin conditions such as acne utilise bioactive compounds, vitamins, and minerals derived from marine plants and animals (Martins et al. 2014). Sources range from sea whips (Munro et al. 1999) to Antarctic and deep-sea bacteria (Martins et al. 2014). For example, the cosmetic product Abyssine contains an anti-inflammatory compound derived from bacteria isolated from a deep-sea polychaete *Alvinella pompejana* found at hydrothermal vents along the East Pacific Rise (Leary et al. 2009).

Food and nutraceuticals

From algae to microbes, there are several examples of the use of natural compounds for purposes such as gelling, thickening, and stabilising (Martins et al. 2014), including from crustacean by-products such as chitin, to antifreeze proteins derived from fish, bacteria, fungi, crustaceans, and microalgae (Kim et al. 2017). Bioactive properties of marine organisms may also be used as sources of food ingredients towards applications in new functional foods. Research into functional food marine-derived ingredients has seen recent rapid growth owing to potential beneficial effects in terms of disease prevention or maintenance of health and well-being, over and above the basic nutritional value of the marine product (Lordan et al. 2011; Freitas et al. 2015). One example is sea cucumbers, which are currently being investigated for this purpose owing to their valuable and wide-ranging nutritional properties (Bordbar et al. 2011; Xu et al. 2018).

Industrial processes and commodity chemicals

The adaptations of organisms to extremes of temperature, pressure, and pH can be utilised to create substances for industrial processes. For example, methane-consuming bacteria can be used for biofuel production to increase the stability and efficiency of bioindustrial processes (Marlow et al. 2019). A deep-sea organism is the source of the key bioactive ingredient in Fuelzyme, an enzyme used for starch liquefication (Leary and Juniper 2013). Other examples include accelerating advanced manufacturing of chemicals for uses ranging from pharmaceuticals (National Academy of Sciences 2015) to agriculture.

Scientific research

Proteins, such as enzymes, from marine organisms are already being used in scientific research processes. A reagent for genetic research uses derivatives sourced from the 'rushing fireball' microbe *Pyrococcus furiosus*, found at hydrothermal vents, on account of its ability to function at high temperature. Another reagent for genetic and biochemical research is green fluorescent protein, derived from the bioluminescent jellyfish *Aequorea victoria*, which absorbs UV light and emits it as green light. Other examples include the use of luciferases derived from deep-ocean organisms to measure cytotoxicity for research relating to the identification of tumour-associated antigens (Matta et al. 2018).

Pharmaceuticals

From aspirin to antibiotics, there are myriad examples of the use of natural products to tackle diseases such as cancer and to relieve pain (Skropeta 2011). Half of drugs used to treat cancer and 75 per cent of drugs used to treat infectious diseases derive from natural products (Skropeta and Wei 2014). Natural products derived from marine organisms; in particular, marine invertebrates are valued for bioactivity and as potential drug leads, with more than 50 per cent showing pharmacological activity (Skropeta 2011). Such compounds can be used to promote or inhibit biological processes and are used for functions ranging from antitumour, antimicrotubule, antiproliferative, antibiotic, or anti-infective (Martins et al. 2014; Newman and Cragg 2016). Of the deep-sea natural products identified between 2009 and 2013, 75 per cent were found to be biologically active, and almost 50 per cent showed anticancer potential (Skropeta and Wei 2014). Since marine-based drug discovery started in the mid-twentieth century, several leads have been identified for treatment of cancer, inflammation, pain, HIV, and Alzheimer disease. There are eight marine-derived drugs currently on the market (five for cancer, one for pain, one for viruses, and one for hypertriglyceridemia). However, while many are in clinical trials, few will ever make it to market. This is due to the long, complex, and costly research and development processes involved in developing and commercialising new drugs (Martins et al. 2014; Oldham et al. 2014; Harden-Davies 2016).

The utilisation of genetic resources to tackle grand challenges of food, health, and energy security and to advance scientific knowledge and boost scientific and technological capacity underpins the societal value of deep-ocean organisms (Arico 2010; Abida et al. 2013). In its twentieth preambular paragraph, the CBD explicitly refers to the essential importance of access and benefit sharing of genetic resources and technologies for the conservation and sustainable use of biodiversity and, in turn, for

meeting food, health, and other requirements of a growing global population. The need to preserve environmental values, both instrumental and inherent, of genetic resources supports the need for biodiversity conservation (van Zonneveld et al. 2018) and approaches that seek to consider the value of genetic resources as far more than economic or financial gains from commercial products.

Inherent value

Inherent value refers to nonuse values such as cultural, aesthetic, or religious, pertaining to individual organisms or ecosystems (Rayfuse 2007). The CBD refers to the 'ecological, social, economic, scientific, educational, cultural, recreational and aesthetic values of biological diversity and its components' in the Preamble, in Article 7(a) and in Annex I. The CBD also refers to the 'social, scientific or economic importance' of genomes and genes; the 'social, economic, cultural or scientific importance' of ecosystems, habitats, species, and communities in the first paragraph of Annex I (CBD 1992). The Forest Genetic Resources Plan (FAO, 2014) even incorporates such value into the definition of genetic resources—considering genetic resources to be genetic material of 'actual or potential economic, environmental, scientific or societal value'. It is increasingly recognised that the issue of benefit sharing from ABNJ is relevant for environmental and social reasons, as well as economic, scientific reasons (Marciniak 2017). Healthy ocean ecosystems harbour sites of cultural importance such as potential World Heritage sites of ABNJ (Freestone et al. 2016).

Table 2 Valuing deep-sea biodiversity

Value	Description
Instrumental	Use values: direct (e.g. pharmaceuticals, materials, scientific research tools, industrial processes, food, and nutraceuticals; biotechnology development to tackle grand challenges such as health or food security) or indirect (e.g. ecosystem services such as biomass, carbon sequestration, and nutrient recycling)
Inherent	Nonuse values of an organism or ecosystem (e.g. cultural and aesthetic)
Intrinsic	Value of an entity in, of, and for itself (e.g. environmental)

Intrinsic value

'Intrinsic value' refers to the value of an entity 'in, of and for itself' (Rayfuse 2007). Unlike instrumental or inherent value, intrinsic values do not depend on the perspective of an external valuer or beneficiary. The 'intrinsic value' of biodiversity is recognised explicitly in the CBD in the first paragraph of the Preamble and in Article 7(a).

7.3 Capacity: capturing benefits

Considering perceptions of value is important to identify potential benefits from deep-sea genetic resources, and to ascertain how those benefits can be derived and shared. In this section, (i) the potential benefits to be derived from deep-sea genetic resources are discussed, and (ii) the scientific and technological capacity requirements or constraints to 'exploit biodiversity' to capture and share such benefits are then considered. The critical role of science and technology in capturing and sharing benefits, particularly from deep-sea biodiversity, is highlighted.

7.3.1 Benefits

Considering the 'actual or potential value' of biodiversity and genetic material as instrumental, intrinsic, and inherent supports an expansive interpretation of the value of marine genetic resources incorporating economic, environmental, cultural, societal, and scientific aspects. However, the CBD definition of genetic resources does not seem to reflect the full value, role in ecosystem function, potential utilisation, or, indeed, the full spectrum of benefits from genetic resources (Deplazes-Zemp 2018). Despite the more expansive description of the instrumental and intrinsic value of biodiversity in the CBD, the only explicit mention of value in the Nagoya Protocol is to the 'economic value of ecosystems and biodiversity' (in the sixth paragraph of the Preamble).

The emphasis on economic value in the Nagoya Protocol appears to be reflected in two broad categories of benefits that are considered therein: monetary or nonmonetary (Table 3). Monetary benefits include, for example, financial or commercial outcomes such as payments (up-front, milestone, or royalties), fees (access, license, or special), research funding, joint intellectual property rights ownership, and patents. Nonmonetary benefits include collaboration and international cooperation in scientific research; technology transfer, and access to samples, data, and knowledge; capacity building and technology transfer including scientific training and access to resources, research infrastructure, and technology; and other social and economic benefits (Table 3). The scope of benefits is another point of contention in the development of the BBNJ negotiations (Broggiato et al. 2014).

Science has played a central role in reframing the conversation about the true value of deep-sea genetic resources and the significance of 'nonmonetary' benefits such as science collaboration, technology transfer, and capacity building. In the development of the BBNJ agreement, an emphasis on economic value and monetary benefits was prominent at the start of the process. However, awareness has grown of the high uncertainty (Oldham et al. 2014), high cost, and high failure rates (Martins et al. 2014) associated with commercialisation of products that are based on genetic resources, as illustrated by the very few examples of commercial products derived from deep-sea BBNJ (Leary et al. 2009). Expectations of the prospects for large financial gains have subsided to some degree. Consequently, there has been a perceptible shift towards a more holistic consideration of benefits, with a prominent focus on scientific and technological capacity building, including access to knowledge, data, training, and equipment (Broggiato et al. 2014; Harden-Davies 2016; Harden-Davies and Gjerde 2019). The scientific community has played an important role in shifting this conversation, for example, as noted by Marlow et al. (2019) 'the true value of the genetic resources of deep-sea ABNJ extends well beyond economic and commercial considerations - as an integral part of the operating system that keeps our planet healthy'. Crucially the scientific research endeavour is, in and of itself, a driver, object, and recipient of sharing benefits from deep-sea genetic resources.

Table 3 Examples of 'nonmonetary' benefits from genetic resources (adapted from Harden-Davies 2018)

Category	Examples of Benefits Provided in Nagoya Protocol Annex
Collaboration in scientific research	Collaboration, cooperation, and contribution in scientific research and development programmes, particularly biotechnological research
Technology transfer, and access to research results/ samples/data/knowledge	Sharing of research and development results
	Admittance to *ex situ* facilities of genetic resources and to databases
	Transfer to the provider of the genetic resources of knowledge and technology that make use of genetic resources or that are relevant to the conservation and sustainable use of biodiversity
	Access to scientific information relevant to conservation and sustainable use of biological diversity, including biological inventories and taxonomic studies
Scientific and technical, human and institutional capacity building	Collaboration, cooperation, and contribution in scientific research and development programmes, particularly biotechnological research
	Collaboration, cooperation, and contribution in education and training
	Institutional and professional relationships
	Training related to genetic resources
	Strengthening capacities for technology transfer
	Institutional capacity building
	Human and material resources to strengthen capacities for the administration and enforcement of access regulations
	Participation in product development
Capturing social and economic outcomes	Research directed towards priority needs, such as health and food security
	Contributions to the local economy
	Food and livelihood security benefits
	Social recognition

7.3.2 Capturing benefits: the role of science and technology

Research and development are critical to the utilisation and derivation of value from genetic resources. This is reflected in the definition of 'Utilisation of genetic resources' in Article 2(c) of the Nagoya Protocol, which is '… to conduct research and development on the genetic and/or biochemical composition of genetic resources, including through the application of biotechnology as defined in Article 2 of the Convention [on Biological Diversity]…'. 'Biotechnology' is defined in Article 2 of the CBD as '… any technological application that uses biological systems, living organisms, or derivatives thereof, to make or modify products or processes for specific use…'. Science and technology are central to accessing, utilising, and exploiting deep-sea biodiversity.

The research and development chain for deep-sea biodiversity begins with deep-ocean investigation, a technology-intensive activity (see Chapter 1).

Although the history of deep-sea exploration has its roots in the expeditions of the *Challenger* in the nineteenth century, it was not until the mid-twentieth century that underwater vehicles (from the first human occupied bathyscaphes in the 1930s to the submersibles, remotely operated vehicles, and autonomous underwater vehicles used today) began to open up deep-sea exploration in earnest (Chapter 1). Deep-sea research remains high-cost and technologically demanding. Research-capacity constraints are the primary limiting factor to exploring deep-sea biodiversity and its biological and genetic and other resources. The fact that so few countries have the scientific and technological potential to access and utilise genetic resources from remote deep-water ABNJ is one of the key challenges facing the development of the BBNJ agreement.

A variety of research infrastructure is needed to access and utilise deep-sea genetic resources. For example, underwater vehicles are usually necessary to access the deep ocean (although there are lower-tech options such as nets) and almost always require research vessels, which carry high capital and operating costs. Just a few countries have sufficient research and development budgets to own and operate such vessels. In addition to off-shore infrastructure for accessing organisms *in situ*, a range of on-shore research infrastructure is needed for research on samples *ex situ* and downstream through research and development pipelines, including for sample storage and curation, data storage, and analysis and modelling capacities.

Several scientific and technological challenges remain in the process of deep-sea biodiscovery. In addition to the barriers to accessing the deep ocean, there are downstream challenges to culturing organisms and recreating high-pressure conditions to keep organisms alive. On the other hand, scientific advances such as the synthesis of compounds in laboratories shows promise. Examples include the chemical Ecteinascidin-743 (trabectedin), the active compound derived from the sea squirt *Ecteinascidia turbinata* used in the anticancer drug Yondelis, which was produced via chemical semi-synthesis (Martins et al. 2014); and the fully synthetic Halaven, modelled on the natural product Halichondrin-B (Newman 2016). However, technical barriers remain (Glaser and Mayer 2009), and there are complex legal and regulatory issues associated with techniques such as synthetic biology (Oldham et al. 2012).

Overcoming technological challenges will depend on the types of international and interdisciplinary (OECD 2013) collaborations that have made deep-sea discoveries possible—from the first discovery of hydrothermal vents to the wreck of the *Titanic* (Ballard 2000). Such collaborations continue to be important today, as illustrated by exceptionally successful programmes such as the Census of Marine Life (Ausubel et al. 2010). International collaboration is crucial to promote scientific advances from genetic resources (Overmann and Hartman Scholz 2017), especially to develop and deploy innovations that decrease cost. One example of this is the rise of DNA sequencing technologies that are increasing rate and decreasing cost (National Academy of Sciences 2015) of genetic research. Such technological advances lower the barrier of entry to research and

development and bring opportunities for capacity building (Harden-Davies and Gjerde 2019). Indeed, the important contribution to sustainable development made by technology transfer and cooperation to build research and innovation capacities for adding value to genetic resources, including for poverty eradication and environmental sustainability, is recognised in the Nagoya Protocol in Article 22.

Opening up the deep ocean to exploration is thought to bring new opportunities for marine biodiscovery. The European Union regulation relating to compliance measures for the Nagoya Protocol (EU 2014) noted that the oceans are widely recognised as the least explored and least well-known realm of the planet and that research on genetic resources is being extended into the deep ocean—'the last great frontier on the planet'.

The value and potential applications of scientific and technological capacity is in no way restricted to the search for compounds of commercial interest, but they can be equally applied to understanding ecosystem function in the deep ocean. As well as potentially opening up opportunities to exploit deep-sea genetic and biochemical attributes, such genetic tools and genomic techniques also contribute to the conservation and sustainable use of marine biodiversity by advancing scientific knowledge of the amount, distribution, and functional significance of genetic variation; and the ability of populations to adapt to change and ecosystems processes like nutrient or energy fluxes. For example, genetic research can also support the design and monitoring of area-based management tools such as marine protected areas, assessments for fish stocks and restoration, and wildlife forensics tools that detect illegal activities such as illegal, unreported, and unregulated (IUU) fishing (Taylor et al. 2017).

7.3.3 Conservation

Given that biological and genetic resources are inextricably linked to biodiversity, it is little surprise that they are considered together and as a whole. Similarly, access and benefit sharing of genetic resources goes hand in hand with conservation and sustainable use of biodiversity in several international legal instruments, such as the CBD and the Nagoya Protocol, and the *2001 International Treaty on Plant Genetic Resources for Food and Agriculture*. The BBNJ agreement is the latest international development following this trend.

The potential to harness the ingenuity of nature to capture a range of potential benefits offers another reason to take action to reverse the current trend of considerable biodiversity loss worldwide (IPBES 2019). Given the significant knowledge gaps regarding the deep ocean, there is a risk that species will be lost entirely before their secrets have been revealed by science, especially given that many species are endemic in deep-ocean habitats. The BBNJ instrument is a historic opportunity to strengthen the framework for the conservation and sustainable use of biodiversity in these remote deep-ocean areas.

The BBNJ instrument is also a critical chance to enable all countries to participate in the exploration and sustainable exploitation of deep-ocean biodiversity, as well as the equitable sharing of benefits arising from their use. Learning from the challenges encountered in the implementation of the Nagoya Protocol will be crucial, including not to place an undue emphasis on economic aspects of instrumental value, but to adopt a more integrated approach to the full value of biodiversity and its genetic resources.

Another key priority should be to avoid unintended consequences of access and benefit sharing that hinder noncommercial research. In ABNJ, deep-sea scientific research is the main actor accessing and collecting marine genetic resources (Oldham et al. 2014) and generating benefits by publishing and sharing knowledge and data, enabling access to deep-sea samples through collections, and promoting international scientific cooperation. The reports of access and benefit-sharing regulations inhibiting scientific research and stifling innovation present a warning to be heeded (Neumann et al. 2018) to ensure that the science needed for biodiversity conservation is not stifled by access and benefit-sharing provisions. Rather, existing scientific best-practices (such as open and facilitated access to data and samples) offer a basis for benefit sharing that builds capacity for conservation and sustainable use of deep-sea biodiversity; these practices can be streamlined and strengthened through the BBNJ agreement to deliver benefit

sharing and improve transparency (Rabone et al. 2019).

The development of the BBNJ agreement, by strengthening the global framework for deep-sea investigation, science, and innovation, could not only support developing countries, but could also be beneficial to developed nations, including by increasing knowledge and promoting enhanced collaboration between funding agencies and disciplines (Oldham et al. 2014; Rogers et al. 2014; Harden-Davies and Gjerde 2019). Continued cross-sectoral engagement will be important, however, to straighten out tangled issues of ownership, proprietary issues and funding barriers. Crucially, such measures are fully consistent with existing rights and responsibilities under UNCLOS Part XIII (marine scientific research) and XIV (development and transfer of marine technology); the strengthened implementation of such measures would support the conservation and sustainable use of biodiversity, including the potential exploitation of its components (Harden-Davies and Gjerde 2019).

7.4 Conclusion

As marine scientific investigation expands and unveils new discoveries of deep-ocean species, the exquisite adaptations of the organisms that dwell there will continue to inspire further research and technological innovation for years to come. The development of the BBNJ agreement is crucial to implement measures for the conservation and sustainable use of biodiversity, to safeguard the survival of species in this vast deep-ocean realm, as well as to equip all nations with the tools and technologies to understand the components and functionality of the ecosystems therein. Given that genetic and biological resources are an integral part of biodiversity, consideration of potential exploitation should continue to go hand in hand with consideration of conservation and sustainable use. Similarly, considerations of the sustainable utilisation of biodiversity and its genetic resources should also continue in conjunction with the development of scientific collaboration and measures to advance research, technology transfer, and capacity building. Doing so will enable the understanding of the instrumental, inherent, and intrinsic value of deep-sea biodiversity needed to guide conservation and sustainable-use measures into the future.

Acknowledgements

The author gratefully acknowledges the support of the University of Wollongong and the Nippon Foundation Nereus Program, the Biodiversity Beyond National Jurisdiction Working Group of the Deep Ocean Stewardship Initiative, and the helpful comments from the two reviewers, which improved the chapter.

References

Abida, H., Ruchaud, S., Rios, L., et al. (2013). Bioprospecting Marine Plankton. *Marine Drugs*, 11(11), 4594–611.

Angel, M. V. (1993). Biodiversity of the Pelagic Ocean. *Conservation Biology*, 7(4), 760–72.

Arrieta, J. M., Arnaud-Haond, S., and Duarte, C. M. (2010). What Lies Underneath: Conserving the Oceans' Genetic Resources. *Proceedings of the National Academy of Sciences USA*, 107(43), 18318–24.

Arico, S. (2010). Marine Genetic Resources in Areas Beyond National Jurisdiction and Intellectual Property Rights, in D. Vidas (ed.) *Law, Technology and Science for Oceans in Globalisation: Iuu Fishing, Oil Pollution, Bioprospecting, Outer Continental Shelf*. Leiden, Netherlands: Martinus Nijhoff, p. 383.

Armstrong, C. W., Foley, C. N. S., Tinch, R., et al. (2012). Services from the Deep: Steps Towards Valuation of Deep Sea Goods and Services. *Ecosystem Services*, 2, 2–13.

Ausubel, J. H., Crist, D. T., and Waggoner, P. E. (eds). (2010). *First Census of Marine Life 2010: Highlights of a Decade of Discovery*. https://pdfs.semanticscholar.org/5d89/00d4b554dd981e3124f69ad98d8c9c8bd6e4.pdf.

Ballard, R. (2000). *The Eternal Darkness: A Personal History of Deep-sea Exploration*. Princeton, NJ: Princeton University Press.

Bergmann, W. and Feeney, R. J. (1951). Contributions to the Study of Marine Products. Xxxii. The Nucleosides of Sponges I. *Journal of Organic Chemistry*, 16(6), 981–7.

Blunt, J. W., Copp, B. R., Keyzers, R. A., Munro, M. H., and Prinsep, M. R. (2014). Marine Natural Products. *Natural Product Reports*, 31(1), 160–258.

Bowler, C., Karl, D. M., and Colwell, R. R. (2009). Microbial Oceanography in a Sea of Opportunity. *Nature*, 459(7244), 180–4.

Bordbar, S., Anwar, F., and Saari, N. (2011). High-value Components and Bioactives from Sea Cucumbers for Functional Foods—A Review. *Marine Drugs*, 9(10), 1761–805. doi:10.3390/md9101761.

Broggiato, A., Arnaud-Haond, S., Chiarolla, C., and Greiber, T. (2014). Fair and Equitable Sharing of Benefits from the Utilization of Marine Genetic Resources in Areas Beyond National Jurisdiction: Bridging the Gaps Between Science and Policy. *Marine Policy*, 49, 176–85.

Carroll, A. R., Copp, B. R., Davis, R. A., Keyzer, R. A., and Prinsep, M. R. (2019). Marine Natural Products. *Natural Product Reports*, 36, 122–73.

CBD 1992. *Convention on Biological Diversity*, opened for signature 5 June 1992, 1760 UNTS 79 (entered into force 29 December 1993).

Danovaro, R., Snelgrove, P. V. R., and Tyler, P. (2014). Challenging the Paradigms of Deep-sea Ecology. *Trends in Ecology & Evolution*, 29(8), 465–75.

Deplazes-Zemp, A. (2018). 'Genetic Resources' an Analysis of a Multifaceted Concept. *Biological Conservation*, 222, 86–94.

Eadie, L. and Ghosh, T. K. (2011). Biomimicry in Textiles: Past, Present and Potential: An Overview. *Journal of the Royal Society Interface*, 8(59). http://doi.org/10.1098/rsif.2010.0487.

FAO (2014). FAO Commission on Genetic Resources for Food and Agriculture, 'Global Plan of Action for the Conservation, Sustainable Use and Development of Forest Genetic Resources'. :FAO, Rome.

Freitas, A., Pereira, L., Rodrigues, D., et al. (2015). Marine Functional Foods. doi: 10.1007/978-3-642-53971-8_42.

Freestone, D., Laffoley, D., Douvere, F., and Badman, T. (2016). World Heritage in the High Seas: An Idea Whose Time Has Come. UNESCO Report, Paris.

Gerwick, W. H. and Fenner, A. M. (2013). Drug Discovery from Marine Microbes. *Microbial Ecology*, 65(4), 800–6.

Glaser, K. B. and Mayer, A. M. S. (2009). A Renaissance in Marine Pharmacology: From Preclinical Curiosity to Clinical Reality. *Biochemical Pharmacology*, 78, 440–8.

Gjerde, K., Clarke, N., and Harden-Davies, H. (2019). Building a Platform for the Future: The Relationship of the Expected New Agreement for Marine Biodiversity in Areas Beyond National Jurisdiction and the UN Convention on the Law of the Sea. *Ocean Yearbook*, 33(1), 1–44.

Harden-Davies, H. (2016). Deep-sea Genetic Resources: New Frontiers for Science and Stewardship in Areas Beyond National Jurisdiction. *Deep-Sea Research II*, 137, 503–13.

Harden-Davies, H. (2018). *Marine Genetic Resources Beyond National Jurisdiction: An Integrated Approach to Benefit-sharing, Conservation and Sustainable Use*. Doctor of Philosophy thesis, Australian National Centre for Ocean Resources and Security. University of Wollongong. https://ro.uow.edu.au/theses1/557/.

Harden-Davies, H. and Gjerde, K. (2019). Building Scientific and Technological Capacity: A Role for Benefit-sharing in the Conservation and Sustainable Use of Marine Biodiversity Beyond National Jurisdiction. *Ocean Yearbook*, 33(1), 377–400.

IPBES 2019. Global Assessment Report on Biodiversity and Ecosystem Services of the Intergovernmental Science-Policy Platform on Biodiversity and Ecosystem Services. E. S. Brondizio, J. Settele, S. Díaz, and H. T. Ngo (eds.). Bonn, Germany: IPBES Secretariat.

Kim, H. J., Lee, J. H., Hur, Y. B., et al. (2017). Marine Antifreeze Proteins: Structure, Function, and Application to Cryopreservation as a Potential Cryoprotectant. *Marine Drugs*, 15(2), 27.

Leary, D., Vierros, M., Hamon, G., et al. (2009). Marine Genetic Resources: A Review of Scientific and Commercial Interest. *Marine Policy*, 33(2), 183–94.

Leary, D. and Juniper, S. K. (2013) Addressing the Marine Genetic Resources Issue: Is the Debate Heading in the Wrong Direction?', in C. H. Schofield, S. Lee, and M-S. Kwon (eds.), *The Limits of Maritime Jurisdiction*. Leiden, Netherlands: Brill, p. 769.

Lordan, S., Ross, R. P., and Stanton, C. (2011). Marine Bioactives as Functional Food Ingredients: Potential to Reduce the Incidence of Chronic Diseases. *Marine Drugs*, 9(6), 1056–100. doi:10.3390/md9061056.

Marciniak, K. J. (2017). Marine Genetic Resources: Do They Form Part of the Common Heritage of Mankind Principle?, in M. Lawrence, C. Salonidis, and C. Hioureas (eds.), *Natural Resources and the Law of the Sea: Exploration, Allocation, Exploitation of Natural Resources in Areas under National Jurisdiction and Beyond*. International Law Institute, p. 373–406.

Marcot, B. G. (2007). Biodiversity and the Lexicon Zoo. *Forest Ecology and Management*, 246(1), 4–13.

Marlow, J., Harden-Davies, H., Snelgrove, P., et al. (2019). The Full Value of Marine Genetic Resources. DOSI Policy Brief. https://www.dosi-project.org/wp-content/uploads/2019/07/Full-value-mgr-March2019.pdf.

Martins, A., Vieira, H., Gaspar, H., and Santos, S. (2014). Marketed Marine Natural Products in the Pharmaceutical and Cosmeceutical Industries: Tips for Success. *Marine Drugs*, 12(2), 1066–101.

Matta, H., Gopalakrishnan, R., Choi, S., et al. (2018). Development and Characterization of a Novel Luciferase Based Cytotoxicity Assay. *Scientific Reports*, 8(1), 199.

Mora, C., Tittensor, D. P., Adl, S., Simpson, A. G. B., and Worm, B. (2011). How Many Species Are There on Earth and in the Ocean? *PLoS Biology*, 9(8), e1001127.

Morris, J. P., Wang, Y., Backeljau, T., and Chapelle, G. (2016). Biomimetic and Bio-inspired Uses of Mollusc Shells. *Marine Genomics*, 27 (1), 85–90.

Mossop, J. (2010). Regulating Uses of Marine Biodiversity on the Outer Continental Shelf, in D. Vidas (ed.) *Law, Technology and Science for Oceans in Globalisation: IUU Fishing, Oil Pollution, Bioprospecting, Outer Continental Shelf*. Leiden, Netherlands: Martinus Nijhoff, p. 319.

Müller, W. E., Wang X., Cui F. Z., et al. (2009). Sponge Spicules as Blueprints for the Biofabrication of Inorganic-organic Composites and Biomaterials. *Applied Microbiology and Biotechnology*, 83(3), 397–413.

Munro M. H., Blunt, J. W., Hickford, S. J., et al. (1999). The Discovery and Development of Marine Compounds with Pharmaceutical Potential. *Journal of Biotechnology*, 70(1–3), 15–25.

National Academy of Sciences (2015). *Industrialization of Biology: A Roadmap to Accelerate the Advanced Manufacturing of Chemicals*. Report. National Academies Press.

Neumann, D., Borisenko, A. V., Coddington, J. A., et al. (2018). Global Biodiversity Research Tied up by Juridical Interpretations of Access and Benefit-sharing. *Organisms Diversity and Evolution*, 18(1). doi: 10.1007/s13127-017-0347-1.

Newman, D. J. (2016). Developing Natural Product Drugs: Supply Problems and How They Have Been Overcome. *Pharmacology and Therapeutics*, 162, 1–9.

Newman, D. J. and Cragg, G. M. (2016). Natural Products as Sources of New Drugs from 1981 to 2014. *Journal of Natural Products*, 79(3), 629–61.

Organisation for Economic Cooperation and Development (OECD) (2013). Marine Biotechnology: Enabling Solutions for Ocean Productivity and Sustainability' (OECD Report).

Oldham, P., Hall, S., Barnes, C., et al. (2014). Valuing the Deep: Marine Genetic Resources in Areas Beyond National Jurisdiction. Report, One World Analytics.

Overmann, J. and Hartman Scholz, A. (2017). Microbiological Research Under the Nagoya Protocol: Facts and Fiction. *Trends in Microbiology*, 25(2), 85–8.

UN (2017). Report of the Preparatory Committee Established by General Assembly Resolution 69/292: Development of an International Legally Binding Instrument Under the United Nations Convention on the Law of the Sea on the Conservation and Sustainable Use of Marine Biological Diversity of Areas Beyond National Jurisdiction. UN doc. A/AC.287/2017/PC.4/2. (31 July). New York: UN. http://www.un.org/ga/search/view_doc.asp?symbol=A/AC.287/2017/PC.4/2.

Pugliano, M., Vanbellinghen, X., Schwinte, P., Benkirane-Jessel, N., and Keller, L. (2017). Combined Jellyfish Collagen Type II, Human Stem Cells and Tgf-β3 as a Therapeutic Implant for Cartilage Repair. *Journal of Stem Cell Research and Therapy*, 7, 4. doi: 10.4172/2157-7633.1000382.

Rabone, M., Harden-Davies, H., Collins, J. E., et al. (2019). Access to Marine Genetic Resources (MGR): Raising Awareness of Best-Practice Through a New Agreement for Biodiversity Beyond National Jurisdiction (BBNJ). *Frontiers in Marine Science*, 6, 520. https://doi.org/10.3389/fmars.2019.00520.

Rayfuse, R. (2007). Biological Resources, in D. Bodansky, J. Brunnee, and E. Hey (eds.) *The Oxford Handbook of International Environmental Law*. Oxford: Oxford University Press, pp. 362–294.

Robison, B. H. (2009) Conservation of Deep Pelagic Biodiversity. *Conservation Biology*, 23(4), 847–58.

Rogers, A., Sumaila, U. R., Hussain, S. S., and Balcomb, C. (2014). The High Seas and Us: Understanding the Value of High Seas Ecosystems. Report. Global Ocean Commission. http://www.oceanunite.org/wp-content/uploads/2016/03/High-Seas-and-Us.FINAL_.FINAL_.high_.spreads.pdf.

Shepherd, R.F., Ilievski, F., Choi, W. et al. (2011). Multigait Soft Robot. *Proceedings of the National Academy of Sciences*, 108 (51), 20400–3.

Skropeta, D. (2008). Deep-sea Natural Products. *Natural Product Reports*, 25(6), 1131–66.

Skropeta, D. (2011). Exploring Marine Resources for New Pharmaceutical Applications, in W. Gullett, C. Schofield, and J. Vince (eds.), *Marine Resources Management*. : LexisNexis Butterworths, p. 211.

Skropeta, D. and Wei, L. (2014) Recent Advances in Deep-sea Natural Products. *Natural Product Reports*, 31(8), 999–1025.

Snelgrove, P., Tunnicliffe, V., Metaxas, A., and Baker, M. (2018) Sustaining Biodiversity Beyond National Jurisdictions: The Major Science Challenges. Deep Ocean Stewardship Initiative (DOSI) Policy Brief. https://www.dosi-project.org/wp-content/uploads/2019/07/009-Policy-Brief-DOSI-final_Nov2018.pdf.

Sogin M. L., et al. (2006). Microbial Diversity in the Deep Sea and the Underexplored 'Rare Biosphere'. *Proceedings of the National Academy of Sciences USA*, 103(32), 12115.

Taylor, H. R., Dussex, N., and van Heezik, Y. (2017). Bridging the Conservation Genetics Gap by Identifying Barriers to Implementation for Conservation Practitioners. *Global Ecology and Conservation*, 10, 231–42.

van Zonneveld M. (2018). Bridging Molecular Genetics and Participatory Research: How Access and Benefit-sharing Stimulate Interdisciplinary Research for Tropical Biology and Conservation. *Biotropica*, 50 (1), 178–86.

Vierros, M., Curtis S., Harden-Davies, H., and Burton, G. (2016). Who Owns the Ocean? Policy Issues Surrounding Marine Genetic Resources. *Limnology and Oceanography Bulletin*, 25(2), 29–35.

Wang, X., Schröder, H. C., Feng, Q., Draenert, F., and Müller, W. E. G. (2013). The Deep-sea Natural Products, Biogenic Polyphosphate (Bio-PolyP) and Biogenic Silica (Bio-Silica), as Biomimetic Scaffolds for Bone Tissue Engineering: Fabrication of a Morphogenetically-Active Polymer. *Marine Drugs*, 11, 718–46.

Whitesides, G. M. (2015). Bioinspiration: Something for Everyone. *Royal Society Interface Focus*, 5: 20150031.

Xu, C., Zhang, R., and Wen, Z. (2018). Bioactive Compounds and Biological Functions of Sea Cucumbers as Potential Functional Foods. *Journal of Functional Foods*, 49, 73–84. https://doi.org/10.1016/j.jff.2018.08.009.

Yao, H., Ming, D., Huang, J., et al. (2010). Protection Mechanisms of the Iron-plated Armor of a Deep-sea Hydrothermal Vent Gastropod. *Proceedings of the National Academy of Sciences USA*, 107(3), 987–92.

Yu C., Li, Y., Zhang, X., et al. (2014). Adaptive optoelectronic camouflage systems with designs inspired by cephalopod skins. *Proceedings of the National Academy of Sciences USA*, 111 (36), 12998–3003.

CHAPTER 8

The deep ocean's link to culture and global processes: nonextractive value of the deep sea

Andrew R. Thurber and Amanda N. Netburn

8.1 Ecosystem services and nonuse values

The natural assets (biological, physical, and geological) of an ecosystem together comprise the overall natural capital of an ecosystem with a subset of these assets leading to ecosystem services, the direct and indirect benefits that people gain from an ecosystem. Ecosystem services have been increasingly factored into management decisions concerning resource use and allocation across a diversity of habitats (reviewed in Costanza et al. 2017). A crux in a holistic evaluation of the benefits an ecosystem provides to society is inclusion of the various kinds of services beyond those derived from resource extraction and exploitation (Chan et al. 2012). This challenge is further complicated by the importance of including facets of an ecosystem that provide nonmonetary societal benefits. In order for ecosystem services to be considered in evaluating management options, the relative value of ecosystem services under the different scenarios must be calculated, and ideally within a framework that links to the diverse components that together form the overall natural capital of an environment.

Some of the most challenging factors when attempting to quantify an ecosystem's value are the nonuse cultural services and the broad categories of regulating and supporting services. Cultural services are 'aesthetic, artistic, educational, spiritual and/or scientific values of ecosystems' (Costanza et al. 1997) that 'provide contributions to the non-material benefits (e.g., capabilities and experiences) that arise from human-ecosystem relationships' (Chan et al. 2012). An example might be a field that inspires a landscape painting which is then enjoyed by many people. Regulating and supporting services are the processes that provide direct benefits to society, in many cases tied to a 'healthy' ecosystem, which do not directly result in an extractive resource in a materialistic sense. An example of a regulating and supporting service is photosynthesis. All people benefit from the oxygen produced by photosynthesis; however, we do not generally provide a monetary value for this service. Numerous classification systems have been developed for this purpose, each with its own merits and weaknesses (Costanza et al. 1997; Carpenter et al. 2006; Chan et al. 2012). Because we do not purchase or trade nonuse services, there is an inherent challenge to estimating the exact value of these services. Translating services to benefits to *values* is an exciting and expanding field. Discourse on this topic reveals many hurdles, especially when moving to the later stages. There are divergent frameworks and concerns of double counting that inflates the values of ecosystems if a supporting service results in multiple benefits (Boyd and Banzhaf 2007; Wallace 2007; Silvertown 2015; Costanza et al. 2017). These challenges have also led to a greater appreciation of the natural capital of the ecosystem, as natural capital is a more foundational treatment of the resources present in an ecosystem independent of if they directly tie to an ecosystem service.

Andrew R. Thurber and Amanda N. Netburn, *The deep ocean's link to culture and global processes: nonextractive value of the deep sea* In: *Natural Capital and Exploitation of the Deep Ocean*. Edited by: Maria Baker, Eva Ramirez-Llodra, and Paul Tyler, Oxford University Press (2020). © Oxford University Press.
DOI: 10.1093/oso/9780198841654.003.0008

One of the most remarkable findings when considering nonuse factors is that often, when monotonised, they can provide a significant proportion of the total value of an ecosystem. For example, regulating, supporting, and cultural services provide approximately 93.6 per cent of the total value that ecosystems provide society globally (Costanza et al. 1997).Within the marine realm, the open ocean alone is estimated to provide 8 trillion (US) dollars a year in ecosystem services, about a quarter of the value of ecosystem services of the entire planet. Only about 0.3 per cent of the ocean's ecosystem services are food resources, including fishing, even though it is the most apparent service provided by the marine realm (Costanza et al. 1997; Kubiszewski et al. 2017). These values help inform the importance of a balanced management portfolio that maintains the continued availability of nonuse benefits and the societal need for the resources extracted from an environment.

Given the projected increase in exploitation of deep-sea resources in coming years, it is critical to estimate the value of the deep-sea ecosystem in order to make sound management decisions. However, it remains especially challenging to quantify the deep sea's nonuse value. The deep sea is unique in how many of the processes cover both vast areas and take a long time (decades to centuries) to result in societal benefit (Thurber et al. 2014). Because the deep sea is so vast, making up 93 per cent of the volume of the planet, management of deep-sea resources frequently crosses national jurisdictions. Even though the nonuse value of this habitat may be significant (Armstrong et al. 2010), there are numerous hurdles to overcome in estimating it. Among the greatest barriers is not (an often-perceived) lack of scientific understanding about the habitat, but instead a lack of knowledge and awareness among the general public (Jobstvogt et al. 2014). Estimating nonuse value of a system is often accomplished by probing people's implicit value for it, for example by finding out the willingness to pay for protecting it (Arin and Kramer 2002; Peters and Hawkins 2009). However, willingness to pay is often driven by knowledge tied to personal experiences (Chan et al. 2012), making a general

lack of knowledge by a person an impediment to an informed estimation of how much they may be willing to pay to conserve a resource. Further, personal values vary across countries and cultures, thus creating significant hurdles for valuation of any ecosystem that spans national jurisdictions (Chan et al. 2012). These challenges are compounded in trying to assign nonuse values in the deep sea because few people even know enough about the deep sea to assign an implicit value to it. At the nexus of economics and ecology, the deep sea is both intriguing and a key area to study, especially as use and extractive approaches are pursued (See Chapters 4–7).

8.2 A diverse and inspiring dark sea

When we each picture the deep sea something different comes to mind: perhaps it is an anglerfish with its bright bioluminescent lure, a giant squid fighting an interminable battle with a sperm whale, or a yeti crab combing its luxurious and bacteria-laden hair gently bathed by toxic and yet nutritious chemicals released from the seafloor. For many, these mental images are inspired by art ranging from documentary (e.g. BBC's Blue Planet II) and fictional films (e.g. Disney's Finding Nemo) to paintings and literature (e.g. Michael Crichton's 'Sphere' and Orson Scott Card's 'The Abyss') to legends based upon the oddities that may occur in the deep, including mythical creatures such as the Kraken (i.e. Lord Tennyson's sonnet 'The Kraken' in Lord Tennyson 2015) The inspiration of the deep sea is not limited to adults: children's cartoons can both inspire and inform (e.g. 'The Octonauts' and 'The Wild Kratts'), translating into lasting impacts on the young public's imagination and psyche (Figure 1; Thurber et al. 2014). These various forms of art and communications create a bridge between the natural world and society, resulting in significant nonuse benefits.

The diversity of habitats and animals in the deep often provides the whimsical inspiration that fuels art. The largest benthic (seafloor) habitats on the planet are the abyssal plains, massive areas of flat, soft, muddy seafloor and gently sloping hills that cover >60 per cent of the globe (Chapter 1). The

Figure 1 Shows such as "The Octonauts" can be a bridge that connects the natural capital of the deep sea to society, creating a lasting non-use benefit. In an example, the show recently partnered with NOAA's Office of Ocean Exploration. In the show Tweak Bunny (Engineer), Peso Penguin (Medical Specialist), Captain Barnacles (Captain and Polar Bear), and Professor Inkling (Scientist and Octopus) explore the ocean and convey scientific discoveries to a wide audience. Image courtesy of the NOAA Office of Ocean Exploration and Research, Discovering the Deep: Exploring Remote Pacific MPAs.

abyssal plains are broken up by mountains, continents, hydrothermal vents, and methane seeps, in addition to massive areas of mineral deposits that can cover over 1.1 million km^2 (Beaulieu et al. 2017). The water column too is an area of habitat diversity. Gentle gradients in pressure and temperature are broken up by hydrothermal vent plumes that can spread thousands of kilometres across ocean basins (Resing et al. 2015), and discrete horizons of particles form thin layers called nepheloid layers, creating vertical microstructure. Gradients in oxygen concentrations including layers with little to no oxygen can create distinct niches in the pelagic (water column) environment. There are vast swarms of animals that swim up and down on a daily basis in a process called diel vertical migration, which is the greatest migration on the planet. In addition to vertical changes, shifts across ocean basins in temperature, oxygen, and food supply all have altered our view of these habitats from monotonous to highly heterogeneous across every dimension. Within this mixture of habitat heterogeneity, biodiversity abounds in life, genes, and resources (Chapters 1 and Chapter 7).

8.2.1 The deep, dark water

The deep pelagic ocean harbours a remarkable diversity of organisms. The animals here are uniquely adapted for life in an environment with no firm, fixed structural habitat. Animals 'hide' in the darkness of their environment with black or red pigmentation (McFall-Ngai 1990). The majority of the organisms living in the deep pelagic realm are small, but the abundance and biomass of midwater assemblages, particularly in the mesopelagic zone (200–1000 m deep), is enormous: biomass of mesopelagic fishes alone is estimated at 10 billion tonnes (Koslow et al. 1997; Kaarrtvelt et al. 2012; Irigoien et al 2014). Other taxa are also found in significant numbers, such as crustaceans (shrimp), cephalopods (squid), tunicates (salps and pyrosomes), and cnidarians (jellyfish). Many of the deep pelagic organisms exhibit bioluminescence, the ability to produce light (Widder 2010). Green fluorescent protein, a luminescent molecule that is produced by a jellyfish (Chalfie et al. 1994), has become a ubiquitous biological marker for biomedical and cellular research (Tsien 1998; Specht et al. 2017). Chapter 7 discusses in depth the potential value of marine natural products from the deep sea, recognising the untapped potential of products derived from the significant unknown biodiversity of the pelagic realm. Though fishing pressure on deep pelagic organisms is still relatively minimal (Lamhauge et al. 2008; Prellezo 2019), these organisms provide numerous hidden ecosystem services (St. John et al. 2016). As we will see in the coming sections, the deep pelagic is a storage centre for carbon and an epicentre of nutrient cycling. The animals transport carbon to the deep sea and are integral to supporting healthy marine food webs.

8.2.2 The expanse of marvellous mud

At the bottom of this global water column is most frequently a swath of mud that extends around the globe, with an average depth around 4000 m. This habitat, the abyssal plain of the deep, is covered by roaming urchins (echinoids) and sea cucumbers (holothurians) that can appear and devour food in dense assemblages before continuing on with their travelling (Billett et al. 2010). Much like the deep water column, the absence of light and thus photosynthesis at these depths defines the life that lives in and on the abyssal plains: organisms are small (Rex et al. 2006; Wei et al. 2010); food limited (Smith et al. 2008a) and unexpectedly diverse (Hessler and Jumars 1974). Life is slow at the seafloor boundary: organisms reach astronomical ages of thousands of years (Jochum et al. 2012); larvae grow slowly (Marsh et al. 2001); and some animals become sexually mature in decades not weeks or months (Drazen and Haedrich 2012). The gentle rain of food that fuels this habitat is broken up by food falls including seasonal rapid deposition of food in the form of 'phytodetritus pulses', dead or dying phytoplankton that rapidly transverse thousands of metres of water before being deposited on the seafloor. Food also falls to the seafloor in the form of wood (Turner 1973; Bernardino et al. 2010; McClain and Barry 2014) and dead carrion that have specialised scavengers adapted to track down this food source (Drazen and Sutton 2017). These food depositions add to the heterogeneity of the deep, and while processes are slow there, the massive area that it covers

helps shape the impact of this realm on the greater function of the ocean.

8.2.3 Habitats that break the global mud belt

Much like the expanses of flat lands that dominate vast sections of Earth's continents, the abyssal plains are broken up by mountains (seamounts and mounds), rift valleys (populated with hydrothermal vents and deep-ocean trenches), and continents whose margins are excised by deep canyons so massive that they dwarf those on land. These features not only create a variety of seafloor depths but also interject hard substrate, areas of boulders, pavements, and stones, into the otherwise soft, muddy expanse. This hard substrate creates habitat diversity that is used by a variety of organisms that do not live on mud: cold-water corals and sponges require hard substrate to attach to for their survival. Corals and sponges in turn serve as habitat for a variety of mobile organisms, such as fishes, crustaceans, and echinoderms. Features like seamounts and canyons also modify the flow of water through the oceans, creating interactions between the deep pelagic and the deep seafloor. As water currents encounter sloped regions of the seafloor, they accelerate, increasing the rate at which plankton and detritus—food for filter-feeding corals and sponges—interact with the seafloor, thus transporting abundant food to support lush cold-water coral beds, epicentres of biomass in the otherwise low-biomass deep sea. Canyons also accelerate water and act as conduits of food from shallow water to the deep sea, further shifting the abundance of organisms and the rate at which the organisms can respire and reproduce, in some cases leading to areas of astronomically high biomass (De Leo et al. 2010; Vetter et al. 2010). Recently we have learned that canyons cut through the subsurface coasts (continental slopes and margins) in a density much greater than was ever thought (De Leo et al. 2010). Seamounts and knolls (smaller features) also accelerate water and shift the biodiversity of the oceans and, like canyons, are incredibly abundant, covering combined more than 20 per cent of the ocean floor (Yesson et al. 2011). Seafloor mapping and biological studies have shown that the deep sea is in fact a patchwork of unique habitats rather than a swath of unrelenting (yet still marvellous) mud.

Most deep-sea habitats ultimately derive their energy from photosynthesis taking place at the surface, often through the rain of detritus and larger particles (e.g. whale falls) from the surface. Hydrothermal vents and methane seeps are exceptions to this. At these environments, chemical energy is harnessed by bacteria and archaea to fuel some of the strangest animals known to date. Both vents and seeps release water from the seafloor containing key chemicals, including methane, sulphide, and sometimes hydrogen: at vents, the temperatures of these fluids can exceed 400 °C, while seep water is extruded at ambient temperatures. Free-living bacteria and archaea, in addition to bacteria living in partnership with worms, clams, and crustaceans, capture this energy and use it to fix carbon into organic molecules and biomass, analogous to photosynthesis in the surface ocean. These epicentres of biomass also create plumes of energy and heat that spread throughout the ocean basins and create the aforementioned diversity in the water column. While the fixation of carbon from chemical energy was long thought to be largely limited to these habitats, it now appears to be a common phenomenon in the dark reaches of the planet from the abyssal plains (Sweetman et al. 2019) to the water column (Wuchter et al. 2003; Ingalls et al. 2006).

Habitat diversity supports biodiversity across all size spectra of life, which together form the basis of many nonuse services provided by the deep sea. Habitat diversity, chemical diversity, and the incredible size of the deep sea allow a plethora of slow processes to result in significant impacts on the functioning of the Earth. Increasingly we are beginning to understand the importance of the ocean as one large connected system with many distinct parts: pelagic systems are connected to benthic systems that influence the pelagic systems in a constant back and forth with epicentres of activity at the habitats that break up the expansive mud and gentle gradients in the water. Further, habitat diversity adds to animal diversity, and those animals are often the focus of inspiration for art and intrigue for people around the world.

8.3 Cultural services

Cultural values include art and the cultural richness that stems from interactions with art. Artists have long sought inspiration from the natural world, and art, in turn, serves to inspire the public at large (Curtis et al. 2014). Nature itself evokes 'environmental aesthetics' which often results in appreciation (e.g. sunsets and mountains), and a scientific understanding of what is evoking this response can even enhance the level of appreciation of nature (Carlson 2010; Harrower et al. 2018). In many ways, science and art are part of a chain of influence that connects society to the world around them (Curtis 2009). As a result, art can be used to galvanise support around environmental issues or protection of areas that people have never physically experienced but are emotionally connected to through artists and their work (Harrison and Harrison 1993). In a demonstration project, 95 per cent of individuals that were shown both a live science talk and a film on the science found the film evoked a stronger sense of urgency than the talk alone; it was also found that the paired format reinforced each other (Harrower et al. 2018).

The diversity of habitat and life in the deep sea also leads to cultural values driven by the discoveries of biological oddities that stretch the imagination of society; the treasure trove of new species constantly uncovered captivate artists and the public alike. In numerous cases we have discovered taxa that have taken decades to decipher even what phylum they belong in (Rouse and Fauchald 1995; McHugh 1997). Even recently it has taken years to tease apart where on the tree of life of certain strange forms belong, such as the now-infamous zombie worms (Rouse et al. 2004). From glow in the dark sharks (Grace et al. 2019) to a giant squid captured on camera for the first time (Widder 2013), discoveries are the catalyst for imagination. An example of this was the discovery of the yeti crab that inspired Lily Simonson, a visual artist, and led her down a line of artistic inquiry into the deep sea. Thousands of people have now interacted with the deep sea through her art (Figure 2). This is an example of the discovery of a

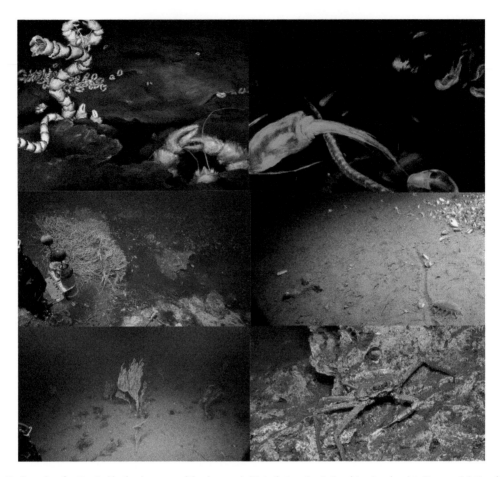

Figure 2 Examples of art inspired by the deep sea and the deep sea habitats that are inspirational to art and society. Top row: Paintings by Lily Simonson inspired by the deep sea. Left: "Party of Yeti's" inspired by the discovery of Yeti Crabs. Right: "To the Lighthouse" inspired by deep water fauna. (Images © Lily Simonson. See Lilysimonson.com). Middle Left: Methane seeps add to the diversity of life through creating structure on the deep seafloor. Middle Right: Sea cucumbers (*Scotoplanes* sp.) found in an aggregation around a methane seep. Bottom Left: Cold water corals also add to the diversity of habitats that break up the mud belt of the deep sea. Bottom Right: Tanner Crabs can occur in dense assemblages at methane seeps providing a mechanism to spread the methane-derived energy around the deep sea. Images courtesy of Ocean Exploration Trust (Cruises NA095 and NA072 aboard the E/V *Nautilus* © of all images belong to Ocean Exploration Trust).

new species that led to the stimulation of the neural soup that leads to beauty.

While discoveries are one aspect of cultural values, the exploration and existence of the unknown within the deep is another stimulant for art and culture. Exploration itself is a strong driver for society, as evidenced by the first human-occupied exploration of the deep sea. William Beebe's dive into the deep sea was the first glimpse of this realm with human eyes (see Chapter 1). His observations initiated numerous new lines of scientific inquiry, such

as the shifting colour spectrum with increasing depth and the oddities that comprise deep-sea fish communities. Yet some argue that the historic first dives were more influential by stimulating the public's imagination for exploration than they were on advancing science (Alaimo 2013). In discussing the cultural impact of Beebe's writings and observations, Alaimo argues that these initial forays into the deep sea by humans marked a point 'in which human knowledge systems and terrestrial horizons are overwhelmed, overtaken, undone' (Alaimo

2013). This example brings to light two key facets of deep-sea cultural values—discoveries that excite the public and the excitement of discovery itself.

We are currently in a new age of exploration with regards to how information about the deep sea is communicated. Exploration is no longer relegated to scientists who make discoveries alone at sea, who then slowly release fuzzy images and sparse data of their discoveries; now programmes exist that allow anyone with an Internet connection to explore along with the scientists. In real time, areas of the seafloor and water column that have never been seen by human eyes are displayed around the globe. In many ways, this has opened up the avenues of arm-chair deep-sea explorers where the public can experience the wonder of the deep sea and be present to observe discoveries as they are made. Two ships are currently the epicentre of this: NOAA Ship *Okeanos Explorer* and the Exploration Vessel (E/V) *Nautilus*. Live video streams from the deep have collectively reached over two million viewers in just one year (2018). Through the cultivation of this video and release on YouTube and through social media channels, more than 13 million viewers are reached annually through these new efforts.

In many ways, our view of the deep sea is shifting due to the newfound and connected ways in which society, beyond a select few scientists, can now experience the deep sea, creating what could almost be described as a tourism aspect to the deep. More traditional tourism has value as a nonuse cultural service, which is an emerging topic of study for the field. Much like there are an elite few people that are anticipated to travel to space as tourists due to their wealth in the coming years, there are currently only select few who have the wealth to own or rent deep-sea submersibles to reach far into the deep sea. Even recreational but technical SCUBA diving can start to reach the upper end of the deep sea; however, the majority of tourism to the deep is through video, or the aforementioned live streams. In many ways, live streaming deep-sea discoveries is a unifying form of tourism as it can be shared among people. The newfound capabilities for anyone with curiosity to experience the deep sea helps alleviate some of the challenges of assigning a universal value on the aesthetic services of the deep

sea. As stated earlier, the perceived value of a system varies significantly between people depending on cultural background, socioeconomic status, and awareness (Le et al. 2017). In that context, sharing the experience of exploration is a way to break down financial barriers, allowing a greater cross section of the populous to experience the deep; it also helps to increase awareness on a massive scale.

Art and artists can be a motivator for social action, and the deep sea is rife with existing and emerging human impacts, such as pollution, deep-sea fishing, oil and gas extraction, and seabed mining (Ramirez-Llodra et al. 2011). In a recent study, an animated short film about loss of biodiversity of trees resulted in 95 per cent of the audience feeling like they were more informed by the film than a science talk, and that the film elicited an emotional response that that they would not have had otherwise (Harrower et al. 2018). Exploration and visibility have brought about increasing awareness of pollution and particularly plastics, with recent studies finding plastic in the deepest areas of the ocean (Anastasopoulou et al. 2013; Van Cauwenberghe et al. 2013; Woodall et al. 2014; Taylor et al. 2016; Jamieson et al. 2019). Appreciation of the global impact of plastic waste has become a realised concern across ocean systems (Lebreton et al. 2018). This begs the question: is the value of inspiration diminished if a newly discovered habitat is already degraded? Or stated differently, are conservation-focused emotional responses as great to an already degraded habitat? As an example, habitat degradation from trawling off New Zealand was discovered when the first methane seeps were documented off the coast there (Baco et al. 2010). Off the Western Coast of North America, an iconic discovery was made of a new vigorous area of methane seepage that had a red plastic 'Solo' cup discovered right in the centre of activity (pers. obs. during E/V *Nautilus* live stream from the deep; Figure 3.) While this brings up the potential depreciation of an ecosystem if it is already impacted, marine pollution and impacts in the deep sea may also be a catalyst for increased visibility resulting in an increase in appreciation by the public and, thus, an increase in the perceived worth of a habitat by society.

Figure 3 Uses of the deep sea often occur prior to discovery, for example waste being present at habitats that have never been previously seen. Top: A red "Solo" cup found during the discovery of a large methane hydrate boulder during the first dive on this methane seep. Lower, additional trash in the deep sea while exploring novel habitats. Cruises NA095 and NA072 aboard the E/V *Nautilus* (images © Ocean Exploration Trust and used with permission).

8.4 Deep-sea science: exploration and research to understand the past, present, and future earth

Included within nonuse services are scientific advances made from a habitat that leverage the natural capital of the region. Research in the deep sea has helped explain the role of the oceans in how the planet functions, the history of the Earth, and how it will change in the future. Some deep-sea ecosystems, such as hydrothermal vents, are even used as proxies for conditions that could support life on other planets. We often think of the deep sea as a place of the unknown, but in many ways it could be

viewed instead as an epicentre of knowledge generation. Monumental discoveries such as plate tectonics (DeMets et al. 1994), the evolution of Earth's atmosphere (Lear et al. 2000; Pagani et al. 2005), or the real-time prediction or reaction to events such as earthquakes, tsunamis (Kawaguchi et al. 2015), and ENSO (El Niño Southern Oscillation) events (McPhaden et al. 2010) are all the results of study of, or in, the deep sea. In many ways, the focus of the deep is part of an observing strategy to understand our globe with exciting and transformative paths ahead (Levin et al. 2019). This is in addition to the potential for medical breakthroughs (Chapter 7). While a book alone could be written just about scientific advances made as the result of deep-sea research, we will focus the remaining sections on how directed study has impacted our understanding of the interaction between deep-sea habitats and the greater functioning of the planet, including its societies.

8.5 Supporting and regulating services

8.5.1 Primer on deep-ocean flow and function

While nuances abound, much of the regulating and supporting services are driven by the flow of water through the ocean and food that falls through that water to the seafloor, being eaten at every step along the way. The water that bathes the deep sea starts largely at the poles and sinks underneath the warmer water that occurs at the surface of the temperate and tropical regions of the earth starting its long, dark journey throughout the deep ocean (see Chapter 1). This initial contact with air sets the initial biogeochemical environment by becoming saturated with gases, including both oxygen and carbon dioxide, in equilibrium with the current atmosphere. This begins the water conveyor belt where it moves slowly throughout the globe and is forced under warmer, more buoyant water, cutting itself off from the atmosphere until it is pulled up to the surface, often 1000 years later in a different ocean basin than it started in. Throughout this time, it interacts with the surface and the seafloor but with the surface only by particles and animals moving through the realm: phytoplankton cells fall from the surface through the deep water, providing food for the diversity of life that lives there. As this food is consumed, carbon dioxide and nutrients are released, and oxygen is consumed; 55 per cent of the food that sinks below 1000 m is consumed prior to its reaching the seafloor (Jahnke 1996). Organisms swim throughout the water, redistributing nutrients, which microorganisms use for energy and growth. As a result, as the water continues along its path, it builds up nutrients and loses oxygen. The amount of food that falls through it depends on the amount and type of photosynthetic production at the surface (Lutz et al. 2002); however, only about 1 per cent of the energy fixed at the surface will ever reach the abyssal deep-sea floor (Ducklow 1995;

Lutz et al. 2007). This overall movement of water and particles falling through it lead to the distribution of animals, their diversity, and the services provided, except in cases where energy from the seafloor augments this gentle rain of food.

8.5.2 Nutrients for the shallows that fuel fisheries and oxygenate the atmosphere

The phytoplankton in the surface oceans of the planet flourish in areas where the accumulated nutrients of the deep are brought to the surface. Through food web interactions, abundant phytoplankton at the surface ultimately fuels fisheries production and generates oxygen to support life across the globe, both critical ecosystem services. Nutrients such as phosphate, nitrate, and ammonium are released as sinking particles are degraded, so after the long, dark path of the deep sea's conveyor belt, the water at depth is enriched in nutrients. Winds drive the water on the surface of the ocean apart, causing the deep, nutrient-rich water to be brought back to the surface fuelling photosynthesis by marine phytoplankton. As marine phytoplankton provide half of the oxygen in the Earth's atmosphere, the impact of the deep-sea nutrients eventually brought to the surface is significant to all life on the planet. This nutrient regeneration has been valued at 5.4 trillion (US) dollars for the surface ocean (Costanza et al. 1997), and it is a dominant deep-sea process that contributes to the natural capital of the oceans.

Nutrients accumulate both through the slow degradation and conversion of organic matter by microbes and by the physical movements and deposition of nutrients by animals that undergo diel vertical migration (DVM), that is, moving either into or away from the surface every night. A large component of deep pelagic animal fauna undergo DVM, to either avoid predation or find food (Pearcy et al. 1977; Watanabe et al. 1999). As there is no light-fuelled primary productivity in the deep water column, most deep pelagic animals that swim to the surface or shallower waters find increased food in the waters where photosynthesis (productivity) has occurred. They then return to the deep as the sun rises to avoid predation by visually oriented predators during the day. In conducting these massive migrations, the deep pelagic biota provides a critical ecosystem service in transferring large amounts of energy from surface waters to the deep ocean (Davison et al. 2013). Through defecation and death (via both predation and disease), the animals bring organic carbon down and release nutrients into the deep pelagic waters, where they are prey to nonmigrating organisms or substrate for microbial decomposition and subsequent nutrient release. The particles that make it through this gauntlet of microbial degradation and animals predation arrive on the seafloor where they are further broken down. While the rates of degradation/respiration are very small compared to what we consider normal for shallow water habitats, the combined time (1000 years) and vast areas that the water moves over concentrates tremendous amounts of nutrients. For example, up to 50 times the concentration of nitrate is present in the deep water in contrast to the value which the water contained when it was last in contact with the atmosphere (upwards of 50 µmol/kg nitrate; NOAA Ocean Atlas).

Carbon sequestration in the deep

Regulating services includes gas storage, and the deep pelagic is integral to this through biological activity and physical processes. The world is shifting due to the carbon that humans have put in the atmosphere through the burning of fossil fuels. The rate that this is happening is actually mitigated by the capture and storage of carbon in the deep sea. Carbon molecules are captured by the surface water before it begins its 1000 year journey through the ocean basins. Phytoplankton, animals, and even coastal algae (Krause-Jensen and Duarte 2016) trap carbon that they transport in their biomass as they sink to deeper depths. As the particles that fall through the water are respired or defecated out by vertically migrating organisms, carbon is also released, further loading up the deep sea with carbon from the atmosphere. Currently there are about 37,700 gigatons of carbon in organic and inorganic forms in the deep pelagic and 6 billion gigatons of carbon captured in deep-sea sediment (Sabine and Feely 2007). These values all dwarf the 700 gigatons of carbon in the atmosphere. Particles of all sizes that make it into the deep sediment and

are not consumed or broken down by microbes and fauna keep the carbon trapped for millennia creating one of the most permanent repositories on the planet. Through these routes, the deep ocean has already captured one third of the carbon that we have released into the atmosphere (Sabine and Feely 2007).

Nitrogen, phosphorous, and carbon in the future oceans

While one may imagine that the vast ocean-scale processes discussed in the previous sections are too large to be impacted by human activity, the ecosystem services derived from them is the result of slow integrative activity making them susceptible to impact and making any impacts long lasting. Ocean warming reduces the amount of carbon that can be captured from the atmosphere. The rate of the deep-ocean conveyor belt is slowing; at reduced speeds, there is less opportunity to take up atmospheric gases, further limiting how much carbon the ocean takes up from the atmosphere (Bryden et al. 2005; Boulton et al. 2014). While remineralisation in the deep ocean releases nutrients, it also uses oxygen, creating vast areas known as oxygen minimum zones. These areas also have low pH (acidic) water as the result of build-up of carbon dioxide caused by organic matter decomposition. Both the acidity of the deep water and the oxygen minimum zones have expanded and are expanding as a result of our changing climate (Stramma et al. 2011). Our warming atmosphere is also warming the ocean (Levitus et al. 2000; Barnett et al. 2005; Domingues et al. 2008); this warmer water causes organisms' metabolic processes to increase, further releasing carbon dioxide into the deep sea and using up oxygen. Complicating this, the majority of the oceans will have shifts in surface production that will result in fewer particles sinking into the deep (Mora et al. 2013; Jones et al. 2017; Sweetman et al. 2017). This reduced flux of food to the deep sea also means less of the carbon dioxide that is captured by the phytoplankton on the surface will sink to the deep sea, further reducing the role of the deep sea in acting as a sink for atmospheric carbon dioxide.

Warmer temperatures, reductions in the number of sinking organic particles, and expanding oxygen minimum zones are likely to have direct impacts on the amount of nutrients available at the surface. Usable forms of nitrogen can be made effectively biologically unavailable through a variety of biological processes, including denitrification and anaerobic ammonium oxidation (annamox). Nitrogen cycling is a complex web of microbial processes composed of microbial winners and losers, and under certain conditions, it can result in a net loss of nitrogen from the system. Denitrification and annamox are both especially abundant in oxygen minimum zones (Lam and Kuypers 2011), which are expanding (Stramma et al. 2011). Increased remineralisation of sinking particles at shallower depths will also decrease the amount of time that the carbon is stored in the oceans, and warming ocean waters may also lead to a greater proportion of the carbon that reaches the seafloor being released. In every one of these scenarios we have altered if not reduced the extent of the ecosystem services provided by the deep ocean and thus reduced their value to society through our actions.

8.5.3 A bottom-up view of vents and seeps

While processes in the deep mud and deep water are slow, building up over centuries, there are also epicentres of rapid activity in the deep sea: hydrothermal vents and methane seep habitats. Vents are areas where superheated fluid that has circulated through the Earth's crusts release massive amounts of energy in both heat and chemical food sources that can fuel biological productivity. Methane seeps are where very small portions of a vast reservoir of methane, buried underneath the seafloor, leak out enough where it can create a cornucopia of chemical food for microorganisms and animals that live either off of or in symbiosis with those bacteria; they are often called cold seeps as they are the same temperature as the surrounding ocean. Hydrothermal vents and seeps are key examples of habitats that further expand the diversity of the deep while also adding regulating and supporting services to the ocean ecosystem (Turner et al. 2019). Complementing the energy that rains down from above, these habitats' organisms harness chemicals released from the seafloor to create biomass and fuel communities based on microbial primary production. Sulphide, methane, hydrogen, and a diversity of metals (iron,

copper, silica, as well as many other forms) all can be harnessed by the microbes at these systems (Peterson et al. 2011). Vent and seep habitats are anomalous in the deep sea; food is no longer limiting and instead heat (at hydrothermal vents) and often hydrogen sulphide (a compound toxic to most life however abundant at vents and seeps) becomes more deterministic in the distribution of taxa who aim to take advantage of the energy, and even competition and predation become increasingly important in structuring the communities (Micheli et al. 2002). While vents (Corliss et al. 1979) and seeps (Paull et al. 1985) were only discovered relatively recently, and once thought to be rare, we now know that these habitats are common and sometimes abundant throughout the oceans (Beaulieu et al 2013; Skarke et al. 2014; Beaulieu et al 2017; Seabrook et al. 2018). The fauna that exist there is a combination of unique animals that are only found in these areas and animals that are found throughout the deep sea (Grassle 1987; Van Dover 2000; Levin et al. 2016; Seabrook et al. 2019). As a result of these fauna, including a unique suite of bacteria and archaea (Knittel and Boetius 2009; Ruff et al. 2015), these areas are significant to the overall functioning of the planet, and they provide a series of benefits to society (Thurber et al. 2014; Levin et al. 2016).

While an energy source for microbes, methane is also a potent greenhouse gas that is 25 time more effective than carbon dioxide at warming our atmosphere (Reeburgh 2007); however, life in the oceans eats the vast majority of the released methane, providing an important regulating service (Knittel and Boetius 2009). A subset of vents (depending on the geologic underpinnings) and methane seeps leak out this methane, in both bubbles and dissolved forms. This release of methane is estimated to be globally on the order of 0.02 GtC annually (Boetius and Wenzhöfer 2013), which is a small fraction of the estimated 1800–10,000 GtC that are stored under marine sediment (Reeburgh 2007; Ruppel and Kessler 2017). If released into the atmosphere, even this slow leak could drastically alter the trajectory of the climate. Massive methane releases are purported to have driven half of the planet's species extinctions throughout Earth's history (Dickens et al. 1997). However, only a small fraction of the methane contained in modern-day reservoirs is released, with most studies estimating that approximately 6 Tg of methane are released annually into the atmosphere from marine methane sources, which is 1 per cent of all of the methane released annually into the atmosphere from all quantified sources (Ruppel and Kessler 2017).

Vents and seeps are increasingly being viewed as conduits for energy from the seafloor to the greater ocean ecosystem (Levin et al. 2016). The energy released also provides a trophic subsidy supporting both seep and nonseep endemic life. Vents and seeps are estimated to provide on the order of 10 per cent of all of the energy that enters the realm bellow the sunlit surface sea (Levin et al. 2016), with cascading effects through the food web further impacting nutrient regeneration and biomass production and even fuelling fisheries production (a provisioning service). Seeps and vents support increased biomass at the seafloor (Van Dover 2000), and in certain cases, there is a strong overlap between the fauna that occur at vents and seeps and the rest of the deep sea, with as much as half of the animals living in the sediment at seeps being found both at these habitats and occurring away from the seep; this overlap is also not limited to the seafloor, as water column bacteria use seep production for energy too (Kelleyl et al. 1998). Commercially harvested species such as crabs are known to utilise vents (Colaço et al. 2002) and seeps (Seabrook et al. 2019) and to derive some of their energy from these systems. In addition Patagonian toothfish, black oreo, and a variety of other deep-sea fishery species are found in greater abundance at seeps (Sellanes et al. 2012; Bowden et al. 2013). Indeed, there has been increasing evidence that vent and seep habitats are important to local fisheries productivity, as postulated years ago (Levy and Lee 1988). While much of this may seem like a highly localised phenomenon, exploration has revealed that vents and especially seeps can be common. On a 830 km stretch of margin habitat offshore the Western United States there are more than 2500 seeps that form a broken line of seepage all along the region (based on updated values from Seabrook et al. 2018). This dense conglomeration of seeps is unlikely to be unique, but we know of its distribution because we have looked there. Methane

seeps are not rare but are instead a common and overlooked feature of the ocean margins (e.g. Skarke et al. 2014) that could provide significant food resources to a diversity of taxa including species that are targets of deep-sea fisheries.

Vents and seeps can also transport their chemical energy throughout large swaths of the ocean basins through their plumes, marauding taxa, and pelagic larval production. There is increased biomass of bacteria, as well as plankton, present in the plumes of hydrothermal vents (Burd and Thompson 1994; Wiebe et al. 1988; Kaartvedt et al. 2012), and these plumes stretch for thousands of kilometres (Resing et al. 2015). Much like animals conducting DVM, transporting energy from surface production into the deep, marauding taxa—mobile taxa that only occur at vents or seeps for a short period of their lives—can move the energy that is released at vents around the oceans as they wander searching for more food (MacAvoy et al. 2008). Many of the animals at vents and seeps also reproduce through the release of gametes that provide an additional conduit for energy to leave these habitats and in certain cases, cross the ocean basins. Energy-rich larvae can transport energy for more than 100 days across the oceans (Marsh et al. 2015) and, if feeding, can cross entire ocean basins (Arellano and Young 2009). The output can also be massive: clams beds in Sagami Bay, Japan, release 5.8×10^8 eggs m^{-2} yr^{-1} (Fujikura et al. 2007). Together this export of production is another way in which these small patches of energy released from the seafloor can seed food throughout ocean basins.

Shifting services of vents and seeps in differing use scenarios

Vents and seeps are also facing differing anthropogenic stressors that may reduce their ecosystem services, including their role in regulating climate change. Methane is trapped below the seafloor as methane hydrate, a form of chemical ice, through a combination of cold temperatures and high pressures. However, ocean warming reduces the stability of the reservoirs of methane in the deep subsurface (Johnson et al. 2015). While still under intense debate whether ocean warming accelerated seepage is now occurring, the simple physics behind reservoir gas stability point to a future where increases in the rate of seepage may challenge the ocean's ability to consume this greenhouse gas prior to its release. If organisms are able to consume the extra methane that will be released into the ocean from destabilising methane reservoirs, there will be increased oxygen stress and lowered pH, as consumption of the increased methane will use oxygen and produce carbon dioxide (Biastoch et al. 2011).

Methane reservoirs, as well as hydrothermal vents, are prime areas for extracting mineral and energy resources (see Chapters 5 and 6 for comprehensive discussion of this). Removal of methane reservoirs may remove the fuel that provides basal energy sources for seep ecosystems, and quantifying and modelling the impacts of this are an emerging area of research. Mining vent structures will also shift their ability to provide a trophic subsidy in the near term, especially if sedimentation from mining vents smothers the biological activity that could otherwise fuel the gamete-producing taxa that spread energy ocean wide. The plumes created by mining may also impact the availability of oxygen, further stressing the marine environment. These stressors may also impact marine genetic resources, as discussed in Chapter 7.

8.6 An overlap of use and nonuse

Integration of nonuse values into management decisions is most key when differing use scenarios impact the nonuse services provided by a habitat (Table 1), especially if those impacts will last a significant period of time. In the deep sea, this concern is great, as many of the resources should be considered nonrenewable, and, thus, impacts can be viewed as indefinite. Hydrothermal vent fields will take decades to centuries to recover from seabed mining (Tunnicliffe et al. 1997; Van Dover 2011), and nodule fields will take on the order of millennia (Thiel 2001; Smith et al. 2008b). Both estimates assume that species or populations are not lost, which would permanently impact the biological diversity of the oceans. Armstrong et al. (2017) argue that due to the slow growth rates of cold-water (deep-sea) corals, they should be considered a nonrenewable resource when identifying what is an acceptable impact to them. Thus impacts, be it

Table 1 Overlap of nonuse factors and anthropogenic impacts on the deep sea.

Nonuse Factor	Overlap with	Impact
Artistic inspiration	Waste disposal	Reduce the inspiration that a pristine environment can evoke
	Deep-sea fishing	Removal of structures that are epicentres of inspiration
	Mining	Removal of structures that are epicentres of inspiration and/or loss of visibility from mining plumes
	Climate change	
Exploration	Waste disposal	Reduced excitement for discovering impacted areas
	Deep-sea fishing	Removal of large inspirational fauna
	Mining	Removal of biodiversity including large fauna that are epicentres of nutrient cycling
	Climate change	Loss of biodiversity from deoxygenation and/or warming ocean etc.
Nutrient regeneration	Deep-sea fishing	Trawling can impact the rate of nutrient regeneration
	Mining	Removal of large inspirational fauna and biodiversity
	Climate change	Expanding oxygen minimum zones shift nutrient cycles
Methane consumption	Deep-sea fishing	Unknown impact
	Methane mining	May reduce methane input into communities and/or decrease oxygen through leaks
	Climate change	Reduce oxygen, which could impact macrofaunal community composition
Scientific discovery or MGR	Deep-Sea fishing	Loss of biodiversity and impact of new habitats prior to discovery
	Methane mining	Unknown Impact
	Climate change	Shift in biodiversity with unknown impact

In addition to negative impacts that are highlighted here, these overlaps also provide an avenue for art and communication to be a mechanism to enact change, which is another use of the natural capital of the deep sea.

from mining or fishing, affect the future available uses of oceans.

A unique facet of the deep sea is that apparently miniscule processes can add up to provide significant ecosystem services over time. Supporting services provide an example of the importance of this view point: the release of nitrogen from sinking particles as they transit from the surface ocean to the deep seafloor is almost immeasurably small; however, when this immeasurably small process occurs unencumbered for a thousand years, enough nitrogen is carried in the slowly flowing water to fuel the most productive fisheries in the world. In fact, most instantaneous rates in the deep sea are slow: fish and microbes alike grow slowly, and yet over hundreds to thousands of years, their existence results in deposited carbon in the deep sea, regenerated nutrients, and areas of scientific discovery and inspiration. The other side of this is that the longevity of impact is also vast in both time and space. Mining impacts extend for decades (Jones et al. 2017), and sediment plumes from mining an area the size of a small continent could impact the biogeochemical environment of water upwelled thousands of kilometres away centuries later. Impacts on deep-ocean function could result in shifting ecosystem services for decades to a thousand years. This all further supports treating many deep-sea extractive resources as nonrenewable with a directed acknowledgement by resource managers on the potential long-term losses in the natural capital that the deep sea contains, including informed balance of both use and nonuse aspects of the ecosystem.

8.7 Current state of valuation of nonuse values in the deep sea

While valuation of nonuse services is a mechanism to facilitate informed management decisions, valuation is challenging and has only been undertaken a few times in the deep sea. While multiple valuation frameworks have been proposed (Armstrong et al. 2010; 2017; Le et al. 2017), they have only been implemented in a few instances. Jobstyogt et al. (2013) performed a valuation focusing on the deep sea and found people were willing to pay £70–77 (US $85–93) per person to protect the deep sea, with increased support provided by those (as they posit) that had seen ocean landscapes before. A significant take-home message from their study was that a lack of experience and/or education is one of the greatest

challenges to conservation of deep-sea resources. There was a much greater willingness to pay by Scottish households when asked about balancing the needs of fish stocks over oil and gas development. People were willing to pay > €200 annually (US $222) per household to offset the costs of fish stock protection (Aanesen et al. 2015). Respondents of this survey were willing to pay more when oil and gas extraction had direct impact on fisheries than when it did not, which highlights a need for further education about the connectedness of the deep sea. As the services are caused by a combination of the slow build-up of integrated activity across the broad (and internationally managed) deep sea, diffuse impacts can be widespread and should be focused on for directed management (Folkersen et al. 2019). A realised example of this was the widespread impact of the Macondo well disaster (*Deep Water Horizon*), which had a widespread and persistent impact on the amount of oxygen in the waters of the Gulf of Mexico (Kessler et al. 2011).

While the services of the deep sea have been called 'the benefit of mankind as a whole' (in the United Nations Convention on the Law of the Sea, 1982), the excitement for the deep sea is a commonality, especially when people are informed of what is present. Zanoli and colleagues found a group of participants that, when exposed to it, fell under a group of 'Ecosystem Functions Supporters' and 'Deep Coral Lovers', with their preference driven by their experience and exposure (Zanoli et al. 2015). Another example of the interest in the excitement for the deep sea is when people were exposed to the deep sea through a learning centre exhibit, 70 per cent of them told other people about what they learned within a month of visiting the centre (Darr et al. 2020). Connecting informed value to the various nonuse benefits of the deep sea may facilitate informed management, especially when harnessing the general interest and excitement that the general public has for the deep sea if they are exposed to it.

8.8. Summary

The deep sea provides a significant suite of services to humankind that falls within the category of nonuse services. Hydrothermal vents are inspiring (Turner et al. 2019); the deep sea is covered with seamounts and canyons; and deep waters are critical to the functioning of the ocean, while enticing the public with their mystery. The greatest challenge to quantifying the wealth within natural capital of the deep sea is a lack of public knowledge about the deep sea, not an often perceived lack of knowledge by scientists. A commonality to nonuse valuations is application of biodiversity as a valued framework, but in the deep sea, biodiversity itself is a point of debate (Higgs and Attrill 2015; Sinniger et al. 2016), and when that is then applied to ecosystem function, a further debate looms concerning the relationship between biodiversity and ecosystem function (Levin and Dayton 2009). While this may be critical to uncover, the clear hurdle for inclusion of nonuse benefits in valuation is informing the public about what is present in the deep sea and what the deep sea does for society. That knowledge is infectious (Darr et al. 2020), and when amplified by art, documentaries, and literature (services in their own right), it can be used to enact change (Harrower et al. 2018). We are also at a new age of sharing research through live streaming of at-sea footage and partnerships with programmes, from the BBC to the Octonauts, where the knowledge and nonuse benefits of the deep can reach far greater and younger audiences than ever before. Together this will allow us to better include both use and nonuse benefits in our quantification of the natural capital of the deep sea.

Acknowledgements

We thank Lily Simonson both for her generosity in allowing us to use her art in this manuscript and through inspiring many people about the mystery of the deep (including the authors.) We also thank Carlie Wiener (Schmidt Ocean Institute), Sarah Seabrook and Katie Darr (Oregon State University), and Nicole Raineault (Ocean Exploration Trust) for facilitating this work through either providing imagery and/or stimulating discussion on nonuse facets of the deep sea. This work was partially supported by NOAA Ocean Exploration and Research Grant NA19OAR0110301 to ART.

References

Aanesen, M., Armstrong, C., Czajkowski, M., et al. (2015). Willingness to Pay for Unfamiliar Public Goods: Preserving Cold-water Coral in Norway. *Ecological Economics*, 112, 53–67.

Alaimo, S. (2013). Violet-Black, in J. J. Cohen (ed.) *Prismatic Ecology: Ecotheory Beyond Green*. Minneapolis: University of Minnesota Press, Chapter 12.

Anastasopoulou, A., Mytilineou, C., Smith, C. J., and Papadopoulou, K. N. (2013). Plastic Debris Ingested by Deep-water Fish of the Ionian Sea (Eastern Mediterranean). *Deep-Sea Research I*, 74, 11–13.

Arellano, S. M. and Young, C. M. (2009). Spawning, Development, and the Duration of Larval Life in a Deep-sea Cold-seep Mussel. *Biological Bulletin*, 216, 149–62.

Arin, T. and Kramer, R. A. (2002). Divers' Willingness to Pay to Visit Marine Sanctuaries: An Exploratory Study. *Ocean and Coastal Management*, 45, 171–83.

Armstrong, C. W., Foley, N., Tinch, R., and van den Hove, S. (2010). Ecosystem Goods and Services of the Deep Sea. *Hotspot Ecosystem Research and Man's Impact on European Seas*. Deliverable D6 2, 68. https://www.pik-potsdam.de/news/public-events/archiv/alter-net/former-ss/2010/13.09.2010/van_den_hove/d6-2-final.pdf.

Armstrong, C. W., Kahui, V., Vondolia, G. K., Aanesen, M., and Czajkowski, M. (2017). Use and Non-use Values in an Applied Bioeconomic Model of Fisheries and Habitat Connections. *Marine Resource Economics*, 32, 351–69.

Baco, A. R., Rowden, A. A., Levin, L. A., Smith, C. R. and Bowden, D. A. (2010). Initial Characterization of Cold Seep Faunal Communities on the New Zealand Hikurangi Margin. *Marine Geology*, 272, 251–9.

Barnett, T. P., Pierce, D. W., Achutarao, K. M., et al. (2005). Penetration of Human-induced Warming into the World's Oceans. *Science*, 309, 284–7.

Beaulieu, S. E., Baker, E. T., and German, C. R. (2013). An Authoritative Global Database for Active Submarine Hydrothermal Vent Fields. *Geochemistry, Geophysics, Geosystems*, 14, 4892–905.

Beaulieu, S. E., Graedel, T. E., and Hannington, M. D. (2017). Should We Mine the Deep Seafloor? *Earth's Future*, 5, 655–8.

Bernardino, A. F., Smith, C. R., Baco, A., Altamira, I., and Sumida, P. Y. G. (2010). Macrofaunal Succession in Sediments Around Kelp and Wood Falls in the Deep NE Pacific and Community Overlap with Other Reducing Habitats. *Deep-Sea Research I*, 57, 708–23.

Biastoch, A., Treude, T., and Rüpke, L. H. (2011). Rising Arctic Ocean Temperatures Cause Gas Hydrate

Destabilization and Ocean Acidification. *Geophysical Research Letters*, 38, 1–5.

Billett, D. S. M., Bett, B. J., Reid, W. D. K., Boorman, B., and Priede, I. G. (2010). Long-term Change in the Abyssal NE Atlantic: The 'Amperima Event' Revisited. *Deep-Sea Research II*, 57, 1406–17.

Boetius, A., Wenzhöfer, F. (2013). Seafloor Oxygen Consumption Fuelled by Methane from Cold Seeps. *Nature Geoscience*, 6, 725.

Boulton, C. A., Allison, L. C., and Lenton, T. M. (2014). Early Warning Signals of Atlantic Meridional Overturning Circulation Collapse in a Fully Coupled Climate Model. *Nature Communications*, 5, 5752.

Bowden, D. A., Rowden, A. A., Thurber, A. R., et al. (2013). Cold Seep Epifaunal Communities on the Hikurangi Margin, New Zealand: Composition, Succession, and Vulnerability to Human Activities. *PLoS ONE*, 8, e76869.

Boyd, J. and Banzhaf, S. (2007). What Are Ecosystem Services? The Need for Standardized Environmental Accounting Units. *Ecological Economics*, 63, 616–26.

Bryden, H. L., Longworth, H. R., and Cunningham, S. A. (2005). Slowing of the Atlantic Meridional Overturning Circulation at 25° N. *Nature*, 438, 655–7.

Burd, B. J. and Thomson, R. E. (1994). Hydrothermal Venting at Endeavour Ridge: Effect on Zooplankton Biomass Throughout the Water Column. *Deep-Sea Research I*, 41, 1407–23.

Carlson, A. 2010. Contemporary Environmental Aesthetics and the Requirements of Environmentalism. *Environmental Values*, 19, 289–314.

Carpenter, S. R., DeFries, R., Dietz, T., et al. (2006). Ecology. Millennium Ecosystem Assessment: Research Needs. *Science*, 314, 257–8.

Chalfie, M., Tu, Y., Euskirchen, G., Ward, W. W., and Prasher, D. C. (1994). Green Fluorescent Protein as a Marker for Gene Expression. *Science*, 263, 802–5.

Chan, K. M. A., Satterfield, T., and Goldstein, J. (2012). Rethinking Ecosystem Services to Better Address and Navigate Cultural Values. *Ecological Economics*, 74, 8–18.

Colaço, A., Dehairs, F., and Desbruyères, D. (2002). Nutritional Relations of Deep-sea Hydrothermal Fields at the Mid-Atlantic Ridge: A Stable Isotope Approach. *Deep-Sea Research I*, 49, 395–412.

Corliss, J. B., Dymond, J., Gordon, L. I., et al. (1979). Submarine Thermal Springs on the Galápagos Rift. *Science*, 203, 1073–83.

Costanza, R., d'Arge, R., de Groot, R. et al. (1997). The Value of the World's Ecosystem Services and Natural Capital. *Nature*, 387, 253–60.

Costanza, R., de Groot, R., Braat, L., et al. (2017). Twenty Years of Ecosystem Services: How Far Have We Come and How Far Do We Still Need to Go? *Ecosystem Services*, 28, 1–16.

Curtis, D. J. (2009). Creating Inspiration: The Role of the Arts in Creating Empathy for Ecological Restoration. *Ecological Management and Restoration*, 10, 174–84.

Curtis, D. J., Reid, N., and Reeve, I. (2014). Towards Ecological Sustainability: Observations on the Role of the Arts. *S.A.P.I.E.N.S.*, 7(1), 15.

Darr, K. D., East, J. L., Seabrook, S., Dundas, S. J., and Thurber, A. R., 2020. The Deep Sea and Me: Using an Exhibit at a Public Science Center to Promote Public Literacy of the Deep Sea. *Frontiers in Marine Science*. 7:159. doi: 10.3389/fmars.2020.00159

Davison, P. C., Checkley, D. M., Koslow, J. A., and Barlow, J. (2013). Carbon Export Mediated by Mesopelagic Fishes in the Northeast Pacific Ocean. *Progress in Oceanography*, 116, 14–30.

De Leo, F. C., Smith, C. R., Rowden, A. A., Bowden, D. A., and Clark, M. R. (2010). Submarine Canyons: Hotspots of Benthic Biomass and Productivity in the Deep Sea. *Proceedings of the Royal Society B*, 277, 2783–92.

DeMets, C., Gordon, R. G., Argus, D. F., and Stein, S. (1994). Effect of Recent Revisions to the Geomagnetic Reversal Time Scale on Estimates of Current Plate Motions. *Geophysical Research Letters*, 21, 2191–4.

Dickens, G. R., Castillo, M. M., and Walker, J. C. (1997). A Blast of Gas in the Latest Paleocene: Simulating First-order Effects of Massive Dissociation of Oceanic Methane Hydrate. *Geology*, 25, 259–62.

Domingues, C. M., Church, J.A., White, N. J., et al. (2008). Improved Estimates of Upper-ocean Warming and Multi-decadal Sea-level Rise. *Nature*, 453, 1090–3.

Drazen, J. C. and Haedrich, R. L. (2012). A Continuum of Life Histories in Deep-sea Demersal Fishes. *Deep-Sea Research I*, 61, 34–42.

Drazen, J. C. and Sutton, T. T. (2017). Dining in the Deep: The Feeding Ecology of Deep-Sea Fishes. *Annual Review of Marine Science*, 9, 337–66.

Ducklow, H. W. (1995). Ocean Biogeochemical Fluxes: New Production and Export of organic Matter from the Upper Ocean. *Review of Geophysics*, 33, 1271–6.

Folkersen, M. V., Fleming, C. M., and Hasan, S. (2019). Depths of Uncertainty for Deep-sea Policy and Legislation. *Global Environmental Change*, 54, 1–5.

Fujikura, K., Amaki, K., Barry, J. P., et al. (2007). Long-term in Situ Monitoring of Spawning Behavior and Fecundity in *Calyptogena* spp. *Marine Ecology Progress Series*, 333, 185–93.

Grace, M. A., Doosey, M. H., Denton, J. S. S., et al. (2019). A New Western North Atlantic Ocean Kitefin Shark (Squaliformes: Dalatiidae) from the Gulf of Mexico. *Zootaxa*, 4619, 109–20.

Grassle, J. F. (1987). The Ecology of Deep-Sea Hydrothermal Vent Communities, in J. H. S. Blaxter and A. J. Southward (eds.) *Advances in Marine Biology*. San Diego, CA: Academic Press, pp. 301–62.

Harrison H. M. and Harrison N. (1993). Shifting Positions Toward the Earth: Art and Environmental Awareness. *Leonardo*, 26:371–7.

Harrower, J., Parker, J., and Merson, M. (2018). Species Loss: Exploring Opportunities with Art–Science. *Integrative and Comparative Biology*, 58, 103–12.

Hessler, R. R. and Jumars, P. A. (1974). Abyssal Community Analysis from Replicate Box Cores in the Central North Pacific. *Deep Sea Research and Oceanographic Abstracts*, 21, 185–209.

Higgs, N. D. and Attrill, M. J. (2015). Biases in Biodiversity: Wide-ranging Species Are Discovered First in the Deep Sea. *Frontiers in Marine Science*, 2, 717.

Ingalls, A. E., Shah, S. R., Hansman, R. L., et al. (2006). Quantifying Archaeal Community Autotrophy in the Mesopelagic Ocean Using Natural Radiocarbon. *Proceedings of the National Academy of Sciences USA*, 103, 6442–7.

Irigoien, X., Klevjer, T. A., Røstad, A., et al. (2014). Large Mesopelagic Fishes Biomass and Trophic Efficiency in the Open Ocean. *Nature Communications*, 5, 3271.

Jamieson, A. J., Brooks, L. S. R., Reid, W. D. K., et al. (2019). Microplastics and Synthetic Particles Ingested by Deep-sea Amphipods in Six of the Deepest Marine Ecosystems on Earth. *Royal Society Open Science*, 6, 180667.

Jahnke, R. A. (1996). The Global Ocean Flux of Particulate Organic Carbon: Areal Distribution and Magnitude. *Global Biogeochemical Cycles, Spec. Publ.*, 10, 71–88.

Jobstvogt, N., Hanley, N., Hynes, S., Kenter, J., and Witte, U. (2014). Twenty-thousand Sterling Under the Sea: Estimating the Value of Protecting Deep-sea Biodiversity. *Ecological Economics*, 97, 10–19.

Jochum, K. P., Wang, X., Vennemann, T. W., Sinha, B., and Müller, W.E.G. (2012). Siliceous Deep-sea Sponge *Monorhaphis chuni*: A Potential Paleoclimate Archive in Ancient Animals. *Chemical Geology*, 300–1, 143–51.

Johnson, H. P., Miller, U. K., Salmi, M. S., and Solomon, E. A. (2015). Analysis of Bubble Plume Distributions to Evaluate Methane Hydrate Decomposition on the Continental Slope. *Geochemistry Geophysics Geosystems*, 16, 3825–39.

Jones, D. O. B., Kaiser, S., Sweetman, A. K., et al. (2017). Biological Responses to Disturbance from Simulated Deep-sea Polymetallic Nodule Mining. *PLoS ONE*, 12, e0171750.

Kaartvedt, S., Staby, A., and Aksnes, D.L. (2012). Efficient Trawl Avoidance by Mesopelagic Fishes Causes Large Underestimation of Their Biomass. *Marine Ecology Progress Series*, 456, 1–6.

Kawaguchi, K., Kaneko, S., Nishida, T., and Komine, T. (2015). Construction of the DONET Real-time Seafloor

Observatory for Earthquakes and Tsunami Monitoring, in P. Favali, L. Beranzoli, and A. De Santis (eds.) *Seafloor Observatories: A New Vision of the Earth from the Abyss*. Berlin and Heidelberg, Germany: Springer Berlin Heidelberg, pp. 211–28.

Kaartvedt, S., Staby, A., and Aksnes, D. L. (2012). Efficient Trawl Avoidance by Mesopelagic Fishes Causes Large Underestimation of Their Biomass. *Marine Ecology Progress Series* 456, 1–6.

Kelleyl, C. A., Cofjin, R. B., and Cifuentes, L. A. (1998). Stable Isotope Evidence for Alternative Bacterial Carbon Sources in the Gulf of Mexico. *Limnology and Oceanography*, 43, 1962–9.

Kessler, J. D., Valentine, D. L., Redmond, M. C., et al. (2011). A Persistent Oxygen Anomaly Reveals the Fate of Spilled Methane in the Deep Gulf of Mexico. *Science*, 331, 312–15.

Knittel, K. and Boetius, A. (2009). Anaerobic Oxidation of Methane: Progress with an Unknown Process. *Annual Review of Microbiology*, 63, 311–34.

Koslow, J. A., Kloser, R. J., and Williams, A. (1997). Pelagic Biomass and Community Structure over the Mid-continental Slope off Southeastern Australia Based upon Acoustic and Midwater Trawl Sampling. *Marine Ecology Progress Series*, 146, 21–35.

Krause-Jensen, D. and Duarte, C. M. (2016). Substantial Role of Macroalgae in Marine Carbon Sequestration. *Nature Geoscience*, 9, 737.

Kubiszewski, I., Costanza, R., Anderson, S., and Sutton, P. (2017). The Future Value of Ecosystem Services: Global Scenarios and National Implications. *Ecosystem Services*, 26, 289–301.

Lam, P. and Kuypers, M. M. M. (2011). Microbial Nitrogen Cycling Processes in Oxygen Minimum Zones. *Annual Review of Marine Science*, 3, 317–45.

Lamhauge S., Jacobsen, J. A., í Jákupsstovu, H., et al. (2008). *Fishery and Utilisation of Mesopelagic Fishes and Krill in the North Atlantic*. Copenhagen, Denmark: Nordic Council of Ministers.

Lear, C. H., Elderfield, H., and Wilson, P. A. (2000). Cenozoic Deep-sea Temperatures and Global Ice Volumes from Mg/Ca in Benthic Foraminiferal Calcite. *Science*, 287, 269–72.

Lebreton, L., Slat, B., Ferrari, F., et al. (2018). Evidence that the Great Pacific Garbage Patch Is Rapidly Accumulating Plastic. *Science Reports*, 8, 4666.

Le, J. T., Levin, L. A., and Carson, R. T. (2017). Incorporating Ecosystem Services into Environmental Management of Deep-seabed Mining. *Deep-Sea Research II*, 137, 486–503.

Levin, L. A., Baco, A. R., Bowden, D. A., et al. (2016). Hydrothermal Vents and Methane Seeps: Rethinking the Sphere of Influence. *Frontiers in Marine Science*, 3, 72.

Levin, L. A., Bett, B. J., Gates, A. R.,. et al. (2019). Global Observing Needs in the Deep Ocean. *Frontiers in Marine Science*, 6, 241.

Levin, L. A. and Dayton, P. K. (2009). Ecological Theory and Continental Margins: Where Shallow Meets Deep. *Trends in Ecology and Evolution*, 24, 606–17.

Levitus, S., Antonov, J. I., Boyer, T. P., and Stephens, C. (2000). Warming of the World Ocean. *Science*, 287, 2225–9.

Levy, E. M. and Lee, K. (1988). Potential Contribution of Natural Hydrocarbon Seepage to Benthic Productivity and the Fisheries of Atlantic Canada. *Canadian Journal of Fisheries and Aquatic Science*, 45, 349–52.

Lord Tennyson, A. (2015). *Poems, Chiefly Lyrical*. CreateSpace Independent Publishing Platform.

Lutz, M., Dunbar, R., and Caldeira, K. (2002). Regional Variability in the Vertical Flux of Particulate Organic Carbon in the Ocean Interior. *Global Biogeochemical Cycles*, 16. doi: 10.1029/2000GB001383.

Lutz, M. J., Caldeira, K., Dunbar, R. B., and Behrenfeld, M. J. (2007). Seasonal Rhythms of Net Primary Production and Particulate Organic Carbon Flux to Depth Describe the Efficiency of Biological Pump in the Global Ocean. *Journal of Geophysical Research*, 112, 1999.

MacAvoy, S. E., Morgan, E., Carney, R. S., and Macko, S. A. (2008). Chemoautotrophic Production Incorporated by Heterotrophs in Gulf of Mexico Hydrocarbon Seeps: An Examination of Mobile Benthic Predators and Seep Residents. *Journal of Shellfish Research*, 27, 153–61.

Marsh, A. G., Mullineaux, L. S., Young, C. M., and Manahan, D. T. (2001). Larval Dispersal Potential of the Tubeworm Riftiapachyptila at Deep-sea Hydrothermal Vents. *Nature*, 411, 77–80.

Marsh, L., Copley, J. T., Tyler, P. A., and Thatje, S. (2015). In Hot and Cold Water: Differential Life-history Traits Are Key to Success in Contrasting Thermal Deep-sea Environments. *Journal of Animal Ecology*, 84, 898–913.

McClain, C. and Barry, J. (2014). Beta-diversity on Deep-sea Wood Falls Reflects Gradients in Energy Availability. *Biology Letters*, 10, 20140129.

McFall-Ngai, M. J. (1990). Crypsis in the Pelagic Environment. *American Zoologist*, 30, 175–88.

McHugh, D. (1997). Molecular Evidence that Echiurans and Pogonophorans Are Derived Annelids. *Proceedings of the National Academy of Sciences USA*, 94, 8006–9.

McPhaden, M. J., Ando, K., Bourles, B., et al. (2010). The Global Tropical Moored Buoy Array. *Proceedings of Ocean Observations*, 9, 668–82.

Micheli, F., Peterson, C. H., Mullineaux, L. S., et al. (2002). Predation Structures Communities at Deep-sea Hydrothermal Vents. *Ecological Monographs*, 72, 365–82.

Mora, C., Wei, C-L., Rollo, A., Amaro, T., et al. (2013). Biotic and Human Vulnerability to Projected Changes in Ocean Biogeochemistry over the 21st Century. *PLoS Biology*, 11, e1001682.

Pagani, M., Zachos, J. C., Freeman, K. H., Tipple, B., and Bohaty, S. (2005). Marked Decline in Atmospheric Carbon Dioxide Concentrations During the Paleogene. *Science*, 309, 600–3.

Paull, C. K., Jull, A. J. T., Toolin, L. J., and Linick, T. (1985). Stable Isotope Evidence for Chemosynthesis in an Abyssal Seep Community. *Nature*, 317, 709–11.

Pearcy, W. G., Krygier, E. E., Mesecar, R., and Ramsey, F. (1977). Vertical Distribution and Migration of Oceanic Micronekton off Oregon. *Deep-Sea Research I*, 24, 223–45.

Peters, H. and Hawkins, J. P. (2009). Access to Marine Parks: A Comparative Study in Willingness to Pay. *Ocean and Coastal Management*, 52, 219–28.

Petersen, J. M., Zielinski, F. U., Pape, T., et al. (2011). Hydrogen Is an Energy Source for Hydrothermal Vent Symbioses. *Nature*, 476, 176–80.

Prellezo, R. (2019). Exploring the Economic Viability of a Mesopelagic Fishery in the Bay of Biscay. *ICES Journal of Marine Science*, 76, 771–9.

Ramirez-Llodra, E., Tyler, P. A., Baker, M. C., et al. (2011). Man and the Last Great Wilderness: Human Impact on the Deep Sea. *PLoS ONE*, 6, e22588.

Reeburgh, W. S. (2007). Oceanic Methane Biogeochemistry. *Chemical Reviews*, 107, 486–513.

Resing, J. A., Sedwick, P. N., German, C. R., et al. (2015). Basin-scale Transport of Hydrothermal Dissolved Metals Across the South Pacific Ocean. *Nature*, 523, 200–3.

Rex, M. A., Etter, R. J., Morris, J. S., et al. (2006). Global Bathymetric Patterns of Standing Stock and Body Size in the Deep-sea Benthos. *Marine Ecology Progress Series*, 317, 1–8.

Rouse, G. W. and Fauchald, K. (1995). The Articulation of Annelids. *Zoologica Scripta*, 24, 269–301.

Rouse, G. W., Goffredi, S. K., and Vrijenhoek, R. C. (2004). *Osedax*: Bone-eating Marine Worms with Dwarf Males. *Science*, 305, 668–71.

Ruff, S. E., Biddle, J. F., Teske, A. P., et al. (2015). Global Dispersion and Local Diversification of the Methane Seep Microbiome. *Proceedings of the National Academy of Sciences USA*, 112(13), 4015–20.

Ruppel, C. D. and Kessler, J. D. (2017). The Interaction of Climate Change And Methane Hydrates. *Reviews of Geophysics*, 55, 126–68.

Sabine, C. L. and Feely, R. A. (2007). The Oceanic Sink for Carbon Dioxide, in D. Reay, C. N. Hewitt, J. Grace, and K. Smith (eds.), *Greenhouse Gas Sinks*. Oxfordshire: CABI Publishing, pp. 31–49.

Seabrook, S., De Leo, F. C., and Thurber, A. (2019). Flipping for Food: The Use of a Methane Seep by Tanner Crabs (*Chionoecetes tanneri*). *Frontiers in Marine Science*, 6, 43.

Seabrook, S., De Leo, F. C., Baumberger, T., Raineault, N., and Thurber, A. R. (2018). Heterogeneity of Methane Seep Biomes in the Northeast Pacific. *Deep-Sea Research II*, 150, 195–209.

Sellanes, J., Pedraza-García, M. J., and Zapata-Hernández, G. (2012). Las áreas de filtración de metano constituyen zonas de agregación del bacalao de profundidad (*Dissostichus eleginoides*) frente a Chile central? *Latin American Journal of Aquatic Research*, 40, 980–91.

Silvertown, J. (2015). Have Ecosystem Services Been Oversold? *Trends in Ecology and Evolution*, 30, 641–8.

Sinniger, F., Pawlowski, J., Harii, S., et al. (2016). Worldwide Analysis of Sedimentary DNA Reveals Major Gaps in Taxonomic Knowledge of Deep-Sea Benthos. *Frontiers in Marine Science*, 3, 57.

Skarke, A., Ruppel, C., Kodis, M., Brothers, D., and Lobecker, E. (2014). Widespread Methane Leakage from the Sea Floor on the Northern US Atlantic margin. *Nature Geoscience*, 7, 657.

Smith, C. R., De Leo, F. C., Bernardino, A. F., Sweetman, A. K., and Arbizu, P. M. (2008a). Abyssal Food Limitation, Ecosystem Structure and Climate Change. *Trends in Ecology and Evolution*, 23, 518–28.

Smith, C. R., Paterson, G., Lambshead, J., et al. (2008b). *Biodiversity, Species Ranges, and Gene Flow in the Abyssal Pacific Nodule Province: Predicting and Managing the Impacts of Deep Seabed Mining*. (ISA Technical Study, 3) Kingston, Jamaica: International Seabed Authority.

Specht, E. A., Braselmann, E., and Palmer, A. E. (2017). A Critical and Comparative Review of Fluorescent Tools for Live-cell Imaging. *Annual Review of Physiology*, 79, 93–117.

St. John, M. A., Borja, A., Chust, G., et al. (2016). A Dark Hole in Our Understanding of Marine Ecosystems and Their Services: Perspectives from the Mesopelagic Community. *Frontiers in Marine Science*, 3, 31.

Stramma, L., Prince, E. D., Schmidtko, S., et al. (2011). Expansion of Oxygen Minimum Zones May Reduce Available Habitat for Tropical Pelagic Fishes. *Nature Climate Change*, 2, 33.

Sweetman, A. K., Smith, C. R., Shulse, C. N., et al. (2019). Key Role of Bacteria in the Short-term Cycling of Carbon at the Abyssal Seafloor in a Low Particulate Organic Carbon Flux Region of the Eastern Pacific Ocean. *Limnology and Oceanography*, 64, 694–713.

Sweetman, A. K., Thurber, A. R., Smith, C. R., et al. (2017). Major Impacts of Climate Change on Deep-sea Benthic Ecosystems. *Elementa: Science of the Anthropocene*, 5, 4. http://doi.org/10.1525/elementa.203.

Taylor, M. L., Gwinnett, C., Robinson, L. F., and Woodall, L. C. (2016). Plastic Microfibre Ingestion by Deep-sea Organisms. *Science Reports*, 6, 33997.

Thiel, H. (2001). *Environmental Impact Studies for the Mining of Polymetallic Nodules from the Deep Sea*. Oxford: Pergamon.

Thurber, A. R., Sweetman, A. K., Narayanaswamy, B. E., et al. (2014). Ecosystem Function and Services Provided by the Deep Sea. *Biogeosciences*, 11, 3941–63.

Tsien, R. Y. (1998). The Green Fluorescent Protein. *Annual Review of Biochemistry*, 67, 509–44.

Tunnicliffe, V., Embley, R. W., Holden, J.F., et al. (1997). Biological Colonization of New Hydrothermal Vents Following an Eruption on Juan de Fuca Ridge. *Deep-Sea Research I*, 44, 1627–44.

Turner, P. J., Thaler, A. D., Freitag, A., and Colman Collins, P. (2019). Deep-sea Hydrothermal Vent Ecosystem Principles: Identification of Ecosystem Processes, Services and Communication of Value. *Marine Policy*, 101, 118–24.

Turner, R. D. (1973). Wood-boring Bivalves, Opportunistic Species in the Deep Sea. *Science*, 180, 1377–9.

Van Cauwenberghe, L., Vanreusel, A., Mees, J., and Janssen, C.R. (2013). Microplastic Pollution in Deep-sea Sediments. *Environmental Pollution*, 182, 495–9.

Van Dover, C. (2000). *The Ecology of Deep-sea Hydrothermal Vents*. Princeton, NJ: Princeton University Press.

Van Dover, C. L. (2011). Mining Seafloor Massive Sulphides and Biodiversity: What Is at Risk? *ICES Journal of Marine Science*, 68, 341–8.

Vetter, E. W., Smith, C. R., and De Leo, F. C. (2010). Hawaiian Hotspots: Enhanced Megafaunal Abundance and Diversity in Submarine Canyons on the Oceanic Islands of Hawaii. *Marine Ecology*, 31, 183–99.

Wallace, K. J. (2007). Classification of Ecosystem Services: Problems and Solutions. *Biological Conservation*, 139, 235–46.

Watanabe, H., Moku, M., Kawaguchi, K., Ishimaru, K., and Ohno, A. (1999). Diel Vertical Migration of Myctophid Fishes (Family Myctophidae) in the Transitional Waters of the Western North Pacific. *Fisheries Oceanography*, 8, 115–27.

Wei, C-L., Rowe, G. T., Escobar-Briones, E., et al. (2010). Global Patterns and Predictions of Seafloor Biomass Using Random Forests. *PLoS ONE*, 5, e15323.

Widder, E. (2013). The Kraken Revealed the Technology Behind the First Video Recordings of Live Giant Squid. *Sea Technology*, 54, 49–54.

Widder, E. A. (2010). Bioluminescence in the Ocean: Origins of Biological, Chemical, and Ecological Diversity. *Science*, 328, 704–8.

Wiebe, P. H., Copley, N., Van Dover, C., and Tamse, A. (1988). Deep-water Zooplankton of the Guaymas Basin Hydrothermal Vent Field. *Deep-Sea Research I*, 35, 985–1013.

Woodall, L. C., Sanchez-Vidal, A., Canals, M., et al. (2014). The Deep Sea Is a Major Sink for Microplastic Debris. *Royal Society Open Science*, 1, 140317.

Wuchter, C., Schouten, S., Boschker, H. T. S., and Sinninghe Damsté, J. S. (2003). Bicarbonate Uptake by Marine Crenarchaeota. *FEMS Microbiology Letters*, 219, 203–7.

Yesson, C., Clark, M. R., Taylor, M. L., and Rogers, A. D. (2011). The Global Distribution of Seamounts Based on 30 Arc Seconds Bathymetry Data. *Deep-Sea Research I*, 58, 442–53.

Zanoli, R., Carlesi, L., Danovaro, R., Mandolesi, S., and Naspetti, S. (2015). Valuing Unfamiliar Mediterranean Deep-sea Ecosystems Using Visual Q-methodology. *Marine Policy*, 61, 227–36.

CHAPTER 9

Climate change cumulative impacts on deep-sea ecosystems

Nadine Le Bris and Lisa A. Levin

9.1 Introduction

It is now recognised that the ocean is experiencing transient regimes of change in its state variables, unprecedented in the Earth's history (IPCC 2014; IPCC 2018, 2019). As a direct consequence of the accumulation of anthropogenic greenhouse gases (GHG) in the atmosphere and resulting climate change, ocean waters acidify and warm, with the later leading to reduced oxygenation (Pörtner et al. 2014; Hoegh-Guldberg et al. 2018). A growing number of surveys show that anthropogenic CO_2 has already entered deep basins and is resulting in significant pH decline in the regions, especially where deep convection and vertical mixing entrain surface waters to the great depths or as a result of reduced ventilation (e.g. North Pacific, Byrne et al. 2010; Subpolar North Atlantic, García-Ibáñez et al. 2016; Mediterranean Sea, Hassoun et al. 2015; Sea of Japan, Chen et al. 2017). Hydrographic-data time series document significant warming in large oceanic deep basins of the Arctic Ocean, North Atlantic, and Southern Ocean (Purkey and Johnson 2010; Wunsch and Heimbach 2014), equivalent to 0.12 °C per decade for the deep water of the Weddell Sea (Robertson et al. 2002) or 0.13 °C per decade for the Deep Greenland Sea (Somavilla et al. 2013). Long-term seafloor observatories have monitored temperature changes in the same range (e.g. 0.1 °C in 8 years at Hausgarten 2500 m depth; Soltwedel et al. 2016). Simultaneously, the ocean has lost 2 per cent of its oxygen globally over the past half century, with the largest loss in the 100–300 m and 1000–2000 m ranges, and surpassing 4 per cent per decade at the periphery of some oxygen

minimum zones (OMZs) (Stramma et al. 2010; Schmidtko et al. 2017; Levin 2018; Oschlies et al. 2018).

Ocean ecosystems are hence exposed to environmental changes occurring at much more rapid rates than over past, historical climate cycles; these changes are associated with significant disturbance of biogeochemical cycles and ecosystem functioning (Doney et al. 2012). Yet, while climate-change impacts on ocean life and ecosystems have mobilised marine scientists for more than a decade, most of the risk assessments and discussion about adaptation and mitigation concerns mainly coastal waters and ocean surface waters (i.e. Gattuso et al. 2015; Gattuso et al. 2018; IPCC 2018). The reasons are twofold. First, the dark ocean (i.e. waters below about 200 m depth) has been traditionally considered to be relatively isolated from the atmosphere and thus mostly sensitive to climate disturbance at the centennial to millennial scale of ocean thermohaline circulation. Second, experts have focused their attention on marine habitats under direct use by humans (e.g. low-lying areas and areas supporting aquaculture and fisheries or subject to harmful algal blooms) (Gattuso et al. 2018). This perspective is changing as we learn more about the interactions between the surface and deep-ocean layers on short timescales (e.g. days to weeks) that contribute to carbon sequestration (Boyd et al. 2019), and as there is a growing need to monitor and mitigate impacts from industrial activities that are rapidly expanding to greater depths (e.g. fisheries, oil and gas exploration and exploitation, and hydrate and mineral mining prospecting—see Chapters 4, 5, and 6; Figure 1; Ramirez-Llodra et al. 2011; Levin and Le

Nadine Le Bris and Lisa A. Levin, *Climate change* In: *Natural Capital and Exploitation of the Deep Ocean*. Edited by: Maria Baker, Eva Ramirez-Llodra, and Paul Tyler, Oxford University Press (2020). © Oxford University Press.
DOI: 10.1093/oso/9780198841654.003.0009

Bris 2015; Sweetman et al. 2017). The importance of raising climate consciousness in assessing impacts on ecosystems at great depth is furthermore supported by the argument that the deep ocean itself is a key component of the climate system, providing a critical regulating service (see Chapter 2, Table 1). Deep waters have absorbed most of the excess heat generated by greenhouse gases and at least 25 per cent of the anthropogenic CO_2 (Le Quéré et al. 2018). Owing to the tremendous volume of deep-ocean waters, impacts on ecosystems are quantitatively massive at the global scale, with potential feedbacks on climate (i.e. reduction of the carbon pump efficiency and enhancement of other GHG fluxes such as methane or nitrous oxide). The propagation of this disturbance along deep basins will be long-lasting and irreversible on the timescale of human societies. For all of these issues, better knowledge of the complex responses of interconnected ecosystems to climate stressors across the water column is critically needed.

In the past 5–10 years, Earth System Model (ESM) projections have been used to assess the future combination of climate stressors according to different GHG emission scenarios. Significant warming, deoxygenation, changing particulate organic carbon fluxes, and acidification are predicted to occur in most deep-ocean regions, distinctively affecting subsurface layers and the seafloor (Bopp et al. 2013; Mora et al. 2013). Projected changes in temperature, oxygen, and pH for 2100 are often greatest at intermediate water depths, compared to abyssal waters, impacting bathyal ecosystems from continental slopes, seamounts, and canyons that already concentrate the highest anthropogenic pressures in the deep sea (Levin and Le Bris 2015; FAO 2019; IPCC 2019). Cumulative impacts are expected to be specific to each type of environment and its associated ecosystem services. Ocean intermediate and deep waters will be particularly susceptible to deoxygenation and acidification (Bopp et al. 2013; Gehlen et al. 2014). These changes are expected to affect biodiversity hotspots even in the deepest

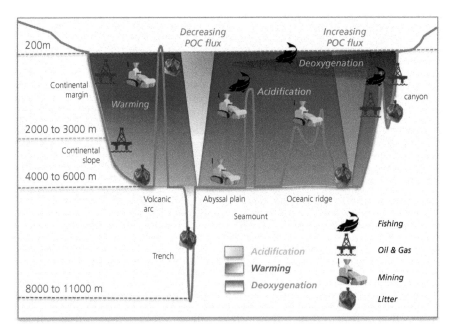

Figure 1 Conceptual scheme of climate change stressor combination and how they can interfere with human imprint in the deep-sea (From Levin and Le Bris 2015).

regions of the seafloor by the end of the twenty-first century (Mora et al. 2013; Gehlen et al. 2014; Sweetman et al. 2017; FAO 2019). Examining the predictions of climate models applied to deep waters and the ocean floor below 200 m in the light of cumulative actual and future anthropogenic pressures, however, requires careful attention as ESM were not designed to represent near-seabed hydrodynamic features on the ocean floor or continental margins. Their spatial resolution is generally too broad to be representative of the typical heterogeneity of many deep-sea habitats (e.g. submarine canyons, seamounts, hydrothermal vents, or methane seeps). In this chapter we address the current understanding of climate-change impact on deep-sea ecosystems and biodiversity, the capacity to anticipate change based on currently available models from the Coupled Model Intercomparison Project Phase 5 (CIMP5), and needs for improved forecasting capacities such as proposed in the future CIMP6 high-resolution models.

The United Nations' (UN) 2030 Agenda for Sustainable Development, adopted in September 2015, identified key interactions among targets that aim at strengthening the health of marine ecosystems and the environmental and societal resilience and adaptive capacities to climate change (International Council for Science 2017). In the context of increasing human impacts in the deep ocean and growing momentum for the exploitation of its resources, it is mandatory to understand the drivers and consequences of climate change on deep-sea ecosystems and to ensure adequate sharing of information. To support relevant strategies in conservation, marine spatial planning and management, including impact assessment and impact monitoring plans in the deep sea, we must address the relevant questions. What do climatic scenarios tell us about change in the deep-ocean layers and the seafloor? What are the main uncertainties of the future climate projections, and what gaps in knowledge remain to be filled to assess species and ecosystem vulnerability? In this perspective, chapter emphasises the need for specific observations to assess cumulative anthropogenic impacts, particularly at temporal and spatial scales representative of

climate-driven changes and deep-sea ecosystem responses to these changes.

9.2 Predicting climate-change impacts: projected changes and species vulnerability

9.2.1 Earth System Model projections and observations at depth

In the framework of CMIP5, projected changes in ocean physical and biogeochemical properties were assessed using ESMs forced by scenarios in the evolution of atmospheric contents in CO_2 and other greenhouse gases (i.e. Representative Concentration Pathways (RCPs), according to IPCC 2014). Simulations performed assess the evolution of ocean variables to the year 2100 for different RCP scenarios (i.e. RCP8.5 for the scenario, leading to more than 4 °C global warming in the year 2100; RCP2.6 with less than 2 °C warming for the 'high-mitigation scenario'; and RCP4.5 being an intermediate scenario. Multimodel averages have proven useful to predict and map expected stresses on marine life resulting from ocean-water warming, deoxygenation, and acidification, down to 600–700 m water depths (Stramma et al. 2010; Cocco et al. 2013) and even to the abyssal layers (Orr et al. 2005; Bopp et al. 2013). These models have been projected to the ocean floor to assess changes in a specific stressor (e.g. Guinotte et al. 2006 for aragonite saturation; Gehlen et al. 2014 for pH), as well as to identify multiple stressor combinations (Mora et al. 2013; Sweetman et al. 2017; FAO 2019). All of these simulations point to significant changes across the entire ocean depth under the RCP4.5 and RCP8.5 scenarios, and more limited changes for RCP2.6, and all help to predict the general exposure of deep-seafloor habitats to these changes.

Predicted shifts in pH, particulate organic carbon (POC) flux, and oxygen concentrations to the end of the century have been useful in pointing out striking differences in the dominant climatic stressors at regional scales, along with strong variations with depth. These changes are generally smaller at great depth than in shallow environments (Mora et al. 2013), but climatic vulnerability depends on the specific properties of deep-sea biomes and their natural variability. As considered by Sweetman et al. (2017), abyssal plains are projected to experience reduced POC export flux in all regions except high-latitude and upwelling areas, with up to 32 to 40 per cent decline in POC flux to seafloor habitats associated with oceanic gyres in the northern and southern Pacific Ocean and southern Indian Ocean. Empirical relationships suggest that POC decline could lead to a reduction in benthic biomass and organism body size in response to declining resource supply (Jones et al. 2014; Yool et al. 2017). Similar impacts of POC decline are expected to occur at bathyal depths, potentially impacting seamount and canyon habitats, which subject to some of the greatest cumulative anthropogenic impacts such as mining cobalt crusts (see Chapter 5) and fishing (see Chapter 4). Bathyal habitats are furthermore overlain by mesopelagic waters, where oxygen and pH are predicted to decline most strongly, particularly in the tropical ocean (Bopp et al. 2013; Sweetman et al. 2017). OMZs that develop in intermediate waters and can reach particularly low O_2 and low pH in some regions are predicted to expand both horizontally and vertically, leading to habitat compression for a whole range of species including migrating micronekton and mesopelagic fishes and cephalopods, as well as for benthic fauna from slopes (Stramma et al. 2010; Gilly et al. 2013; Sato et al. 2017; Levin and Gallo 2019). Additionally, shoaling of the aragonite-saturation horizon (ASH) has been identified as a major threat to ecosystems and biodiversity relying on calcifying species, most of the deep waters of the Pacific and a large part of the Atlantic being undersaturated with respect to aragonite at the end of the twentieth century (Orr et al. 2005). In the deep Pacific Ocean there already a substantial reduction of areas displaying noncorrosive conditions for aragonite, the calcium carbonate forming cold-water coral skeletons (Guinotte et al. 2006). In the North Atlantic, seafloor areas where the pH decrease will exceed interglacial variability are predicted to expand significantly (Gehlen et al. 2014).

Simultaneous changes in temperature, oxygen, pH, and POC fluxes are expected to occur in all types of deep-sea benthic and pelagic environments, the largest cumulated changes being predicted for continental slope and submarine canyons (Mora et al. 2013; FAO 2019). There is general scientific concern that the co-occurrence of these changes could create synergistic effects and significantly impair marine biodiversity and ecosystem services (Gunderson et al. 2016; Boyd et al. 2018). Exploring the regional difference in the abyssal and bathyal seafloor areas for the year 2100, Sweetman et al. (2017) depicted contrasting situations linked to the overturning circulation, with the largest absolute changes in pH, oxygen, and temperature occurring at the high latitudes of the North Atlantic, Arctic Ocean, and Southern Ocean (i.e. in regions of deep-water formation), while relative reduction in the POC export flux would be highest in the mid and low latitude in the Atlantic, Pacific, and Indian Oceans. The co-occurrence of stressors is expected to create a wide range of impacts that cannot be appreciated from the additive changes of individual factors (e.g. as done in Mora et al. 2013). Instead synergistic (or antagonistic) effects on species and the cascade of interactions that lead them to impair deep-sea biodiversity and ecosystem services have to be explored on a case-by-case basis (Gunderson et al. 2016). For instance, food limitation may exacerbate effects of warming and acidification in areas such as the North Atlantic (Sweetman et al. 2017).

The use of these climate-model projections at great depth for local forecasting purposes should, however, be considered with caution to avoid over interpretation. Multimodel averages available as part of CMIP5 offer the best available predictions for different climate stressors, but the spatial resolution (i.e. 1° grid) and the accuracy of models are still limited for the deep sea. Such limitations are partly due to the relative lack of observing data at depth to constrain these models and to the insufficient knowledge of mechanisms driving organic carbon transport to the ocean interior, particularly oxygen in tropical regions (Bopp et al. 2013; Oschlies et al. 2017; Boyd et al. 2019). ESMs are built to simulate the global heat budget and carbon cycle, and are based on physical and biological processes that require a certain degree of simplification to be parameterised. The biogeochemical component of these models generally accounts for organic matter export and carbon transformation processes that are well described in the first few hundreds of metres below the surface of the ocean, where the largest fraction of the photosynthetic organic carbon produced is being remineralised. The CIMP5 projections used so far to assess deep-sea POC fluxes are not expected to be representative of the complexity of biological processes involved across the entire ocean-depth range. Nor are physical sub-mesoscale processes (e.g. eddies) that develop in surface, intermediate, or abyssal waters and strongly influence organic matter transformations and nutrient export suitably represented in the former predictions. Upgraded high-resolution modelling of such ocean internal variability as proposed in the Ocean Model Intercomparison Project (OMIP) of the sixth phase of the Couple Intercomparison Model Project (CIMP6) will greatly improve this capacity (Orr et al. 2017). The new OMIP biogeochemical component should also more specifically consider deep-sea biogeochemical processes and their feedbacks to climate. In particular, the 1D-sedimentation model (Martin's curve) used in previous models has been recently challenged by observations and tracer measurements that suggest more complex processes are at play than have been considered so far (Smith et al. 2018; Boyd et al. 2019). Better parameterisation and spatial resolution of the contributions of lateral nutrient transfer (e.g. Frischknecht et al. 2018), particle-transport and carbon-remineralisation rates (Guidi et al. 2015), and water column chemosynthetic production (Middelburg 2011) will improve POC flux predictions. Parameterisation and ground-truthing of models for animal diel migration and *in situ* carbon fixation will be more challenging (e.g. De Leo et al. 2018), as they involve complex ecological couplings. Upgraded CIMP6 models are also expected to reduce uncertainties in the regional patterns of oxygen change that to reproduce observations in areas of oxygen deficiency (Stramma et al. 2012; Oschlies et al. 2017, 2018). The limited understanding of fundamental processes that govern the availability of oxygen is one of the most critical limitations in our ability to anticipate the impacts of climate

change on bathyal and abyssal ecosystems, independent of whether low-oxygen water forms in high-productivity oceanic regions or (Levin 2018; Breitburg et al. 2018).

9.2.2 Species sensitivity to change in natural abiotic conditions

Warming has been hypothesised as a major threat to deep-sea biodiversity (Yasuhara and Danovaro 2016; Sweetman et al. 2017). The vast majority of deep-sea habitats are thermally much more stable than shallow-water ones on daily to seasonal timescales seafloor temperatures strongly with depth and latitude. Temperature influences on deep-sea species are illustrated by unimodal or bimodal relationships linking diversity and temperature along these gradients in different oceanic basins (Yasuhara and Danovaro 2016). On this basis, important changes in species distribution are predicted from the warming of deep waters (Yasuhara and Danovaro 2016; Sweetman et al. 2017), and are consistent with past-extinction records of deep-sea fauna (Yasuhara et al. 2008). As temperature directly influences metabolic rates, both positive and negative responses are expected. Warming of deep waters, for instance, has allowed the invasion of the Antarctic shelf by the lithodid crab *Neolithodes yaldwyni* with major consequences for the resident benthic communities (Smith et al. 2012). Beyond the extension of the adult thermal niche, potential influences of food availability for the development and dispersal of larvae were also hypothesised (Thatje et al. 2005). The thermal tolerances of deep-sea species are, however, poorly known (Yasuhara and Danovaro 2016), with the exception of a few taxa thriving in extreme hydrothermal conditions or some cold-water coral species living at the edge of their thermal regime (Brooke et al. 2013). The sensitivity of deep-sea species to warming is thus expected to reflect more complex regional patterns, combining temperature increase with changes in resource supply and other climate stressors (Sperling et al. 2016).

Emerging issues in marine ecology relate to the synergistic effects of multiple climate stressors that could result in irreversible impacts on biodiversity and ecosystem functions (Gunderson et al. 2016; Boyd et al. 2018). How combined changes in temperature, oxygen level, and pH affect individual behaviour and physiological responses at individual or population level and how they impact species abundance and interaction networks within communities has been mostly studied for shallow-water species (Pörtner and Farrell 2008; Pörtner 2012; Somero 2010). Conceptual approaches to multistressor interactions, linking thermal tolerance to sensitivity to hypoxia, hypercapnia, or acidification, have been proposed from common physiological bases and field observations (Pörtner 2012). The applicability of these concepts to deep-sea species still needs validation. The assessment of deep-sea species vulnerability is still constrained by the limited knowledge of their physiology, genetic diversity, and ecology (Danovaro et al. 2014). Early studies addressed the thermal-tolerance range and physiological adaptations of deep-sea lineages in oxygen-depleted waters (e.g. Childress 1995; Somero 2010). To the question of specific adaptation of deep-sea fauna to low oxygen, early physiological studies on deep-sea taxa from OMZs, however, concluded that the response is far from intuitive and reflects a complex combination of factors (Childress 1995). Brown and Thatje (2015) identify depth as a key factor defining the fundamental ecological niches of species and adaptation to low-oxygen conditions, but their work mostly concerns Arctic shallow-water species.

To compensate for the lack of physiological studies, natural gradients can be used to assess species sensitivity to abiotic factors. A large set of studies concerning the habitat distribution of deep-sea species in oxygen-depleted waters is shedding light on this question. Gradients in oxygen are known to constrain benthic-species distribution and biodiversity severely along OMZ slopes (Levin 2003; Gooday et al. 2010; Sperling et al. 2016; Levin and Gallo 2019). The same is true for mesopelagic zooplankton in these regions, where very small changes in oxygenation can yield sharp changes in abundance and composition (Wishner et al. 2013, 2018).

Calcifying species structuring benthic habitats and, particularly, cold-water corals have benefited from specific research efforts in the recent years. Scleractinian corals considered particularly vulner-

able to changes in pH (Turley et al. 2007). Guinotte et al. (2006) and Guinotte (2008) reported that, on a global scale, 95 per cent of deep-water corals develop in water supersaturated with respect to aragonite, and when present below this threshold, they not form massive structures. The reduced abundance of scleractinians in the Pacific deep waters, and their replacement by octocorals in the available niche, support the assumption these corals would be particularly sensitive to the shoaling of the ASH (Guinotte and Fabri 2008). Yet, some scleractinian populations can thrive below this threshold, as reported on seamount flanks of the South (Thresher et al. 2011) and North Pacific (Baco et al. 2017) and in a Patagonian fjord (Fillinger and Richter 2013). Temperature is another key variable defining the deep-sea coral niche that constitutes a potential factor of vulnerability. Similarly to aragonite saturation, deep-sea scleractinian corals can occur above the limit of their common thermal niche (e.g. up to 15 °C in the North-west Atlantic and 20 °C in the Red Sea) (Brooke et al. 2013; Roder et al. 2013). These studies suggest the existence of physiological responses that permit corals to survive corrosive conditions or thermal extremes, providing that nutritional resources are available. However, dissolution and increased bioerosion of the nonliving components of cold-water coral reefs is a likely consequence of continued ocean acidification (IPCC 2019).

Running physiological and ecological experiments with deep-sea species, directly *in situ* or in high-pressure aquaria on-board research vessels, is difficult. Most long-term experiments are still performed at shore-based facilities at atmospheric pressure, limiting studies to species that can be acclimatised to depressurisation. Despite this limitation, progress has been made in the understanding of species responses to stressors. Progressively, the stress-exposure conditions challenging cold-water coral growth and survival are being clarified, revealing more complex responses than first expected. Physiological studies of deep-water scleractinian corals confirmed their capacity to maintain skeleton growth in corrosive conditions over 6 months in aquaria (Form and Riebesell 2012; Maier et al. 2013), despite a short-term decrease in growth rate observed in the first days (Lunden et al. 2014).

The combination of climate stressors may lead to more dramatic changes than acidification alone (Lunden et al. 2014; Gori et al. 2016). While acidification only reduced the calcification rate, an increased temperature and low-oxygen conditions have been shown to affect the survivorship of corals directly in 2 weeks (Lunden et al. 2014). Several studies also pointed to high variability in individual responses and suggested that inherent genetic diversity could lead to divergence in the fate of coral populations among regions (e.g. North Atlantic, Gulf of Mexico and Mediterranean Sea; Lunden et al. 2014; Kurman et al. 2017). Most of these studies are performed in stable mesocosm conditions that differ from fluctuating natural conditions (Roberts et al. 2009; Mienis et al. 2014). Growth-rate variability at seasonal to interannual scales has been revealed by *in situ* experiments and highlights distinct sensitivity of different coral species to environmental fluctuations (Lartaud et al. 2014, 2017). Concurrent changes in their food supply are likely to influence this response (Maier et al. 2019). The energetic cost of skeleton growth in corrosive natural conditions has been documented from experimental studies (Hennige et al. 2014; Maier et al. 2016). According to multifactorial studies, fitness would be more influenced by resource availability than exposure to low-pH conditions (Büscher et al. 2017). Increased metabolic rates and oxygen consumption by *Lophelia pertusa* (now as *Desmoplyllum pertusum*) as a result of temperature increase suggest that food supply could also greatly influence the response to thermal or hypoxic stress (Roberts et al. 2009). Key questions thus remain concerning the ability of corals to counter acidification, warming, and deoxygenation in areas combining these changes with increasing food limitation.

9.3 Identifying the drivers and impacts of climate change in deep-sea ecosystems

9.3.1 Export of organic resources at depth

The delivery of food resources to deep-sea biota has a tremendous influence on ecosystem functions (Smith et al. 2008). From an ecological perspective, not only the flux of POC delivered at the seafloor but also the quality of the food supply should be considered. More than the amount of carbon itself, the composition in lipids, sugars, proteins, and amino-acids determines the nutritional resource supplied at depth (Campanyà-Llovet et al. 2017). While the seasonality in organic carbon export can be parameterised from empirical models and satellite primary production data (Lutz et al. 2007), regional patterns of primary productivity are affected by climate change in a complex manner (Boyd et al. 2015), resulting in contrasted patterns in the alteration of diversity and activity of the phytoplankton community. Among the processes driving changes in particulate sinking rate is the size of dominant phytoplankton cells, and particularly the increase in picophytoplankton with the warming of surface waters (as observed from long-term time series in the North Atlantic Moran et al. 2010) or an unusual warm period in surface waters of a Pacific sub-Arctic region (Peña et al. 2019). Also important is their incorporation into aggregates (so-called mineral ballasting effect), relying on complex trophic pathways (Richardson and Jackson 2007). Major changes in the structure of marine ecosystems at shallow depths will therefore affect deep-sea ecosystems. The processes that propagate this disturbance to the deep sea are complex and multifactorial. Primary and secondary producers and all trophic interactions across the water column have to be accounted for. There are still many unknowns, but some key drivers are recognised that contribute to biological carbon sequestration (i.e. 'the biological carbon pump') (Boyd et al. 2019). Warming at shallow depths is expected to reduce the rate of primary production and enhance the degradation of the sinking organic particles (Henson et al. 2012), while acidification of surface waters affects dominant secondary producers in surface waters like pteropods (Orr et al. 2005). The expansion of oxygen-depleted waters will alter vertical migration and disrupt trophic networks by favouring hypoxia-tolerant predatory species (Stewart et al. 2014; Koslow et al. 2011).

Seasonal fluctuations in organic resources have been recognised in abyssal habitats for many years (Gooday 2002; Beaulieu 2002; Lampitt et al. 2010); this variability is often driven by episodic pulses occurring on daily to weekly scales (Smith et al. 2014, 2016, 2018; Pusceddu et al. 2013; Thomsen et al. 2017). Abyssal observatories have first shed light on the predominant role of episodic events supplying high amounts of organic material and their influence on the diversity and activities of deep-sea communities. Change in the intensity and frequency of these phenomena are reported as main drivers of climatic disturbance in abyssal plains of the Pacific and Atlantic (Smith et al. 2008; Smith et al. 2009, 2013, 2018). Submarine canyons are also experiencing resource pulses on short timescales with major ecological consequences (Pusceddu et al. 2010; Thomsen et al. 2017). Primary production and export flux from the euphotic zone similarly depends on complex mechanisms that are highly dependent on season and latitude. The importance of complex mechanisms of organic matter export at depth over space and time has been only recently for integration in models of carbon export at depth (Boyd et al. 2019). The diverse communities of heterotrophs driving transfer of resource to deep waters, the degradation of organic materials during sinking across the water column, introduce additional sensitivity of the export to changes in environmental factors such as temperature, oxygen, nitrate, and pH.

In this context, the role of chemoautotrophy (e.g. organic carbon production via bacteria and archaea using reduced chemicals such as ammonia, hydrogen, iron, methane, or sulphide) raises more questions about the nature and direction of changes in food supply to deep-sea ecosystems under climate change. Geological features hosting chemosynthetic productivity hotpots such as hydrothermal vents or methane seeps have the capacity to supply resources to oligotrophic deep-sea via the lateral export of particle organic carbon (Levin et al. 2016). In the context of potential cumulative impacts on these ecosystems from industrial activities and climate change, quantifying the influence of this production on the surrounding habitats is essential. This task, however, remains challenging as knowledge is largely lacking about the processes that govern the rates of production and transport of this production (Boetius and Wenzhöfer 2013; Le Bris et al. 2019). Beyond the importance of these particular areas of the seabed, the coupling of heterotrophy and chemoautotrophy in the conversion of poorly degradable organic material, such as wood and bones, into labile organic (Treude et al. 2009;

Kalenitchenko et al. 2016, 2018) and the recent account of archaeal and bacterial chemoautotrophic production in oxygenated deep waters and sediments (e.g. Sweetman et al. 2019) would also deserve specific attention.

9.3.2 Combination of climate stressors in space and time

The vulnerability of species to climate change is intimately linked with the range of abiotic environmental conditions that define the species niche (Doney et al. 2012; Bates et al. 2018). The natural variability characteristic of a species' environment is considered a valuable indicator of its capacity to acclimatise (individual scale) or adapt (over generations). This idea is captured in the concepts of time of emergence (when the climate signal emerges from natural variability) (Henson et al. 2017) and climate hazard (projected change from baseline conditions divided by 'natural variability' defined as the standard deviation of baseline (Jones and Cheung 2015; FAO 2019). The prevailing paradigm of a slowly changing deep-sea environment has been challenged by numerous observations in the past decades. Deep-sea-habitat abiotic properties are imprinted by diurnal, tidal, and seasonal cycles, but much less is known about deep-sea taxon exposure to variable environmental conditions than for their shallow relatives. The dynamics of climate stressors vary along gradients of depth, latitude, and longitude, as well as along geomorphological features. As a result of hydrodynamic forcing, the spatial and temporal heterogeneity of environmental conditions is intimately linked to complex seafloor topographies (e.g. seamounts and volcanoes, canyons, fracture zones, trenches, and ridges) (Turnewitsch et al. 2013; St Laurent and Thurnherr 2007). The combination of stressors also varies at 'landscape' scales because of the topographic features that generate specific hydrodynamic conditions driving local environmental variability (e.g. canyon heads and flanks summit and flanks of seamount or abyssal hill; Morris et al. 2016).

Additionally, the variability of physico-chemical factors (pH, oxygen, and POC) in deep-sea benthic habitats is frequently associated the biogenic structures formed by ecosystem engineers

(Buhl-Mortensen et al. 2010). These habitats are characterised by local fluid-flow constraints and high biological activity inducing fluctuations of abiotic factors over a large spectrum of frequencies. The biogenic habitats supporting energy-hotspot ecosystems such as vents and seeps, cold-water coral reefs, and sponge grounds often associated with tidal microclimatic conditions that depart from the environmental background (e.g. Davies et al. 2009; Nedoncelle et al. 2015; Roberts et al. 2018). All of these fluctuations are superimposed on seasonal and long-term trends and have profound influence on faunal growth rates and community structure (e.g. Davies et al. 2009; Soetaert et al. 2016; Nedoncelle et al. 2015; Cuvelier et al. 2017; Chauvet et al. 2018).

The prevalence of episodic extreme events as drivers of environmental variability has been recently emphasised. At abyssal depths in the Mediterranean, important changes in the diversity and abundance of nematodes in response to an abrupt warming were described following a deep convection event (Danovaro et al. 2004). At bathyal depths, submesoscale hydrodynamic fluctuations generated by weather extremes frequently underlie patterns of diversity and ecosystem function (Company et al. 2008; Davies et al. 2009; Pusceddu et al. 2013; Khripounoff et al. 2014; Chatzievangelou et al. 2016; Romano et al. 2017; Doya et al. 2017). Short-term events hence directly connect changes in the climate regime at the surface ocean to deep-sea ecosystems. The intensity of these events is expected to drive substantial seasonal to decadal variability in regional patterns. For instance, short-term oxygen decreases at depth up to 450 m have been described as a result of the transport of eddies across OMZs and their propagation off-shore (Stramma et al. 2014), while simulations confirm that seasonal modulation of oxygenation within the Peru OMZ can result from the formation of eddies extending at least down to 300 m (Vergara et al. 2016).

9.4 Required monitoring to forecast vulnerability

In many areas, particularly on continental margins and seamounts, deep-sea ecosystems are threatened by extensive habitat alteration by destructive fishing practices, leading to important efforts to develop protection measures and ensure sustain-

ability this activity (Huvenne et al. 2014; Roger et al. 2018, see Chapter 4). Extractive activities have also strengthened their regulations to minimise impacts, and industry is committed to monitoring these impacts over the complete exploration/exploitation cycle (Cordes et al. 2016, this volume). The development of all types of industrial activities at greater depths still challenges the capacity to suitably monitor these impacts. Potential cumulative threats with climate-change stressors further support the need for advanced monitoring strategies (Levin 2019). In the context of an emerging deep-sea mining industry (Chapter 5), this need is even more critical, as synergistic effects are expected. For example, deep coral and sponge gardens affected by warming may be more vulnerable to sedimentation or toxic-metal exposure, and the habitats impacted by fishing disturbance may be less amenable to restoration. Even chemosynthetic ecosystems, like hydrothermal vents and methane seeps, may be vulnerable to climate change through effects of warming, acidification, deoxygenation, and POC flux declines on mixotrophic adults, planktonic larvae, or symbiont–host interactions supported by high oxygen-consumption rates (Levin and Le Bris 2015; FAO 2019). Long-range impacts of mining activities may be further amplified by climate-related changes in water-mass properties (e.g. remobilisation of metals in an oxygen-depleted and more acidic water column), and these impacts themselves could further enhance climate stressors (e.g. acidification and deoxygenation due to metal sulphide dissolution).

So far, the resolution of ESM developed to address global or regional ocean dynamics is too limited to accurately reflect the effects of submesoscale phenomena on seafloor habitats; thus, the assessment of drivers of change for a particular deep-sea ecosystem cannot be directly derived from these projections. Prediction of climate-change impacts at depth will be improved by the monitoring of short-term components of deep-sea ecosystem dynamics. Not only should meteorological extremes like hurricanes, storms, or shelf-water cascades be accounted for, but also episodes of intense surface productivity such as salp and jellyfish blooms (Smith et al. 2014; Sweetman et al. 2014) associated with massive export of organic matter to the deep sea. Monitoring habitat sensitivity to climate change is of key importance to establish indicators of ecological status and their evolution, and to identify refugia or vulnerable ecosystems in the context of growing human pressure (FAO, 2019; Dunn et al. 2019; Levin 2019). Monitoring strategies should, however, ensure that, in addition to direct impact assessments, trends in climatic stressors are monitored at representative spatial and temporal scales (Gunderson et al. 2016). The impact of climate stressors on deep-sea ecosystem function and services typically multiple spatial scales, with typical distances ranging from a kilometre to thousands of kilometres (Figure 2). Deep-sea fisheries usually target high-productivity areas associated with small-scale oceanographic features, such as fronts and the resulting eddies, or with seafloor topographic heterogeneities on the order of few kilometres, such as seamounts or submarine canyons. Hydrocarbon-seepage areas and hydrothermal vent fields associated with fossil fuel reservoirs and metal resource accumulations typically span over the same scale. Even large abyssal areas considered for nodule exploitation display a certain degree of fragmentation, as shown by the influence of abyssal hills on organic resource supply (Morris et al. 2017). Assessing how climate change impacts deep-sea biodiversity and ecosystems

affected by human activities requires downscaling of climate predictions (generally 1° in resolution, i.e. about 100 km) to the scale of settings with critical biodiversity and ecosystem services: seamounts, canyons, abyssal hills, seepage areas, and hydrothermal vent fields, fractures, faults, or axial valleys. While some high-resolution ocean models have started to include surface processes into global climate modelling and should be increasingly accounted for in future large multimodel assessments in CIMP6 (Haarsma et al. 2016), the processes driving ocean internal variability at great depth are still far from being fully described and parameterised.

The habitat temporal fluctuations that are playing a key role in the functional responses of species also require specific attention (Bates et al. 2018). Ecological responses to disturbance, more generally rely on complex and dynamic interaction among species and between species and their environment, for which we lack mechanistic understanding. Migration of predators and disturbance of trophic guilds as a result of changes in temperature (Smith et al. 2012) or oxygen (Stewart et al. 2014), synergistic effects on habitat builders (e.g. the combination of corrosive conditions and warming or decrease in resources, deoxygenation, and acidification on

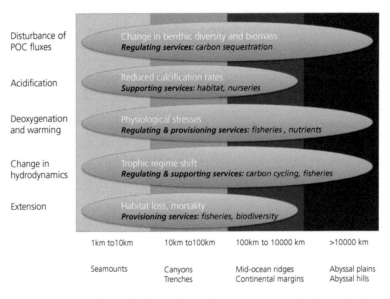

Figure 2 Scales of major climate stressors on deep-sea habitats and related impacts on ecosystem functions and services.

cold-water corals) are directly related to regional and local oceanographic features (e.g. coastal upwelling and dense-water cascading). Early studies have documented rapid changes in benthic community structure and functions combined with changes in carbon-export flux and temperature (Danovaro et al. 2004; Smith et al.). The mechanisms driving deep-sea ecosystem responses to continuous changes in abiotic conditions and resource fluxes generated by GHG accumulation in the atmosphere are still largely unknown, but the role of habitat builders such as coral reefs and gardens, sponge grounds, and carbonate bioherms are of particular concern as species local extinction and turnover is expected to have major ecological consequences. A recent study of cold-water coral cover conducted over 4 years on the Porcupine Seabight off Ireland has shown a significant decrease in abundance of scleractinian corals and increase in sponge cover, pointing to the potential disappearance of *Madrepora oculat*, a dominant coral species from the area (Boolukos et al. 2019). Local mesoscale and submesoscale (1–10 km) hydrodynamic processes associated with seafloor topography should also be better monitored, particularly in areas hosting habitat-structuring species such hydrothermal vent and methane seep invertebrates, cold-water coral, and sponge grounds. Understanding the 'microclimatic' conditions they create should be considered one particular research priority (Buhl-Mortensen et al. 2010), as the biogenic structures that drive hydrodynamic and biogeochemical processes have the potential to amplify or attenuate climatic threats (FAO 2019).

Long-term ecological surveys and process-based studies are needed to provide quantitative information to ground-truth and improve model parameterisation. Beside long-term observatories sustained by large infrastructures, a globally distributed strategy for the monitoring of ecosystems exposed to multiple anthropogenic pressures over all climate-sensitive areas is critically needed for management of seafloor activities (Levin et al. 2019). Areas targeted for impact management and protection measures can contribute to this monitoring effort and enhance regional or species-oriented assessments, though they should not replace the value of model ecosystems where long-term obser-

vation and mechanistic understanding is available (Smith et al. 2018). Cost-effective low-energy devices integrating chemical sensors (Nedoncelle et al. 2015; Kalenitchenko et al. 2018), cameras (Cuvelier et al. 2017; Chauvet et al. 2018), and onboard robotics and/or data treatment systems are increasing our capacity to build miniaturised observatories to characterise benthic process dynamics (Smith et al. 2013; Thomsen et al. 2017; Aguzzi et al. 2019). These data will complement repeated seafloor surveys and help us to better understand the drivers of change by bridging the expertise of deep-sea physical oceanographers, biogeochemists, and ecologists. This momentum benefits from the development of floats and gliders. Even in the most remote environments, miniaturised sensors on marine mammals can be used as oceanographic mobile platforms informing on ecological features (e.g. prey concentration) linked with oceanographic data (temperature, salinity, and more recently, oxygen) (e.g. up to a maximum of 2000 m depth for elephant seals in the Southern Ocean; Bras et al. 2017). The vast majority of observing networks are, however, focusing on interfaces with the atmosphere and land where photosynthetic productivity is confined. A significant effort is needed to expand long-term observations down to intermediate and abyssal depths, enhancing monitoring efforts on biodiversity hotspots (Danovaro et al. 2017). Long-term oceanographic times series are still rare at great depths, and it is particularly important to better document the high-frequency dynamics of deep-sea environments (Levin et al. 2019). Integration of oceanographic-ecological parameters both on and above the seafloor are furthermore required. Future observation networks on the seafloor will benefit from the innovative observing strategies combining current observing infrastructure with new technologies and advances in data management that are becoming available (e.g. long-range under-ice AUV and crawlers; Smith et al. 2016; Doya et al. 2017; Thomsen et al. 2017; Aguzzi et al. 2019). Emphasis by the Deep Ocean Observing Strategy on the integrated monitoring of physical, biogeochemical, and biological essential ocean variables offers promise for improved mechanistic understanding of climate-change impacts in the deep sea (Levin et al. 2019).

9.5 Climate policy and the deep sea

The deep ocean exists within territorial seas (e.g. around island nations and coastal regions with narrow shelves or incised by canyons), within most exclusive economic zones, and in vast areas beyond national jurisdiction. Each realm is governed by different laws and policies which are relevant to climate change (see Chapter 3). The United Nations Framework Convention on Climate Change (UNFCCC) was created to address 'dangerous' human interference with the climate system. Adopted at the Rio Earth Summit in 1992, it entered into force in March 1994 with 197 parties (ratified by 184 to date). The Paris Agreement (2015) builds upon this convention, to provide a global response to combat climate change and adapt to its effects by reducing CO_2 emissions and keeping temperature increases below 2 °C, with assistance for developing countries to undertake the effort. The Paris Agreement requires nationally determined contributions (NDCs), allowing each country to set how and when it will achieve climate mitigation and adaptation.

The ocean is mentioned and considered only minimally in these legal instruments, despite the fact that the ocean is a major climate mitigator. The ocean has taken up 26–30 per cent of excess CO_2 from the atmosphere, and 93 per cent of the excess heat (IPCC 2014, 2018), with accelerating rates of warming and redistribution of heat into the ocean interior (Cheng et al. 2019). Nineteen per cent of this heat increase resides below 2000 m (Talley et al. 2016; Glecker et al. 2016). The Paris Agreement notes the importance of ensuring the integrity of all ecosystems, including oceans, and the protection of biodiversity. The ocean appears in Subsidiary Body for Scientific and Technical Advice discussions, but until recently, it was not discussed in UNFCCC texts or at Conference of Parties (COP) plenaries or formal side events. This has changed over the past 5 years, with attention being drawn to the importance of the ocean in climate mitigation and adaptation, through states' initiatives (e.g. Because the Ocean, Oceans Pathway), high-level statements in COP 24 plenaries by environment ministers (2018), and workshops, Ocean Day events, and a myriad of side events and exhibits by sponsored states, academic institutions, and nongovernmental organisations (NGOs). Over 70 per cent of NDCs mention the ocean, although some of the large developed-nation NDCs do not (Gallo et al. 2017). However, the deep ocean is almost never mentioned. Intergovernmental panel on climate change (IPCC) assessment reports (AR1 to AR5), produced every 6 years, have provided increasing attention to the ocean over the years, but the deep ocean has received little emphasis. The IPCC Special Report on Oceans and Cryosphere (2019), however, evaluates climate impacts on a host of deep-ocean ecosystems and, to a lesser extent, addresses consequences for ecosystem services. Many of findings of this report concerning impacts of climate change on deep-sea ecosystems have 'low confidence', highlighting the need for expanded observation and research on this topic.

Technically the jurisdiction of the UNFCCC and the Paris Agreement includes only national lands and waters. This means that no political entity is responsible for climate change in the high seas or the international deep seabed—either the regulation of climate-altering emissions, the impacts on the ecosystems, or adaptation of human activities to the impacts. The United Nations Convention on the Law of the Sea (1982) includes a provision for states *to protect and preserve the marine environment* (Article 192) throughout the entire ocean, and to *prevent, reduce and control marine pollution* (Article 194). These could apply to harmful effects of greenhouse gases on the marine environment, with prosecution of violations possible under the binding dispute settlement process (Doelle 2006). Currently the International Maritime Organization addresses emissions from international shipping, as well as dumping linked to climate geoengineering through the London Convention and but other international activities are unregulated with respect to climate change.

There are, however, a number of UN resolutions and current policy developments unfolding internationally that have or can bring attention to climate influences on various sectoral activities in the deep ocean (Figure 3). For example the United Nations General Assembly resolution 71/123 (2016; para185) calls upon States, individually and through regional fisheries management organisations and arrange-

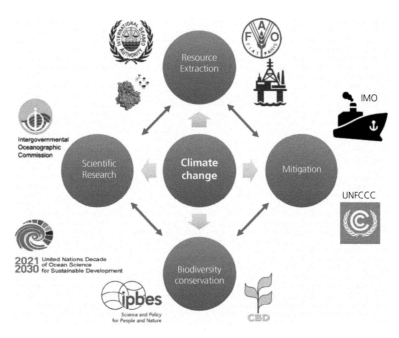

Figure 3 Different organisations that could incorporate climate into their management of the deep sea.

ments, *'to take into account the potential impacts of climate change and ocean acidification in taking measures to manage deep-sea fisheries and protect vulnerable marine ecosystems'*. This encouraged the Food and Agricultural Organization of the United Nations (FAO) to work with the Deep-Ocean Stewardship Initiative (DOSI) Climate Working Group to conduct a study and develop recommendations regarding potential climate change impacts on bottom habitats, fish and fisheries at depths of 500–2500 m, focusing on Regional Fisheries Management Organization regions (FAO 2019).

In advance of the eighteenth meeting of the UN open-ended Informal Consultative Process on Oceans and the Law of the Sea in May 2017, titled *'The Effects of Climate Change on Oceans'*; the Division for Ocean Affairs; and the Law of the Sea (DOALOS) called for input. The UNFCCC submission acknowledged the importance of deep-ocean warming in sea-level rise and ecosystem health, and of deep circulation on nutrient supply, heat, and CO_2 uptake. The response of the International Seabed Authority notes potential for climate impacts to interact with seabed mining, and the importance of distinguishing climate change from mining-induced perturba-

tions. They acknowledge that seabed-mining disturbances (sedimentation, plume effects, and metal release) could act cumulatively with climate-induced warming, acidification, and deoxygenation, affecting resilience and recovery rates (see Chapter 5). The designation of no-mining (protected) areas and design of mining configurations could incorporate the need for corridors and refugia. A recently proposed framework for design and evaluation of protected-area networks to support conservation of benthic ecosystems on ridges introduces climate-induced change in evaluation of different spatial scenarios (Dunn et al. 2018).

Within the Biodiversity realm, the Convention on Biodiversity has identified Ecologically or Biologically Significant Marine Areas (EBSAs) through a series of regional workshops. The seven criteria do not include vulnerability to climate change; however, a single EBSA (High Aragonite Saturation State Zone Western South Pacific) was described based on projected high aragonite-saturation levels under future climate change. It focused on providing refugia for shallow-water corals. Nowhere do vulnerabilities of deep-sea ecosystems to climate change emerge in the EBSA deliberations, but Johnson and Kenchington (2019) have proposed climate refugia as an eighth EBSA criterion. Climate change may also reduce habitat suitability for species and assemblages that underlie EBSA designation, or shift the location for those EBSAs that rely on mobile oceanographic features such as fronts and the associated eddies. In the North Atlantic, projected changes in climate are shown to impact nearly all existing Marine Protected Areas in national and international waters by 2050 (Johnson et al. 2018). The spatially variable EBSA designation may be expanded to incorporate this concept.

This is important because EBSA workshops may provide the starting point for spatial planning within a broader international treaty on the conservation and sustainable use of biodiversity beyond national jurisdiction (BBNJ process—see Chapter 7). Negotiations have begun at the UN, and will address four issues, most of which need to include climate considerations: area-based management tools, environmental impact assessment, marine genetic resources, and capacity building and technology transfer. Observational and modelled climate projections can identify areas most resistant to climate change that can act as climate refugia, identify the need for migratory corridors between suitable habitats, or identify areas most likely to experience climate stress (Dunn et al. 2019). Climate considerations in spatial planning may involve protection of ecosystems with high carbon (CO_2 or CH_4)–sequestration capacity, or reducing activities where they most exacerbate climate change (Levin 2019). Impact assessment and impact monitoring need to incorporate parameters that can detect the effects of climate on ecological baseline data, assess the climate footprint of human activities and their environmental impacts, and identify synergistic and cumulative impacts. Open, public, and transparent access to climate-relevant data is fundamental to building climate into sustainable use of natural capital in the deep ocean (see Chapter 1 for definition of 'natural capital in the deep ocean').

The large repository of genetic resources in the deep sea, as acknowledged in the BBNJ process, may come under threat from climate change. Many see establishment of large protected areas in international waters as one pathway to conserve biodiversity in the face of multiple threats, but this approach faces many challenges (De Santo 2018). Many West Pacific island nations are considering developing their deep-sea mineral resources within national jurisdictions (e.g. Papua New Guinea, Tonga, Cook Islands, Solomon Islands, Kiribati, etc.), although Fiji have just declared a 10-year moratorium on deep-seabed mining within its waters to allow time for robust scientific research. Deep-sea fishing rights are also transferred between developing and developed countries as a source of income.

There is a need to grow capacity in, and transfer technology to, developing countries to monitor, mitigate, and adapt to climate change, as these deep-blue economic activities accelerate, in order to achieve UN Sustainable Development Goal 14 to 'Conserve and sustainably use the oceans, seas and marine resources for sustainable development'. Creating linkages between industrial-sector activity, other stakeholders, and climate mitigation will be especially challenging in developing nations. Increased access to green (and blue) climate funds to support predictive modelling and deployment and maintenance of observing system platforms and sensors, as well as the adoption of climate-conscious best practices, can help (Levin 2019).

9.6 Conclusion

The ocean is the largest biome on Earth. Alteration of the circulation regime and hydrologic properties of waters and organic resources may lead to massive loss of biodiversity, as revealed by past large-scale extinctions affecting marine biota. The diversity and patchy distribution of habitats, including abyssal plains, continental slopes, ridges, trenches, volcanoes, seamounts, and canyons, within which ecosystem diversity is far from being completely described, serve as home for a large variety of fauna and unicellular organisms including bacteria and archaea that have evolved in the deep ocean over 4 billion years. In about 2 centuries of deep-sea exploration, scientists have only scratched the surface of this unique biological heritage (Ramirez-Llodra et al. 2010; Danovaro et al. 2014). This diversity challenges our capacity to understand and predict the responses to environmental change and how these responses will differ from those in shallow water.

The rising economic momentum towards the exploitation of marine resources (i.e. European concept of 'Blue Economy'), expanding far away from land to the deep ocean, creates an imperative for improved monitoring and forecasting of climate-change impacts on deep-sea and their feedbacks on greenhouse gases fluxes to the atmosphere. Climate change is a multistressor disturbance that affects nearly all aspects of life in the ocean. Accounting for the effects of stressor combinations, including warming, deoxygenation, acidification, and change in nutritional resources, in addition to direct impacts (e.g. habitat destruction, turbidity, overfishing, and exposure to toxic compounds) is essential to achieve any conservation, mitigation, or protection goals. The combination of climate stressors at depth is heterogeneously delineated, with specific links to change in surface productivity superimposed on changes in water-mass properties. To this should be added the local hydrodynamic properties associated with topographic features that characterise the ocean floor. Accurate prediction of changes at relevant scales and modelling of complex ecosystem responses still remain a significant challenge. But the nature and intensity of climate disturbance can be explored by analysing the combination of factors predicted from 'climate models' both on a mechanistic and observational basis. Modelling responses arising from the understanding of the complex interaction network of organisms and their habitats are needed. This new great challenge in ocean science deserves a decade of research effort that expands well beyond the short timelines of most research projects in the deep sea.

Acknowledgements

LL's contribution was supported by grants OCE 1634172 and OCE1829623, and she benefited from a workshop supported by the JM Kaplan Foundation. Would like to acknowledge the Ocean and Climate Platform and the DOSI Climate Change working group.

References

Aguzzi, J., Chatzievangelou, D., Marini, S., et al. (2019). New High-tech Flexible Networks for the Monitoring of Deep-Sea Ecosystems. *Environmental Science and Technology*. doi: 10.1021/acs.est.9b00409.

Baco, A. R., Morgan, N., Roark, E. B., et al. (2017). Defying Dissolution: Discovery of Deep-Sea Scleractinian Coral Reefs in the North Pacific. *Scientific Reports*, 7(1). doi: 10.1038/s41598-017-05492-w

Bates, A. E., Helmuth, B., Burrows, M. T., et al. (2018). Biologists Ignore Ocean Weather at Their Peril. *Nature*, 560(7718), 299–301. doi: 10.1038/d41586-018-05869-5.

Beaulieu, S. (2002). Accumulation and Fate of Phytodetritus on the Sea Floor. *Oceanography and Marine Biology, An Annual Review*, 40, 171–232. doi: 10.1201/9780203180594.ch4

Boetius, A. and Wenzhöfer, F. (2013). Seafloor Oxygen Consumption Fuelled by Methane from Cold Seeps. *Nature Geoscience*, 6, 725–34. doi: 10.1038/ngeo1926.

Boolukos, C. M., Lim, A., O'Riordan, R. M., and Wheeler, A. J. (2019). Cold-water Corals in Decline—A Temporal (4 Year) Species Abundance and Biodiversity Appraisal of Complete Photomosaiced Cold-water Coral Reef on the Irish Margin. *Deep-Sea Research I*, 146, 44–54. doi: 10.1016/j.dsr.2019.03.004.

Bopp, L., Resplandy, L., Orr, J. C., et al. (2013). Multiple Stressors of Ocean Ecosystems in the 21st Century: Projections with CMIP5 Models. *Biogeosciences*, 10, 6225–45. doi: 10.5194/bg-10-6225-2013.

Boyd, P. W., Lennartz, S. T., Glover, D. M., and Doney, S. C. (2015). Biological Ramifications of Climate-change-mediated Oceanic Multi-stressors. *Nature Climate Change*, 5, 71–9. doi: 10.1038/nclimate2441.

Boyd, P. W., Collins, S., Dupont, S., et al. (2018). Experimental Strategies to Assess the Biological Ramifications of Multiple Drivers of Global Ocean Change: A Review. *Global Change Biology*, 24, 2239–61. doi: 10.1111/gcb.14102.

Boyd, P. W., Claustre, H., Levy, M., Siegel, D. A., and Weber, T. (2019). Multi-faceted Particle Pumps Drive Carbon Sequestration in the Ocean. *Nature*, 568, 327–35. doi: 10.1038/s41586-019-1098-2.

Bras, Y. L., Jouma'a, J., and Guinet, C. (2017). Three-dimensional Space Use During the Bottom Phase of Southern Elephant Seal Dives. *Movement Ecology*, 5(1). doi: 10.1186/s40462-017-0108-y.

Brooke, S., Ross, S. W., Bane, J. M., et al. (2013). Temperature Tolerance of the Deep-sea Coral *Lophelia pertusa* from the Southeastern United States. *Deep-Sea Research II*, 92, 240–8. doi: 10.1016/j.dsr2.2012.12.001.

Brown, A. and Thatje, S. (2015). The Effects of Changing Climate on Faunal Depth Distributions Determine Winners and Losers. *Global Change Biology*, 21, 173–80. doi: 10.1111/gcb.12680.

Buhl-Mortensen, L., Vanreusel, A., Gooday, A. J., et al. (2010). Biological Structures as a Source of Habitat

Heterogeneity and Biodiversity on the Deep Ocean Margins: Biological Structures and Biodiversity. *Marine Ecology*, 31, 21–50. doi: 10.1111/j.1439-0485.2010.00359.x.

Büscher, J. V., Form, A. U., and Riebesell, U. (2017). Interactive Effects of Ocean Acidification and Warming on Growth, Fitness and Survival of the Cold-Water Coral *Lophelia pertusa* under Different Food Availabilities. *Frontiers in Marine Science*, 4. doi: 10.3389/fmars.2017.00101.

Byrne, R. H., Mecking, S., Feely, R. A., and Liu, X. (2010). Direct Observations of Basin-wide Acidification of the North Pacific Ocean: pH Changes in the North Pacific. *Geophysical Research Letters*, 37(2), L02601. doi: 10.1029/2009GL040999.

Campanyà-Llovet, N., Snelgrove, P. V. R., and Parrish, C. C. (2017). Rethinking the Importance of Food Quality in Marine Benthic Food Webs. *Progress in Oceanography*, 156, 240–51. doi: 10.1016/j.pocean.2017.07.006.

Chatzievangelou, D., Doya, C., Thomsen, L., Purser, A., and Aguzzi, J. (2016). High-Frequency Patterns in the Abundance of Benthic Species Near a Cold-Seep—An Internet Operated Vehicle Application. *PLoS ONE*, 11(10), e0163808. doi: 10.1371/journal.pone.0163808.

Chauvet, P., Metaxas, A., Hay, A. E., and Matabos, M. (2018). Annual and Seasonal Dynamics of Deep-sea Megafaunal Epibenthic Communities in Barkley Canyon (British Columbia, Canada): A Response to Climatology, Surface Productivity and Benthic Boundary Layer Variation. *Progress in Oceanography*, 169, 89–105. doi: 10.1016/j.pocean.2018.04.002.

Chen, C.A., Lui, H., Hsieh, C., et al. (2017). Deep Oceans May Acidify Faster than Anticipated Due to Global Warming. *Nature Climate Change* 7, 890–4 doi:10.1038/s41558-017-0003-y.

Cheng, L., Abraham, J., Hausfather, Z., and Trenberth, K. E. (2019). How Fast Are the Oceans Warming? *Science*, 363(6423), 128–9. doi: 10.1126/science.aav7619.

Childress, J. J. (1995). Are There Physiological and Biochemical Adaptations of Metabolism in Deep-sea Animals? *Trends in Ecology and Evolution*, 10, 30–6. doi: 10.1016/S0169-5347(00)88957-0.

Cocco, V., Joos, F., Steinacher, M., et al. (2013). Oxygen and Indicators of Stress for Marine Life in Multi-model Global Warming Projections. *Biogeosciences*, 10, 1849–68. doi: 10.5194/bg-10-1849-2013.

Company, J. B., Puig, P., Sardà, F., et al. (2008). Climate Influence on Deep Sea Populations. *PLoS ONE*, 3(1), e1431 doi: 10.1371/journal.pone.0001431.

Cuvelier, D., Legendre, P., Laës-Huon, A., Sarradin, P-M., and Sarrazin, J. (2017). Biological and Environmental Rhythms in (Dark) Deep-sea Hydrothermal Ecosystems. *Biogeosciences*, 14, 2955–77. doi: 10.5194/bg-14-2955-2017.

Davies, A. J., Duineveld, G. C. A., Lavaleye, M. S. S., et al. (2009). Downwelling and Deep-water Bottom Currents as Food Supply Mechanisms to the Cold-water Coral *Lophelia pertusa* (Scleractinia) at the Mingulay Reef Complex. *Limnology and Oceanography*, 54, 620–9. doi: 10.4319/lo.2009.54.2.0620.

Danovaro, R., Dell'Anno, A., and Pusceddu, A. (2004). Biodiversity Response to Climate Change in a Warm Deep Sea: Biodiversity and Climate Change in the Deep Sea. *Ecology Letters*, 7, 821–8. doi: 10.1111/j.1461-0248.2004.00634.x.

Danovaro, R., Snelgrove, P. V. R., and Tyler, P. (2014). Challenging the Paradigms of Deep-sea Ecology. *Trends in Ecology and Evolution*, 29, 465–75. doi: 10.1016/j.tree.2014.06.002.

De Leo, F. C., Ogata, B., Sastri, A. R., et al. (2018). High-frequency Observations from a Deep-sea Cabled Observatory Reveal Seasonal Overwintering of *Neocalanus* spp. in Barkley Canyon, NE Pacific: Insights into Particulate Organic Carbon Flux. *Progress in Oceanography*, 169, 120–37. doi: 10.1016/j.pocean.2018.06.001.

De Santo, E. M. (2018). Implementation Challenges of Area-based Management Tools (ABMTs) for Biodiversity Beyond National Jurisdiction (BBNJ). *Marine Policy*, 97, 34–43. doi: 10.1016/j.marpol.2018.08.034.

Doelle, M. (2006). Climate Change and the Use of the Dispute Settlement Regime of the Law of the Sea Convention. *Ocean Development and International Law*, 37, 319–37. doi: 10.1080/00908320600800945.

Doney, S. C., Ruckelshaus, M., Emmett Duffy, J., et al. (2012). Climate Change Impacts on Marine Ecosystems. *Annual Review of Marine Science*, 4, 11–37. doi: 10.1146/annurev-marine-041911-111611.

Doya, C., Chatzievangelou, D., Bahamon, N., et al. (2017). Seasonal Monitoring of Deep-sea Megabenthos in Barkley Canyon Cold Seep by Internet Operated Vehicle (IOV). *PLoS ONE*, 12(5), e0176917. doi: 10.1371/journal.pone.0176917.

Dutkiewicz, S., Scott, J. R., and Follows, M. J. (2013). Winners and Losers: Ecological and Biogeochemical Changes in a Warming Ocean. *Global Biogeochemical Cycles*, 27, 463–77. doi: 10.1002/gbc.20042.

Dunn, D. C., Van Dover, C., Etter, R. J., et al. (2018). A Strategy for the Conservation of Biodiversity on Mid-ocean Ridges from Deep-sea Mining. *Science Advances* 4(7), eaar4313n. doi: 10.1126/sciadv.aar4313.

Fillinger, L. and Richter, C. (2013). Vertical and Horizontal Distribution of *Desmophyllum dianthus* in Comau Fjord, Chile: A Cold-water Coral Thriving at Low pH. *PeerJ*, 1, e194. doi: 10.7717/peerj.194.

FAO (2019). Deep-ocean Climate Change Impacts on Habitat, Fish and Fisheries, in L. Levin, M. Baker, and

A. Thompson (eds.) FAO Fisheries and Aquaculture Technical Paper No. 638. Rome: FAO, p. 186. Licence: CC BY-NC-SA 3.0 IGO.

Form, A. U. and Riebesell, U. (2012). Acclimation to Ocean Acidification During Long-term CO_2 Exposure in the Cold-water Coral *Lophelia pertusa*. *Global Change Biology*, 18, 843–53. doi: 10.1111/j.1365-2486.2011.02583.x.

Frischknecht, M., Münnich, M., and Gruber, N. (2018). Origin, Transformation, and Fate: The Three-Dimensional Biological Pump in the California Current System. *Journal of Geophysical Research: Oceans*, 123, 7939–62. doi: 10.1029/2018JC013934.

Gallo, N. D., Victor, D. G., and Levin, L. A. (2017). Ocean Commitments Under the Paris Agreement. *Nature Climate Change*, 7, 833–8. doi: 10.1038/nclimate3422.

García-Ibáñez, M. I., Zunino, P., Fröb, F., et al. (2016). Ocean Acidification in the Subpolar North Atlantic: Rates and Mechanisms Controlling pH Changes. *Biogeosciences*, 13, 3701–15. doi: 10.5194/bg-13-3701-2016.

Gattuso, J-P., Magnan, A., Bille, R., et al. (2015). Contrasting Futures for Ocean and Society from Different Anthropogenic CO_2 Emissions Scenarios. *Science*, 349(6243). doi: 10.1126/science.aac4722.

Gattuso, J-P., Magnan, A. K., Bopp, L., et al. (2018). Ocean Solutions to Address Climate Change and Its Effects on Marine Ecosystems. *Frontiers in Marine Science*, 5. doi: 10.3389/fmars.2018.00337.

Gehlen, M., Séférian, R., Jones, D. O. B., et al. (2014). Projected pH Reductions by 2100 Might Put Deep North Atlantic Biodiversity at Risk. *Biogeosciences*, 11, 6955–67. doi: 10.5194/bg-11-6955-2014

Georgian, S. E., Dupont, S., Kurman, M., et al. (2016). Biogeographic Variability in the Physiological Response of the Cold-water Coral *Lophelia pertusa* to Ocean Acidification. *Marine Ecology*, 37, 1345–59. doi: 10.1111/maec.12373.

Gilly, W. F., Beman, J. M., Litvin, S. Y., and Robison, B. H. (2013). Oceanographic and Biological Effects of Shoaling of the Oxygen Minimum Zone. *Annual Review of Marine Science*, 5, 393–420. doi: 10.1146/annurev-marine-120710-100849.

Gleckler, P. J., Durack, P. J., Stouffer, R. J., Johnson, G. C., and Forest, C. E. (2016). Industrial-era Global Ocean Heat Uptake Doubles in Recent Decades. *Nature Climate Change*, 6(4), 394–8. doi: 10.1038/nclimate2915 (Accessed 21 February 2016).

Glover, A. G., Gooday, A. J., Bailey, D. M., et al. (2010). Temporal Change in Deep-Sea Benthic Ecosystems. *Advances in Marine Biology*, 58, 1–95. doi: 10.1016/B978-0-12-381015-1.00001-0.

Gooday, A. J. (2002). Biological Responses to Seasonally Varying Fluxes of Organic Matter to the Ocean Floor: A Review. *Journal of Oceanography*, 58, 305–32. doi: 10.1023/A:1015865826379

Gooday, A. J., Bett, B. J., Escobar, E., et al. (2010). Habitat Heterogeneity and Its Influence on Benthic Biodiversity in Oxygen Minimum Zones: Habitat Heterogeneity and Its Influence on Biodiversity in OMZs. *Marine Ecology*, 31, 125–47. doi: 10.1111/j.1439-0485.2009.00348.x.

Gori, A., Ferrier-Pagès, C., Hennige, S. J., et al. (2016). Physiological Response of the Cold-water Coral *Desmophyllum dianthus* to Thermal Stress and Ocean Acidification. *PeerJ*, 4, e1606. doi: 10.7717/peerj.1606.

Guidi, L., Legendre, L., Reygondeau, G., Uitz, J., Stemmann, L., and Henson, S. A. (2015). A New Look at Ocean Carbon Remineralization for Estimating Deepwater Sequestration. *Global Biogeochemical Cycles*, 29, 1044–59. doi: 10.1002/2014GB005063.

Guinotte, J. M., Orr, J. C., Cairns, S., Freiwald, A., Morgan, L., and George, R. (2006). Will Human-induced Changes in Seawater Chemistry Alter the Distribution of Deep-sea Scleractinian Corals? *Frontiers in Ecology and the Environment*, 4, 141–6.

Gunderson, A. R., Armstrong, E. J., and Stillman, J. H. (2016). Multiple Stressors in a Changing World: The Need for an Improved Perspective on Physiological Responses to the Dynamic Marine Environment. *Annual Review of Marine Science*, 8, 357–78. doi: 10.1146/annurev-marine-122414-033953.

Hassoun, A. E. R., Gemayel, E., Krasakopoulou, E., et al. (2015). Acidification of the Mediterranean Sea from Anthropogenic Carbon Penetration. *Deep-Sea Research I*, 102, 1–15. doi: 10.1016/j.dsr.2015.04.005.

Hennige, S. J., Wicks, L. C., Kamenos, N. A., et al. (2014). Short-term Metabolic and Growth Responses of the Cold-water Coral *Lophelia pertusa* to Ocean Acidification. *Deep-Sea Research II*, 99, 27–35. doi:10.1016/j.dsr2.2013.07.005.

Henson, S. A., Sanders, R., and Madsen, E. (2012). Global Patterns in Efficiency of Particulate Organic Carbon Export and Transfer to the Deep Ocean. *Global Biogeochemical Cycles*, 26(1), doi: 10.1029/2011GB004099.

Henson, S. A., Beaulieu, C., Ilyina, T., et al. (2017). Rapid Emergence of Climate Change in Environmental Drivers of Marine Ecosystems. *Nature Communications*, 8(1). doi: 10.1038/ncomms14682

Hoegh-Guldberg, O., Jacob, D., Taylor, M., et al. (2018) Impacts of 1.5°C Global Warming on Natural and Human Systems, in V. Masson-Delmotte, P. Zhai, H-O. Pörtner, et al. (eds.) *Global Warming of 1.5°C. An IPCC Special Report on the Impacts of Global Warming of 1.5°C Above Pre-industrial Levels and Related Global Greenhouse Gas Emission Pathways, in the Context of Strengthening the Global Response to the Threat of Climate*

Change, Sustainable Development, and Efforts to Eradicate Poverty. Geneva, Switzerland: World Meteorological Organization, pp. 175–313.

International Council for Science (2017). A Guide to SDG Interactions: From Science to Implementation, International Council for Science (ICSU). Paris: International Council for Science. doi: 10.24948/2017.01.

IPCC (2014). *Climate Change 2014: Impacts, Adaptation, and Vulnerability. Contribution of Working Group II to the Fifth Assessment Report of the Intergovernmental Panel on Climate Change.* C. B. Field, V. R. Barros, D. J. Dokken, et al. (eds.). Cambridge: Cambridge University Press.

IPCC (2018). Global Warming of 1.5°C, in V. Masson-Delmotte, P. Zhai, H. O. Pörtner, et al. (eds.) *An IPCC Special Report on the Impacts of Global Warming of 1.5°C Above Pre-industrial Levels and Related Global Greenhouse Gas Emission Pathways, in the Context of Strengthening the Global Response to the Threat of Climate Change, Sustainable Development, and Efforts to Eradicate Poverty.* : IPCC.

IPCC (2019). Summary for Policymakers, in H. O. Pörtner, D. C. Roberts, V. Masson-Delmotte, et al. (eds.) IPCC.

Johnson, D. E. and Kenchington E. L. (2019). Should Potential for Climate Change Refugia Be Mainstreamed into the Criteria for Describing EBSAs? *Conservation Letters*, 11 February, e12634. https://doi.org/10.1111/conl.12634.

Johnson, D. E., Adelaide Ferreira, M., and Kenchington, E. (2018). Climate Change Is Likely to Severely Limit the Effectiveness of Deep-sea ABMTs in the North Atlantic. *Marine Policy*, 87, 111–22. doi: 10.1016/j.marpol.2017.09.034.

Jones, D. O. B., Yool, A., Wei, C-L., et al. (2014). Global Reductions in Seafloor Biomass in Response to Climate Change. *Global Change Biology*, 20, 1861–72. doi: 10.1111/gcb.12480.

Jones, M. C. and Cheung, W. W. L. (2015). Multi-model Ensemble Projections of Climate Change Effects on Global Marine Biodiversity. *ICES Journal of Marine Science*, 72, 741–52. doi: 10.1093/icesjms/fsu172.

Kalenitchenko, D., Dupraz, M., Le Bris, N., et al. (2016). Ecological Succession Leads to Chemosynthesis in Mats Colonizing Wood in Sea Water. *The ISME Journal*, 10, 2246–58. doi: 10.1038/ismej.2016.12.

Kalenitchenko, D., Péru, E., Contreira Pereira, L., et al. (2018). The Early Conversion of Deep-sea Wood Falls into Chemosynthetic Hotspots Revealed by *in Situ* Monitoring. *Scientific Reports*, 8(1). doi: 10.1038/s41598-017-17463-2.

Khripounoff, A., Caprais, J-C., Le Bruchec, J., et al. (2014). Deep Cold-water Coral Ecosystems in the Brittany Submarine Canyons (Northeast Atlantic): Hydrodynamics, Particle Supply, Respiration, and

Carbon Cycling. *Limnology and Oceanography*, 59, 87–98. doi: 10.4319/lo.2014.59.1.0087.

Koslow, J., Goericke, R., Lara-Lopez, A., and Watson, W. (2011). Impact of Declining Intermediate-water Oxygen on Deepwater Fishes in the California Current. *Marine Ecology Progress Series*, 436, 207–18. doi: 10.3354/meps09270.

Kurman, M. D., Gómez, C. E., Georgian, S. E., Lunden, J. J., and Cordes, E. E. (2017). Intra-Specific Variation Reveals Potential for Adaptation to Ocean Acidification in a Cold-water Coral from the Gulf of Mexico. *Frontiers in Marine Science*, 4. doi: 10.3389/fmars.2017.00111.

Lampitt, R. S., Salter, I., de Cuevas, B. A., et al. (2010). Long-term Variability of Downward Particle Flux in the Deep Northeast Atlantic: Causes and Trends. *Deep-Sea Research II*, 57, 1346–61. doi: 10.1016/j.dsr2.2010.01.011.

Lartaud, F., Pareige, S., de Rafelis, M., et al. (2014). Temporal Changes in the Growth of Two Mediterranean Cold-water Coral Species, in Situ and in Aquaria. *Deep-Sea Research II*, 99, 64–70. doi: 10.1016/j.dsr2.2013.06.024.

Lartaud, F., Meistertzheim, A. L., Peru, E., and Le Bris, N. (2017). In Situ Growth Experiments of Reef-building Cold-water Corals: The Good, the Bad and the Ugly. *Deep-Sea Research I*, 121, 70–8. doi: 10.1016/j.dsr.2017.01.004.

Le Bris, N., Yücel, M., Das, A., et al. (2019). Hydrothermal Energy Transfer and Organic Carbon Production at the Deep Seafloor. *Frontiers in Marine Science*, 5. doi: 10.3389/fmars.2018.00531

Le Quéré, C., Andrew, R. M., Friedlingstein, P., et al. (2018). Global Carbon Budget 2018. *Earth System Science Data*, 10, 2141–94. doi: 10.5194/essd-10-2141-2018.

Levin, L. A. (2003). Oxygen Minimum Zone Benthos: Adaptation and Community Response to Hypoxia. *Oceanography and Marine Biology: An Annual Review*, 41, 1–45.

Levin, L. A. (2018). Manifestation, Drivers, and Emergence of Open Ocean Deoxygenation. *Annual Review of Marine Science*, 10, 229–60. doi: 10.1146/annurev-marine-121916-063359.

Levin, L. (2019). Sustainability in Deep Water: The Challenges of Climate Change, Human Pressures, and Biodiversity Conservation. *Oceanography*, 32, 170–80. doi: 10.5670/oceanog.2019.224.

Levin, L. A., Bett, B. J., Gates, A. R., et al. (2019). Global Observing Needs in the Deep Ocean. *Frontiers in Marine Science*, 6. doi: 10.3389/fmars.2019.00241.

Levin, L. A. and Gallo, N. (2019). IUCN Ch. 8.5 Continental Margin Benthic and Demersal Biota, in D. Laffoley and J. Baxter (eds.) IUCN, *Ocean Deoxygenation: Everyone's Problem. Causes, Impacts, Consequences and Solutions.* In press.

Levin, L. A. and Le Bris, N. (2015). The Deep Ocean Under Climate Change. *Science*, 350(6262), 766–8.

Levin, L. A., Whitcraft, C. R., Mendoza, G. F., Gonzalez, J. P., and Cowie, G. (2009). Oxygen and Organic Matter Thresholds for Benthic Faunal Activity on the Pakistan Margin Oxygen Minimum Zone (700–1100m). *Deep-Sea Research II*, 56, 449–71. doi: 10.1016/j.dsr2.2008.05.032.

Lunden, J. J., McNicholl, C. G., Sears, C. R., Morrison, C. L., and Cordes, E. E. (2014). Acute Survivorship of the Deep-sea Coral *Lophelia pertusa* from the Gulf of Mexico Under Acidification, Warming, and Deoxygenation. *Frontiers in Marine Science*, 1. doi: 10.3389/fmars.2014.00078.

Lutz, M. J., Caldeira, K., Dunbar, R. B., and Behrenfeld, M. J. (2007). Seasonal Rhythms of Net Primary Production and Particulate Organic Carbon Flux to Depth Describe the Efficiency of Biological Pump in the Global Ocean. *Journal of Geophysical Research*, 112(C10). doi: 10.1029/2006JC003706.

Maier, C., Watremez, P., Taviani, M., Weinbauer, M. G., and Gattuso, J. P. (2012). Calcification Rates and the Effect of Ocean Acidification on Mediterranean Cold-water Corals. *Proceedings of the Royal Society B: Biological Sciences*, 279, 1716–23. doi: 10.1098/rspb.2011.1763.

Maier, C., Popp, P., Sollfrank, N., et al. (2016). Effects of Elevated pCO_2 and Feeding on Net Calcification and Energy Budget of the Mediterranean Cold-water Coral *Madrepora oculata*. *Journal of Experimental Biology*, 219, 3208–17. doi: 10.1242/jeb.127159.

Maier, S. R., Kutti, T., Bannister, R. J., et al. (2019). Survival Under Conditions of Variable Food Availability: Resource Utilization and Storage in the Cold-water Coral *Lophelia pertusa*: Cold-water Coral Resource Utilization and Storage. *Limnology and Oceanography*, 64, 1651–71. doi: 10.1002/lno.11142.

Middelburg, J. J. (2011). Chemoautotrophy in the Ocean. *Geophysical Research Letters*, 38 (24). doi: 10.1029/2011GL049725.

Middelburg. J. J. (2018). Reviews and Syntheses: To the Bottom of Carbon Processing at the Seafloor. *Biogeosciences*, 15, 413–27. https://doi.org/10.5194/bg-15-413-2018.

Mienis, F., Duineveld, G. C. A., Davies, A. J., et al. (2014). Cold-water Coral Growth Under Extreme Environmental Conditions, the Cape Lookout Area, NW Atlantic. *Biogeosciences*, 11, 2543–60. doi: 10.5194/bg-11-2543-2014.

Mora, C., Wei, C-L., Rollo, A., et al. (2013). Biotic and Human Vulnerability to Projected Changes in Ocean Biogeochemistry over the 21st Century. *PLoS Biology*, 11(10), e1001682. doi: 10.1371/journal.pbio.1001682.

Morán, X. A. G., López-Urrutia, Á., Calvo-Díaz, A., and Li, W. K. W. (2010). Increasing Importance of Small Phytoplankton in a Warmer Ocean. *Global Change Biology*, 16, 1137–44. doi: 10.1111/j.1365-2486.2009.01960.x.

Morris, K. J., Bett, B. J., Durden, J. M., et al. (2016). Landscape-scale Spatial Heterogeneity in Phytodetrital Cover and Megafauna Biomass in the Abyss Links to Modest Topographic Variation. *Scientific Reports*, 6(1). doi: 10.1038/srep34080.

Nedoncelle, K., Lartaud, F., Contreira Pereira, L., et al. (2015). *Bathymodiolus* Growth Dynamics in Relation to Environmental Fluctuations in Vent Habitats. *Deep-Sea Research I*, 106, 183–93. doi: 10.1016/j.dsr.2015.10.003.

Orr, J. C., Fabry, V. J., Aumont, O., et al. (2005). Anthropogenic Ocean Acidification over the Twenty-first Century and Its Impact on Calcifying Organisms. *Nature*, 437, 681–6. doi: 10.1038/nature04095.

Oschlies, A., Duteil, O., Getzlaff, J., et al. (2017). Patterns of Deoxygenation: Sensitivity to Natural and Anthropogenic Drivers. *Philosophical Transactions of the Royal Society A: Mathematical, Physical and Engineering Sciences*, 375(2102), 20160325. doi: 10.1098/rsta.2016.0325.

Oschlies, A., Brandt, P., Stramma, L., and Schmidtko, S. (2018). Drivers and Mechanisms of Ocean Deoxygenation. *Nature Geoscience*, 11, 467–73. doi: 10.1038/s41561-018-0152-2.

Peña, M. A., Nemcek, N., and Robert, M. (2019). Phytoplankton Responses to the 2014–2016 Warming Anomaly in the Northeast Subarctic Pacific Ocean. *Limnology and Oceanography*, 64, 515–25. doi: 10.1002/lno.11056.

Pörtner, H. O. and Farrell, A. P. (2008). Ecology: Physiology and Climate Change. *Science*, 322(5902), 690–2. doi: 10.1126/science.1163156.

Pörtner, H. (2012). Integrating Climate-related Stressor Effects on Marine Organisms: Unifying Principles Linking Molecule to Ecosystem-level Changes. *Marine Ecology Progress Series*, 470, 273–90. doi: 10.3354/meps10123.

Pörtner, H. O., Karl, D. M., Boyd, P. W., et al. (2014) Ocean Systems, in C. B. Field, V. R. Barros, D. J. Dokken, et al. (eds.) *Climate Change 2014: Impacts, Adaptation, and Vulnerability. Part A: Global and Sectoral Aspects. Contribution of Working Group II to the Fifth Assessment Report of the Intergovernmental Panel on Climate Change*. Cambridge: Cambridge University Press, pp. 411–84.

Purkey, S. G. and Johnson, G. C. (2010). Warming of Global Abyssal and Deep Southern Ocean Waters Between the

1990s and 2000s: Contributions to Global Heat and Sea Level Rise Budgets. *Journal of Climate*, 23, 6336–51.

Ramirez-Llodra, E., Brandt, A., Danovaro, R., et al. (2010). Deep, Diverse and Definitely Different: Unique Attributes of the World's Largest Ecosystem. *Biogeosciences*, 7, 2851–99. doi: 10.5194/bg-7-2851-2010.

Ramirez-Llodra, E., Tyler, P. A., Baker, M. C., et al. (2011). Man and the Last Great Wilderness: Human Impact on the Deep Sea. *PLoS ONE*, 6(8), e22588. doi: 10.1371/journal.pone.0022588.

Richardson, T. L. and Jackson, G. A. (2007). Small Phytoplankton and Carbon Export from the Surface Ocean. *Science*, 315, 838–40. doi: 10.1126/science.1133471.

Roberts, J., Davies, A., Henry, L., Dodds, L., et al. (2009). Mingulay Reef Complex: An Interdisciplinary Study of Cold-water Coral Habitat, Hydrography and Biodiversity. *Marine Ecology Progress Series*, 397, 139–51. doi: 10.3354/meps08112.

Roberts, E. M., Mienis, F., Rapp, H. T., et al. (2018). Oceanographic Setting and Short-timescale Environmental Variability at an Arctic Seamount Sponge Ground. *Deep-Sea Research I*, 138, 98–113. doi: 10.1016/j.dsr.2018.06.007.

Robertson, R., Visbeck, M., Gordon, A. L., and Fahrbach, E. (2002). Long-term Temperature Trends in the Deep Waters of the Weddell Sea. *Deep-Sea Research II*, 49, 4791–806. doi: 10.1016/S0967-0645(02)00159-5.

Roder, C., Berumen, M. L., Bouwmeester, J., et al. (2013). First Biological Measurements of Deep-sea Corals from the Red Sea. *Scientific Reports*, 3. doi: 10.1038/srep02802.

Rogers, A. D. (2018). The Biology of Seamounts: 25 Years On. *Advances in Marine Biology*, 79, 137–224. doi: 10.1016/bs.amb.2018.06.001.

Romano, C., Flexas, M. M., Segura, M., et al. (2017). Canyon Effect and Seasonal Variability of Deep-sea Organisms in the NW Mediterranean: Synchronous, Year-long Captures of 'Swimmers' from Near-bottom Sediment Traps in a Submarine Canyon and Its Adjacent Open Slope. *Deep-Sea Research I*, 129, 99–115. doi: 10.1016/j.dsr.2017.10.002.

St Laurent, L. C. and Thurnherr, A. M. (2007). Intense Mixing of Lower Thermocline Water on the Crest of the Mid-Atlantic Ridge. *Nature*, 448, 680–3. doi: 10.1038/nature06043.

Sato, K. N., Levin, L. A., and Schiff, K. (2017). Habitat Compression and Expansion of Sea Urchins in Response to Changing Climate Conditions on the California Continental Shelf and Slope (1994–2013). *Deep-Sea Research II*, 137, 377–89. doi: 10.1016/j.dsr2.2016.08.012.

Schmidtko, S., Stramma, L., and Visbeck, M. (2017). Decline in Global Oceanic Oxygen Content During the Past Five Decades. *Nature*, 542, 335–9. doi: 10.1038/nature21399.

Smith, C., Bernardino, A., Sweetman, A., and Arbizu, P. (2008). Abyssal Food Limitation, Ecosystem Structure and Climate Change. *Trends in Ecology and Evolution*, 23, 518–28. doi: 10.1016/j.tree.2008.05.002.

Smith, C. R., Grange, L. J., Honig, D. L., et al. (2012). A Large Population of King Crabs in Palmer Deep on the West Antarctic Peninsula Shelf and Potential Invasive Impacts. *Proceedings of the Royal Society B: Biological Sciences*, 279, 1017–26. doi: 10.1098/rspb.2011.1496.

Smith, K. L., Ruhl, H. A., Bett, B. J., et al. (2009). Climate, Carbon Cycling, and Deep-ocean Ecosystems. *Proceedings of the National Academy of Sciences USA*, 106, 19211–18. doi: 10.1073/pnas.0908322106.

Smith, K. L., Ruhl, H. A., Kahru, M., Huffard, C. L., and Sherman, A. D. (2013). Deep Ocean Communities Impacted by Changing Climate over 24 y in the Abyssal Northeast Pacific Ocean. *Proceedings of the National Academy of Sciences USA*, 110, 19838–41. doi: 10.1073/pnas.1315447110.

Smith, K. L. Jr, Sherman, A. D., Huffard, C. L., et al. (2014). Large Salp Bloom Export from the Upper Ocean and Benthic Community Response in the Abyssal Northeast Pacific: Day to Week Resolution. *Limnology and Oceanography*, 59, 745–57. doi: 10.4319/lo.2014.59.3.0745.

Smith, K. L., Huffard, C. L., Sherman, A. D., and Ruhl, H. A. (2016). Decadal Change in Sediment Community Oxygen Consumption in the Abyssal Northeast Pacific'. *Aquatic Geochemistry*, 22, 401–17. doi: 10.1007/s10498-016-9293-3.

Smith, K. L., Ruhl, H. A., Huffard, C. L., Messié, M., and Kahru, M. (2018). Episodic Organic Carbon Fluxes from Surface Ocean to Abyssal Depths During Long-term Monitoring in NE Pacific. *Proceedings of the National Academy of Sciences USA*, 115, 12235–40. doi: 10.1073/pnas.1814559115.

Soetaert, K., Mohn, C., Rengstorf, A., Grehan, A., and van Oevelen, D. (2016). Ecosystem Engineering Creates a Direct Nutritional Link Between 600-m Deep Cold-water Coral Mounds and Surface Productivity. *Scientific Reports*, 6. doi:10.1038/srep35057.

Soltwedel, T., Bauerfeind, E., Bergmann, M., et al. (2016). Natural Variability or Anthropogenically-induced Variation? Insights from 15 Years of Multidisciplinary Observations at the Arctic Marine LTER Site HAUSGARTEN. *Ecological Indicators*, 65, 89–102. doi: 10.1016/j.ecolind.2015.10.001.

Somavilla, R., Schauer, U., and Budéus, G. (2013). Increasing Amount of Arctic Ocean Deep Waters in the

Greenland Sea. *Geophysical Research Letters*, 40, 4361–6. doi: 10.1002/grl.50775.

Somero, G. N. (2010). The Physiology of Climate Change: How Potentials for Acclimatization and Genetic Adaptation Will Determine 'Winners' and 'Losers'. *Journal of Experimental Biology*, 213, 912–20. doi: 10.1242/jeb.037473.

Sperling, E. A., Frieder, C. A., and Levin, L. A. (2016). Biodiversity Response to Natural Gradients of Multiple Stressors on Continental Margins. *Proceedings of the Royal Society B: Biological Sciences*, 283, 20160637. doi: 10.1098/rspb.2016.0637.

Stewart, J. S., Hazen, E. L., Bograd, S. J., et al. (2014). Combined Climate- and Prey-mediated Range Expansion of Humboldt Squid (*Dosidicus gigas*), a Large Marine Predator in the California Current System. *Global Change Biology*, 20, 1832–43. doi: 10.1111/gcb.12502.

Stramma, L., Schmidtko, S., Levin, L. A., and Johnson, G. C. (2010). Ocean Oxygen Minima Expansions and Their Biological Impacts. *Deep-Sea Research I*, 57, 587–95. doi: 10.1016/j.dsr.2010.01.005.

Stramma, L., Oschlies, A., and Schmidtko, S. (2012). Mismatch Between Observed and Modeled Trends in Dissolved Upper-ocean Oxygen over the Last 50 Yr. *Biogeosciences*, 9, 4045–57. doi: 10.5194/bg-9-4045-2012.

Sweetman, A. K., Chelsky, A., Pitt, K. A., et al. (2016). Jellyfish Decomposition at the Seafloor Rapidly Alters Biogeochemical Cycling and Carbon Flow Through Benthic Food-webs: Jelly-fall Effects on the Benthos. *Limnology and Oceanography*, 61, 1449–61. doi: 10.1002/lno.10310.

Sweetman, A. K., Thurber, A. R., Smith, C. R., et al. (2017). Major Impacts of Climate Change on Deep-sea Benthic Ecosystems. *Elementa, Science of the Anthropocene*, 5, doi: 10.1525/elementa.203.

Sweetman, A. K., Smith, C. R., Shulse, C. N., et al. (2019). Key Role of Bacteria in the Short-term Cycling of Carbon at the Abyssal Seafloor in a Low Particulate Organic Carbon Flux Region of the Eastern Pacific Ocean. *Limnology and Oceanography*, 64, 694–713. doi: 10.1002/lno.11069.

Talley, L. D., Feely, R. A., Sloyan, B. M., et al. (2016). Changes in Ocean Heat, Carbon Content, and Ventilation: A Review of the First Decade of GO-SHIP Global Repeat Hydrography. *Annual Review of Marine Science*, 8, 185–215. doi: 10.1146/annurev-marine-052915-100829.

Thatje, S., Anger, K., Calcagno, J. A., et al. (2005). Challenging the Cold: Crabs Reconquer the Antarctic. *Ecology*, 86, 619–25. doi: 10.1890/04-0620.

Thomsen, L., Aguzzi, J., Costa, C., et al. (2017). The Oceanic Biological Pump: Rapid Carbon Transfer to Depth at Continental Margins During Winter. *Scientific Reports*, 7(1). doi: 10.1038/s41598-017-11075-6).

Thresher, R., Tilbrook, B., Fallon, S., Wilson, N., and Adkins, J. (2011). Effects of Chronic Low Carbonate Saturation Levels on the Distribution, Growth and Skeletal Chemistry of Deep-sea Corals and Other Seamount Megabenthos. *Marine Ecology Progress Series*, 442, 87–99. doi: 10.3354/meps09400.

Treude, T., Smith, C., Wenzhöfer, F., et al. (2009). Biogeochemistry of a Deep-sea Whale Fall: Sulfate Reduction, Sulfide Efflux and Methanogenesis. *Marine Ecology Progress Series*, 382, 1–21. doi: 10.3354/meps07972.

Turley, C. M., Roberts, J. M., and Guinotte, J. M. (2007). Corals in Deep-water: Will the Unseen Hand of Ocean Acidification Destroy Cold-water Ecosystems? *Coral Reefs*, 26, 445–8. doi: 10.1007/s00338-007-0247-5.

Turnewitsch, R., Falahat, S., Nycander, J., et al. (2013). Deep-sea Fluid and Sediment Dynamics—Influence of Hill- to Seamount-scale Seafloor Topography. *Earth-Science Reviews*, 127, 203–41. doi: 10.1016/j.earscirev.2013.10.005.

Turnewitsch, R., Dumont, M., Kiriakoulakis, K., et al. G. (2016). Tidal Influence on Particulate Organic Carbon Export Fluxes Around a Tall Seamount. *Progress in Oceanography*, 149, 189–213. doi: 10.1016/j.pocean.2016.10.009.

UNCLOS (1982). *United National Convention on the Law of the Sea. Part IX*. Montego Bay, Jamaica: UNCLOS.

Vergara, O., Dewitte, B., Montes, I., et al. (2016). Seasonal Variability of the Oxygen Minimum Zone off Peru in a High-resolution Regional Coupled Model. *Biogeosciences*, 13, 4389–410. doi: 10.5194/bg-13-4389-2016.

Wishner, K. F., Outram, D. M., Seibel, B. A., Daly, K. L., and Williams, R. L. (2013). Zooplankton in the Eastern Tropical North Pacific: Boundary Effects of Oxygen Minimum Zone Expansion. *Deep-Sea Research I*, 79, 122–40. doi: 10.1016/j.dsr.2013.05.012.

Wishner, K. F., Seibel, B. A., Roman, C., et al. (2018). Ocean Deoxygenation and Zooplankton: Very Small Oxygen Differences Matter. *Science Advances*, 4(12), eaau5180. doi: 10.1126/sciadv.aau5180.

Wunsch, C. and Heimbach, P. (2014). Bidecadal Thermal Changes in the Abyssal Ocean. *Journal of Physical Oceanography*, 44, 2013–30. doi: 10.1175/JPO-D-13-096.1.

Yasuhara, M., Cronin, T. M., deMenocal, P. B., Okahashi, H., and Linsley, B. K. (2008). Abrupt Climate Change and Collapse of Deep-sea Ecosystems. *Proceedings of the*

National Academy of Sciences USA, 105, 1556–60. doi: 10.1073/pnas.0705486105.

Yasuhara, M. and Danovaro, R. (2016). Temperature Impacts on Deep-sea Biodiversity: Temperature Impacts on Deep-sea Biodiversity. *Biological Reviews*, 91, 275–87. doi: 10.1111/brv.12169.

Yool, A., Martin, A. P., Anderson, T. R., et al. (2017). Big in the Benthos: Future Change of Seafloor Community Biomass in a Global, Body Size-resolved Model. *Global Change Biology*, 23, 3554–66. doi: 10.1111/gcb.13680.

CHAPTER 10

Space, the final resource

S. Kim Juniper, Kate Thornborough, Paul Tyler, and Ylenia Randrianarisoa

10.1 Introduction

The ocean water column below continental shelf depths represents well over 1 billion km^3, and the floor of the deep sea occupies 47 per cent of the surface of our planet (Chapter 1). In terms of physical dimensions alone, these deep-sea spaces merit consideration as a capital asset. This chapter will review some of the ways in which global society has been utilising space in the deep ocean. We have divided our survey into three sections, beginning with an examination of the use of the deep sea as a reservoir for solid, liquid, and hazardous waste produced by terrestrial-based societies. We will consider organised deliberate disposal of waste, as well as unintentional waste streams and accidental disposal events. Next is an exploration of our dependence on the deep sea as a buffer that absorbs nonsolid, nonliquid industrial waste streams, specifically CO_2 from fossil fuel and biomass burning; heat from a warming atmosphere; and noise produced by maritime traffic. Finally, we will consider examples of infrastructure installations in the deep sea that are simply making use of freely available space and convenient locations. Valuations are beyond the scope of our consideration (see Chapter 2), but we will conclude with a consideration of whether or not humans are making high-value use of this component of the 'natural capital of the deep sea' (as defined in Chapter 1).

The deep sea beyond national jurisdictions is a space of legal jeopardy, where there is a lack of precise regimes, and occasional puzzling implementation of legal instruments. There are some substantial, binding and nonbinding Conventions, Protocols, Agreements, and Treaties that can frame or guide the use of space in the deep sea. Broadest in scope is the United Nations Convention on the Law Of the Sea (UNCLOS), the 'Constitution of the Ocean', that consists of 320 Articles and nine Annexes, governing all aspects of ocean space from delimitations to control scientific research, economic and commercial activities, technology, and settlement of ocean-related disputes (see Chapter 3). International legal regimes for controlling high-seas pollution have been mainly developed through the International Maritime Organisation London Convention and London Protocol (IMO) (Mankabady 1986), and the UNCLOS. Measures for the protection of marine species are found in the Convention on Biological Diversity, the Convention on Wetlands, and the Convention on Migratory Species. The relevance of these and other legal instruments to the use of space in the deep sea is summarised in Table 1. See Chapter 3 for detailed discussion of legal regimes relevant to the deep sea.

10.2 Organised, deliberate waste disposal

There is no location in the ocean unaffected by anthropogenic waste (UNEP 2009; Miyake et al. 2011; Galgani et al. 2015), and the deep sea is probably the largest marine waste repository on Earth (Pham et al. 2014). Recent publications emphasise the fact that the remoteness of the deep seafloor has promoted the disposal of residues and litter for decades, using the sea as a dump site (Ramirez-Llodra et al. 2011; Angiolillio 2019).

S. Kim Juniper, Kate Thornborough, Paul Tyler, and Ylenia Randrianarisoa, *Space, the final resource* In: *Natural Capital and Exploitation of the Deep Ocean*. Edited by: Maria Baker, Eva Ramirez-Llodra, and Paul Tyler, Oxford University Press (2020). © Oxford University Press. DOI: 10.1093/oso/9780198841654.003.0010

Others, such as Goldberg (1981), writing in a marine pollution–themed issue of the journal *Oceanus*, have argued for the use of the ocean as a receptacle for waste when 'land or atmospheric dissemination of waste becomes scientifically, economically, or socially unjustifiable'. The reality is that the combination of the deep sea's huge assimilative capacity and remoteness has allowed practices such as those described below to continue. In some cases, national or international regulations (Table 1) emerged in response to evidence that assimilative capacity was being approached or surpassed.

10.2.1 Particulate waste: sewage sludge, dredge spoils, and mining tailings

For coastal nations, the marine environment has been a long-favoured disposal site for dredge spoils and sewage sludge. Ocean dumping of dredge spoils began in the nineteenth century as rapidly developing ports required deeper channels to be dredged and maintained to cope with ever-larger ships (Nihoul 1991). Marine sands represent the largest proportion of dredge spoils and are relatively uncontaminated, compared with silts and muds, which can be highly contaminated, as a result of flocculation at the fresh-and-saltwater interface that causes deposition of pollutants (Nihoul 1991). Sewage sludge is the solid end-product of primary and secondary sewage treatment processes. It is rich in organic matter and nutrients such as nitrogen and phosphorous, and contains elevated levels of heavy metals, a large number of low-concentration poorly biodegradable organic compounds, and potentially pathogenic organisms.

Globally, the volume of ocean disposal of dredge-spoils and sewage-sludge disposal in offshore waters increased throughout the twentieth century, as economic and population growth drove the expansion of shipping and increases in vessel size, and the adoption of sewage treatment by urban centres. Data for European waters indicate the volume of dredge-spoil disposal at sea to be in the order of 1.3×10^9 to 1.5×10^9 tonnes per year from 2008 to 2014 (OSPAR 2019). Most of the disposal mentioned here has traditionally been directed into

shallow water. The deep has been the target when in close proximity or when other uses of the nearshore take priority.

The immediate impacts of dredge-spoil disposal are the smothering of the seabed, potential suffocation of the nonmobile infauna, and deoxygenation of the nearby water column if the spoil contains a high organic load. The classic study was conducted at Deepwater Dumpsite 106 at 2500 m depth in the NW Atlantic. Now known as Ocean Disposal Sites and managed by the US Environmental Protection Agency, the numbered dumpsites were historically used for disposal of liquid and solid wastes and now primarily receive dredge spoils only (EPA 2019). Site 106 is located 106 miles off the New Jersey coast, with depths ranging from 1550 m to 2750 m (Greig et al. 1976). Between 1986 and 1992, 26×10^6 tonnes of dredge spoils were dumped at site 106 (Thiel 2003). Grassle (1991) reported increased macrofaunal abundance in the impacted zone, whilst stable isotope analysis of sea urchin tissues (Van Dover et al. 1992) found indications of the incorporation of sewage-derived organic matter.

Documented ecosystem effects of dredge-spoil disposal provide some insight into the impact of current and future ocean mining operations. The current footprint of ocean mining operations is very small compared with disturbances from dredge-spoil disposal, but it is likely to grow in the near future and, more importantly, it will extend anthropogenic seabed disturbance to areas far offshore that have never been affected by dredge-spoil disposal or bottom-contact fishing (see Chapter 5 for description of potential mining sites). Studies of marine disposal of mining tailings from land have found effects on deep-water organisms that are similar to those of dredge-spoil disposal (Ramirez-Llordra et al. 2015). Mevenkamp et al. (2017) demonstrated that a 1 mm layer of mining tailings (rich in iron) in a Norwegian fjord compromised the function of the community to assimilate ^{13}C labelled algae, whilst a 3 cm layer resulted in high mortality of the local meiofaunal community as a result of food loss. In an example from an actual ocean mining operation, Rogers and Li (2002) documented irreversible reorganisation of sediment stratigraphy and increased substratum patchiness in seabed areas exploited for offshore diamond mining on the Namibian continental shelf.

10.2.2 Marine litter: shipping and commercial fishing sources

'Litter' (or 'debris') is a generic term used here to define shipping and commercial fishing waste formed by any persistent, manufactured or processed solid material and deliberately discarded, disposed of, or abandoned in the marine environment during shipping or fishing operations (adapted from UNEP 2009). The most common waste types found in the ocean today were dumped mostly during the last 50 years (Gregory and Ryan 1997; Derraik 2002; Tabau et al. 2015; Garcia-Rivera et al. 2017). This subsection will consider sources of deliberate, at-sea disposal, in keeping with the focus of Section 10.2. Such a singular focus is not without its challenges. For categories such as 'dumped or lost fishing gear', it is often easier to identify sources than determine whether the waste was disposed of deliberately or accidentally lost. For 'plastics', even distinguishing between at-sea and terrestrial sources can be difficult. The primary sources of waste discarded at sea are commercial vessels and fishing fleets (Puig et al. 2012; Tabau et al. 2015). As technological advances have increased access to the open sea and driven the development of commercial shipping and fishing industries, so too have the volume and spatial extent of waste disposed of at sea.

Ship-sourced waste has been entering the ocean for as long as sea-going vessels have existed. The first significant commercial vessel waste type reaching the deep seabed was unburnt coal and coal clinker, the residue of burnt coal used by steam-powered vessels, dumped overboard along shipping routes. Clinker remains the most abundant debris material beneath shipping lanes in the North Atlantic (Kidd and Huggett 1981; Briggs et al. 1996; Ahrens and Morrisey 2005; Pham et al. 2014, Chapter 1), at depths of 3000–5000 m in the Venezuelan Basin (Briggs et al. 1996) and off British Columbia (Ramirez-Llodra et al. 2013). Coal-powered steam ships only operated for approximately 150 years from the late eighteenth century

until around 1940 when internal combustion engines and gas turbines replaced them.

Globally, plastics are the dominant benthic waste type in the deep sea, followed by discarded fishing gear and metal objects (Galgani et al. 1996, 2000, 2010; Koutsodendris et al. 2008; Watters et al. 2010; Mordecai et al. 2011; Schlining et al. 2013; Strafella et al. 2015; Tabau et al. 2015; Garcia-Rivera et al. 2017; Chapter 1, Figure 3D). Plastics alone account for 60–80 per cent of total marine litter (Gregory and Ryan 1997; Derraik 2002; Tabau et al. 2015; Garcia-Rivera et al. 2017). Benthic waste in submarine canyons of the North Mediterranean Sea was composed of plastics equivalent to 72 per cent of the marine waste component, followed by 17 per cent of discarded fishing gear (Tabau et al. 2015). Lost and dumped fishing gear (FAO 2010–2019) has been identified in almost every single study of seabed waste around the world (Galgani et al. 1995a, 2000; Lee et al. 2006; Watters et al. 2010; Bergmann and Klages 2012; Pham et al. 2014; Vieira et al. 2014; Chapter 1, Figure 3C).

Discarded fishing gear has been identified as a significant source of benthic marine litter in Monterey Canyon and Nazaré Canyon (Lee et al. 2006; Mordecai et al. 2011; Schlining et al. 2013), and on heavily fished continental shelves, such as the Celtic Sea and off California (Galgani et al. 2000; Watters et al. 2010). The distribution of discarded/lost fishing gear has been linked to differences among regional fishing traditions, accessibility of fishing grounds, distance from the coast, exploited depth range, fishing efforts, gear type, and benthic community composition (Angiolillo 2019). Significant amounts of fishing gear on canyon floors at depths greater than the maximum limit of the fishing activity is evidence of deliberate dumping of old, damaged, or tangled nets and cables from fishing vessels (Tabau et al. 2015).

Despite the long history of marine waste disposal, studies of waste in the deep sea only began in the 1970s as international regulatory regimes emerged (Jewett 1976). These were primarily trawl surveys and later video surveys. Identifying and quantifying litter in the deep sea are challenging but crucial to understanding sources of waste. Approximately 15 per cent of marine litter is floating at the sea surface, 15 per cent remains in the water column, and 70 per cent lies on the seafloor (UNEP 2005), where it has been estimated (not directly measured) to equal or exceed the biomass of megafauna, in some areas (Ramirez-Llodra et al. 2013). Waste already on the seafloor can be remobilised and transported to more distant, deeper waters by passive sinking or cascading (Canals et al. 2006; Puig et al. 2008; Durrieu de Madron et al. 2013; Tabau et al. 2015). Conversely, 'heavy' waste (e.g. coal clinker, metal waste, and glass) found in the deep sea is expected to predominantly result from direct sinking to the seafloor following dumping or accidental loss from vessels (Tabau et al. 2015).

The geographic distribution of litter amongst different deep-sea environments is variable, although it tends to accumulate on the deep-sea floor close to main population centres and along shipping lanes (Ramirez-Llodra et al. 2013; Garcia-Rivera et al. 2017). Accumulation rates appear to be higher in the Northern Hemisphere than the Southern Hemisphere (Angiolillo 2019). In addition, geomorphic and hydrodynamic processes result in certain areas of the seafloor being more prone to capture and accumulate wastes (Galgani et al. 1996; Watters et al. 2010; Mordecai et al. 2011; Schlining et al. 2013; Pham et al. 2014; Tabau et al. 2015; Garcia-Rivera et al. 2017). These deep-sea waste concentrations tend to include a mixture of land-based and marine-based litter. Light litter, particularly plastics and fishing gear (e.g. nets, wire tangles, and nylon line), are susceptible to physical forces and often form waste aggregations on the deep seabed. Tabau et al. (2015) found that, in general, the highest reported concentrations of seafloor waste occur in submarine canyons and on seamounts, banks, mounts, and ocean ridges. There have been fewer comparable surveys in abyssal plain environments. The highest mean concentration of litter ever recorded on the deep-sea floor was in two submarine canyons in the North Mediterranean Sea (La Fonera and Cap de Creus canyons), with 15,057 and 8090 items per km^2, respectively (Tabau et al. 2015).

The ecological impact of shipping and fishing waste is related to waste type and the method of disposal. Heavy waste, such as coal clinker and metal waste, causes substrate disturbance as it arrives on the seafloor. It can also entrap deep-sea benthic organisms, produce contaminants, and provide a hard substratum for settlement of sessile species. Lighter-weight waste, such as plastics and nylon fishing line, can be transported over long distances and is more likely to entangle and/or entrap sessile deep-sea species (e.g. cold-water corals and sponges) and habitats and result in the release of contaminants (Orejas et al. 2009; Madurell et al. 2012). Long-line fishing in the deepsea uses nylon lines which, when discarded or lost, can pose a risk of entanglement for up to 600 years before they degrade (Bollmann et al. 2010; Tabau et al. 2015; Chapter 1, Figure 3C).

Some authors have noted that marine litter may provide additional substrata that may result in an apparent increase in biodiversity, but this has not been studied in detail (Briggs et al. 1996; Watters et al. 2010; Mordecai et al. 2011; Tabau et al. 2015). This waste material may provide shelter for vagile fauna including fish, echinoderms, crabs, and other crustaceans, and/or suitable larval settlement substrata for encrusting or sessile organisms (Briggs et al. 1996; Watters et al. 2010; Miyake et al. 2011; Mordecai et al. 2011; Angiolillo et al. 2015). The presence of the litter can increase the species diversity and abundance of organisms in a single habitat (Katsanevakis et al. 2007), for example, suspension-feeding anemones (Briggs et al. 1996), or can extend the potential range of aquatic invasive species (Barnes and Milner 2005; Barnes et al. 2009; Mordecai et al. 2011).

10.2.3 Radioactive waste

With the coming of the atomic age in the post–World War II period, radioactive waste rapidly accumulated. European nations discarded low-level radioactive waste in metal drums in the NE Atlantic from 1949 to 1982 (Thiel 2003). Initially the waste was mixed with bitumen or concrete and sealed within the metal drum. Later techniques used a second enclosing drum, and finally, in the later stages of dumping, the drums were encased in a concrete jacket (Thiel 2003). The sites chosen for disposal in the NE Atlantic ranged from NW of Ireland south through the Bay of Biscay to Madeira.

Unfortunately, only approximate disposal locations are known, in part because of the limited accuracy of available maritime navigational instruments. The total number of drums discarded was 122,732, equivalent to 142,284 tonnes of low-level radioactive waste (see Table 13.1 Thiel 2003). Radioactive waste disposal by other countries also took place off Japan, Korea, and New Zealand. To this catalogue must be added any disposal of radioactive waste by the former Soviet Union (Vartanov and Hollister 1997) and the loss of nuclear-powered submarines (see Section 10.3.1).

From 1972 the disposal of radioactive waste has been controlled by the London Dumping Convention and its subsequent amendment in 1994. The parties banned ocean disposal of low-level radioactive wastes as of February 1994 and the dumping of industrial waste after 1995. The amendment also stated that the Contracting Parties would review the protocol after 25 years (i.e. in 2020), based on scientific studies. The dumping of high-level radioactive waste was discussed in the latter years of the twentieth century but was abandoned with the preference for land-based disposal.

Thiel (2003) reviewed the ecological impact of radioactive waste disposal in deep water. The general opinion appeared to be that there was little or no radioactive leakage from the dumped drums. When some radioactive elements were found, they were considered to have arisen from natural processes related to particle flux from surface waters (Nies and Simmons 1989) or were related to biogeochemical processes such as those that concentrate elements at redox boundaries (Feldt et al. 1989). However, there has been no interest in examining these waste deposits for over 25 years, during which time the drums may have corroded.

10.2.4 Chemical and pharmaceutical waste

The chemical and pharmaceutical contamination of the ocean has been more the result of diffuse sources such as release from treated sewage rather than deliberate dumping of chemical or pharmaceutical waste. Much chemical waste arrives in the ocean via rivers that flow through industrial areas, and the various contaminants are advected across the

shelf by hydrodynamic processes, some entering the deep sea downslope or through submarine canyons. Soluble chemical waste can also be transferred to the deep sea by high-energy processes such as dense shelf water cascading (Canals et al. 2006; Salvadó et al. 2012a, b).

Some xenobiotics, such as persistent organic pollutants, toxic metals, radioelements, pesticides, herbicides, and pharmaceuticals are resistant to degradation, such that deep waters and sediments have been suggested as sites of accumulation for these pollutants (Ramirez-Llodra et al. 2011). High concentrations of a pollutant can kill animals immediately, but more insidiously, sublethal concentrations may impact physiological processes such as respiration or reproduction, decreasing reproductive output such that the effect at the population level is not recognised for a long period of time.

Synthetic organic contaminants include chlorofluorocarbons and organochlorine compounds. Many of these organic compounds are highly toxic. The use of tributyltin (TBT) as an antifouling paint gave rise to imposex disorder in shallow water gastropods, and it has been found at 980 m in Surugu Bay in a variety of invertebrates and fish (Takahashi et al. 1997; Thiel 2003). Deep-sea fish collected at depths of 1000 to 1800 m in the Gulf of Lyon carried as much as 175 ng.l^{-1} (wet weight) total butyltin residues in their tissues, comparable to contamination levels in coastal fish collected along the Catalan coast, and attesting to the exposure of deep-sea biota to TBT (Borghi and Porte 2002). Rotllant et al. (2006) reported high concentrations of dioxins from persistent organic pollutants of industrial origin in the deep-sea rose shrimp, *Aristeus antennatus*, an important human food resource fished between 600 and 2500 m in the Mediterranean.

The inorganic chemical composition of seawater is generally conservative except for nutrients such as nitrate and phosphate, and for trace metals (Riley and Chester 1971). As for other contaminants, trace metals enter the ocean from rivers and are advected over the continental slope into the deep sea. An additional pathway is through the atmosphere in the form of dust, which enters the surface waters and is scavenged by organic particles or phytoplankton and enters the deep sea through the downward vertical flux of particulate organic matter (see Chapter 1). It was by this pathway that radionuclide particles from atomic and thermonuclear tests in the 1960s were transported to the abyssal seafloor and rapidly ingested by holothurians (Oesterberg et al. 1963).

Deliberate pharmaceutical waste disposal is best documented in the Puerto Rico Trench, where 387,000 tonnes of waste were dumped between 1973 and 1978. Not unexpectedly, analysis of microbial communities in surface waters showed significant shifts in composition and metabolic activity (Peele et al. 1981), whilst Grimes et al. (1984) noted a substantial decrease in the *Pseudomonas* spp. population. Invertebrate populations including amphipods were also affected, particularly their growth, fecundity, and survival. At present, there is no bulk disposal of antibiotics into the deep ocean, but there are concerns that the flushing of animal antibiotics, birth control medications, pain killers, and antidepressants into coastal waters may ultimately impact the deep sea (Ramirez-Llodra 2011).

10.2.5 Munitions

The disposal of munitions at sea has been extant since the beginning of the nineteenth century and has increased as the firepower and size of munitions has increased. A peak arrived at the conclusion of the Second World War, when most of the unwanted munitions were dumped in relatively shallow water. Conventionally, the primary munitions disposed of at sea are bombs, grenades, mines, torpedoes, but also incendiary devices and chemical munitions. It should be noted that military aviation security rules prohibit bombers from landing with their bombs aboard; thus, unused bombs were often dropped at sea on the return from an unsuccessful mission (Maritime Executive 2015). Researchers at the James Martin Center for Nonproliferation Studies in Monterey estimate that 127 sites of immersion of these munitions are in the ocean. The Chemical Munition Convention (CWC), which entered into force in April 1997, prohibits the manufacture of chemical weapons and requires the destruction of their stockpiles. The London Convention on the Prevention of Marine Pollution by Dumping of Wastes and Other Matter banned ocean disposal of chemical weapons from 1972 onward.

The disposal of munitions in the deep sea has a more limited history (Thiel 2003). In 1994 the South African government disposed of 14,000 tonnes of munitions at 3500 m depth off South Africa, whilst in the same year, the Portuguese scuttled a vessel containing 2000 tonnes of various munitions (some from the Gulf War) at 4000 m off the Portuguese coast. The UK has used various deep-water sites off Northern Ireland, the Hebrides, and at 4500 m in the Southwest Approaches for munitions disposal, some of which contain poisonous gases. Britain ceased this practice is 1992. The main problem in determining the extent of munition dumping is that most data are confidential, and most military forces have 'sovereign immunity' (Thiel 2003). Carton and Jagusiewicz(2009) advocate for research into the nature and location of sea-disposed munitions in order to manage risk from detonation and exposure to munition constituents.

There are no published detailed reports on the ecological impact of munitions at the deep-sea bed. Effects both locally and further afield will depend on release of toxic contents, dilution and dispersal, and rate of chemical degradation.

10.3 Inadvertent disposal

10.3.1 Shipwrecks and maritime accidents

The iconic shipwreck of all time is the *Titanic*. Although this has generated long-lasting considerable interest in the public mind, the loss of ships at sea has always been one of the hazards of maritime exploration and commerce. There are many reasons for the demise of a vessel such as extreme weather conditions, armed conflicts, and navigational or other human errors (Rogowska and Namieśnik 2009). Such losses are significantly augmented in the times of war when the sinking of vessels is a deliberate act.

Accidental losses between 1971 and 1990 amounted to 3701 ships, equivalent to 12,861,975 tonnes, most being lost due to fire, foundering, explosion, or unknown disappearances (Thiel 2003). Thiel (2003) suggests that if 10 per cent of this loss is over the high seas, some eighteen ships or 65,000 tonnes of shipping are sinking to the deep-sea floor each year.

The majority of deliberate, for-disposal sinkings have been in shallow or shelf-edge environments, particularly in the last 30 years. In many cases this is because the wreck will be used as an artificial reef, or as a target ship for training or testing weapons. Currently scuttling in deep waters is primarily restricted to military operations, or for economic disposal of assets such as offshore platforms. Controversy exists around scuttling for economic reasons, in relation to its environmental impact, but this also has to be weighed against the environmental cost of shore-based ship breaking and recycling facilities (Devault et al. 2016).

The initial ecological impact of wrecks arriving at the seabed are physical, including the crushing of any organisms on which the ship settles, the formation of deep plough marking depending on the orientation of the ship as it hits the seabed, and the suspension of sediment created by the turbulent nature of the impact. Not all wrecks end up on soft sediment (although in deeper water the chances are greater), but some could land on biological structures such a deep-water corals, causing crushing-type damage.

The most significant ecological outcome of a vessel sinking is that a 'new' substratum may be introduced into the natural environment. The metal and wood *Titanic* is surrounded by soft sediments which contain a fauna not requiring a hard substratum. Whilst animals living in sediment are mainly deposit feeders, those that live on hard substrata are often filter feeders, and it is groups such as Anthozoa and crinoids that are found settling on metallic surfaces of wrecks. The wood associated with wrecks is colonised by species of the wood-boring genus *Xylophaga*, resulting in the ultimate disintegration of the wood. The establishment of a faunal community on the wreck is dependent on the source of suitable larvae and the prevailing physical oceanography. This may explain why so few organisms have been found on the *Titanic* and wrecks such as the *Kumanovo* off Morocco at 2570 m depth (Chapter 1, Figure 3E).

The introduction of a wreck not only produces a new substratum but may also introduce contamination from its constituent materials, cargo, and fuel. Chemical contamination is potentially significant

(Jones 2007; Dimitrakakis et al. 2014). In a passenger ship, heavy metals are present almost everywhere (in the antifouling paints, in the electric and electronic equipment, in its hull, and in the sacrificial anodes). The corrosion of all the aforementioned parts of a ship and consequently metals under seawater is a multifactorial process, influenced by both physicochemical and biological parameters (Dimitrakakis et al. 2014). Prego and Cobelo-Garcia (2004) and Santos-Echandia et al. (2005) reported contamination from Pb to Cu in the Galician water column near the point of the '*Prestige*' shipwreck. Other scientists studied heavy metal contamination from benzene or oil spills (Mirlean et al. 2001; Lin and Hu 2007; Bu-oolayan and Al-Yakoob 1998). Jones (2007) revealed contamination by Cu and Zn in sediments near the '*Norwegian Crown*' wreck, mainly attributed to the leaching of antifouling paints (Dimitrakakis et al. 2014).

A not insignificant accidental loss is that of nuclear submarines, as they represent both a hard substratum entering the deep sea and a source of radioactive materials. In the 1960s the United States lost two nuclear submarines in deep water, the USS *Thresher* (1963) and USS *Scorpion* (1968) (Naval Technology 2014). In 1989 the Russians lost the *Konsomolets*, which still contains two nuclear warheads, as well as its reactor, in the Barents Sea (Høibråten et al. 1997), and the *Kursk* in 2000, most of which was later recovered (Davidson et al. 2003). In 1993 Russia admitted the disposal of up to seventeen nuclear reactors near Novaya Zemlya in the deep Arctic Ocean. It should be noted that radioactive elements are also found naturally in deep-sea marine organisms. Cherry and Heyraud (1982) recognised naturally elevated levels of ^{210}Po in deep midwater shrimp in the NE Atlantic attributable to atmospheric weapons testing, whilst Charmasson et al. (2009) described elevated levels of radioactive elements, especially U and Po-Pb, in hydrothermal organisms as a result of the fluids emanating from vent orifices, reflecting high levels of radioactive elements in discharging hydrothermal fluids.

10.3.2 Microplastics

Plastics can enter the marine environment through a variety of land-based or marine-based pathways but particularly through rivers (Lebreton et al. 2017). Most of the marine debris captured in surveys appears to be land-based, having been washed or discarded into the marine environment and transported off the continental shelf through physical processes (Tabau et al. 2015; Garcia-Rivera et al. 2017). The term 'plastic' refers to a family of organic polymers derived from petroleum sources, including polyvinyl chloride, nylon, polyethylene (PE), polystyrene (PS), polypropylene (PP), low-density polyethylene (LDPE), and polyacrylates (Hidalgo-Ruz et al. 2012; Vert et al. 2012; Depledge et al. 2013; Frias et al. 2014). Different plastic polymers can be buoyant, neutral, or sink, depending on their composition, density, and shape (Browne et al. 2007; Cole et al. 2011; Anderson et al. 2016), determining how the plastic will be vertically distributed throughout the high-seas water column (Watters et al. 2010; Vieira et al. 2014). Once in the ocean, plastic can degrade into micro-sized or even nano-sized particles (Galgani et al. 2010; Cole et al. 2011) through biological, photo, thermal, mechanical, thermo-oxidative, and hydrolytic processes (Browne et al. 2007; Andrady 2011; Anderson et al. 2016). Plastic in the ocean is usually classified into macro- or meso-plastics (>5 mm) or microplastics. The definition of 'microplastic' varies, but the upper limit of 5 mm is generally agreed upon in the literature, and many researchers use 0.5 or 1 mm as the cut-off between macro- or meso-plastic and microplastic (Andrady 2011; Cole et al. 2011; Ryan 2015; Van Cauwenberghe et al. 2013; Anderson et al. 2016).

The presence of small plastic particles in the open ocean was first documented in the 1970s (Carpenter and Smith 1972), with the term 'microplastics' coined in 2004 (Thompson et al. 2004). Microplastics have only been sampled a few times in the deep sea (Thompson et al. 2009; O'Brine and Thompson 2010; Woodall et al. 2014; Van Cauwenberghe et al. 2013), but results from these studies indicate a widespread distribution in deep sediments and deep-sea organisms. Taylor et al. (2016) and Courtney-Jones et al. (2019) found evidence that microplastics are ingested and internalised by members of at least three major phyla with different feeding mechanisms. Woodall et al. (2014) collected sediments and coral samples from depths down to 3500 m in the Mediterranean Sea, SW Indian Ocean,

and NE Atlantic Ocean in 2001 and 2012 (Woodall et al. 2014). Microplastics were detected in all sediment samples, at an average abundance of 13.4 pieces/50 mL of sediment. Peng et al. (2020) report the detection of microplastics in sediments from the deepest part of the Challenger Deep. Given their areal extent, deep-sea sediments could represent a considerable sink for microplastics, and might account for plastics 'missing' from mass balance calculations (Woodall et al. 2014; Anderson et al. 2016). The time required to completely mineralise plastics is estimated to be on the order of hundreds to thousands of years (Barnes et al. 2009).

Potential ecological impacts of microplastics are from two primary pathways: ingestion and contamination. Ingestion of microplastics can occur through filter feeding, suspension feeding, consumption of prey exposed to microplastics, or via direct ingestion (Moore et al. 2001; Depledge et al. 2013; Baulch and Perry 2014; Desforges et al. 2014; Besseling et al. 2015; Lusher et al. 2015; Romeo et al. 2015; Taylor et al. 2016). The ingestion of microplastics can lead to a reduction in fitness and in some cases mortality (e.g. physical blockage, damaged stomach lining, energy expenditure for egestion, and a reduction in feeding). Bioavailability or likelihood of uptake of microplastics depends upon particle size, colour, and abundance in water, sediment, or biota. Currently, many of these factors are not well understood, and need to be better characterised in various environments (Anderson et al. 2016).

While there is a rapidly growing body of literature focusing on microplastics in the marine environment, the lack of standardised size definitions for microplastics, mesh, and filter sizes used for collecting and analysing samples can make comparisons across studies difficult (Hidalgo-Ruz et al. 2012). There is a need for standardised methods to improve comparisons between studies as this area of research grows. This is especially true for nano-plastics, particles less than 1.0 μm in diameter that can behave as colloids and are potentially more easily ingested and assimilated by marine organisms (Mattsson et al. 2017; Al-Sid-Cheikh et al. 2018). Nanoplastics in the ocean could represent a considerable future challenge for monitoring and risk assessment (Galgani et al. 2010; Hidalgo-Ruz et al. 2012; Norwegian Institute for Water Research 2014; Koelmans et al. 2015).

10.4 Buffer space

10.4.1 Noise

In recent decades, human activities such as urban development, expansion of transport networks, and resource extraction have added considerable noise to both terrestrial and aquatic environments across the globe, and led to major changes in the acoustic landscape (see McDonald et al. 2006; Watts 2007; Barber et al. 2009). Consequently, anthropogenic noise is now recognised as a major pollutant of the twenty-first century on a global scale.

For marine mammals, it is generally accepted that received levels greater than 120 dB (re 1 µPa - water standard) may cause behavioural changes (Richardson et al. 1995; Moore et al. 2002), and levels greater than 150 dB can lead to effects ranging from severe behavioural disruption to 'Temporary Threshold Shift', a temporary lowering of hearing sensitivity. Levels greater than 170 to 180 dB are considered enough to cause 'Permanent Threshold Shift', which means permanent hearing loss, deafness, and physical damage, including death in some circumstances. These numbers are only guidelines; they may vary according to environmental context, behavioural context, and species, as demonstrated by the deep-diving Cuvier's beaked whale strandings that occurred after repeated exposure to levels believed to be safe.

Two main sources of noise may have an impact on deep-sea residents and deep-diving animals. The first is the sound made by ships whilst steaming. Ambient noise levels in the ocean increased at a rate of more than 1 dB every 2 years from 1960 to 1980 as transoceanic shipping boomed, and continue to increase today albeit at a slower rate (Southall 2005; McDonald et al. 2006; Chapman and Price 2011). Militaries have invested heavily in researching noise-reducing technologies on ships to minimise their detectability, and there is growing interest in adapting these technological advances to make shipping vessels quieter (Mitson 1995; Southall 2005; NOAA 2007). A recent voluntary commercial vessel slowdown trial in shipping lanes

overlapping critical foraging for killer whales in the Salish Sea achieved 22 to 40 per cent reduction in potential foraging time lost to noise disturbance (Joy et al. 2019). Currently, controls on shipping to protect marine mammals are limited to shifting traffic lanes to reduce ship strikes on large whales (Agardy et al. 2007).

The second major source of noise in the ocean is active acoustic operations. Sonar is a critical naval military and deep-sea scientific tool. These high-power anthropogenic sound sources (up to 250 bD re 1 µPa at 1 m distance) radiate low- to high-frequency sound, and individual animals are exposed to high sound levels (above 160 dB re 1 µPa) over relatively short periods of time (acute exposure), as in some military sonar operations. As well as producing high sound levels close to the source, seismic surveys and low-frequency naval sonar may radiate low-frequency sounds over very large areas, thereby exposing populations to lower sound levels (below 160 dB re 1 µPa) over relatively long periods of time (chronic exposure). Continuous exposure to low-frequency sound is also an effect of distant shipping noise, multiple distant seismic surveys, or construction work (Tyack et al. 2003; Nieukirk et al. 2004; Borsani et al. 2007).

There is an increasing body of scientific evidence (e.g. McDonald et al 2006; Chapman and Price 2011; Kunc et al. 2014; Weilgart 2018) documenting the impact of anthropogenic ocean noise pollution on fish, invertebrates, and ecosystem services. Most marine animals use sound for vital life functions, and reported impacts include permanently damaged sensory organs, development delays and malformations, decreased feeding and reproduction, increased stress, and mortality. Impacts extend beyond individual species to include communities of species and how they interact, compromising ecosystem productivity and ecological services.

10.4.2 Heat absorption and transfer and CO_2 uptake

The global pattern of deep-ocean thermohaline circulation is well established and described. The densest waters are formed along the Antarctic Peninsula in winter by a combination of cold surface water and slightly warmer but more saline subsurface layers to form the Weddell Sea bottom water (Gage and Tyler 1991). However, most of the abyssal areas of the World Ocean are covered with waters formed in the Norwegian Sea, where salty water flows in as the North Atlantic Drift, then cools in winter forming an unstable water mass that flows out of the Norwegian Sea along the Denmark Strait in to the deep North Atlantic, subsequently flowing throughout the world's deep ocean. Intermediate water flows originate at the Antarctic convergence and at outflows from the Mediterranean into the Atlantic, and from the Red Sea and Persian Gulf into the North Indian Ocean. With minor perturbations this system had appeared to be stable and from an ecological perspective, is essential to maintain the flow of oxygenated water throughout the world's deep oceans.

The first indication that all was not well came in a paper by Roemmich and Wunsch (1984). Data from their 1981 survey provided evidence for significant warming in an ocean-wide band from 700 to 3000 m in the deep North Atlantic. Subsequent research has added a huge number of temperature measurements in all the main ocean basins. It is beyond the scope of this chapter to review all the literature that has be published in the subsequent 30+ years, but we will select one important study.

Purkey and Johnson (2010) conducted twenty-eight full-depth hydrographic sections throughout the World Ocean to assess temperature trends in twenty-four deep-water basins. Three basins furthest south showed a significant abyssal warming trend. Although the warming signal decreased as the sections moved northwards in the central Pacific, western Atlantic, and eastern Indian Ocean, the sections north to the eastern Atlantic and western Indian Ocean showed no significant cooling. These data were interpreted as a heat-flux rate increase between the 1990s and 2000s of 0.027 W m^{-2} below 4000 m applied over the entire earth, and 0.068W m^{-2} between 1000 and 4000 m south of the Subantarctic Front in the Antarctic Circumpolar Current.

The ecological impact of this increase in heat flux is difficult to assess, particularly in the water column. Of concern is a case where the surface waters involved in large-scale thermohaline circulation fail to cool, and the circulation stops. Over time this would lead to deoxygenation of the water column and subsequent lethal effects on the fauna.

The buffering capacity of the ocean volume with respect to climate change is strikingly illustrated by the recent estimation that 90 per cent of the warming of the Earth over the past 50 years has occurred in the ocean (Rhein et al. 2013). While the atmosphere and the climate have been temporarily buffered from the full extent of global warming, this heat stored in the ocean will eventually be released, committing Earth to additional warming in the future (Dahlman and Lindsey 2018).

Carbon dioxide absorption and ocean acidification are intimately tied up with climate change and are addressed in detail in Chapter 9. As is the case for excess heat absorption, the huge volume of the ocean, most of which is deep ocean, is absorbing anthropogenic CO_2 emissions and thereby slowing the rate of atmospheric increase. According to a recent global carbon budget, between 2002 and 2011, the ocean absorbed approximately 26 per cent of the carbon released as CO_2 from fossil fuel burning, cement production, and land-use changes (Le Quéré et al. 2012). In addition to potential for future release of absorbed CO_2 back to the atmosphere, a more immediate consequence of this planetary system buffering is the acidification of the ocean. Ocean acidification has been expected to be more rapid in surface waters than in the deep, since surface waters are in direct contact with the atmosphere. A recent study in the Sea of Japan (Chen et al. 2017) suggests that the deep ocean may acidify more quickly than projected, as global warming slows thermohaline circulation, thereby reducing ventilation of the deep ocean. In this scenario, CO_2 released from organic-matter decomposition in deep waters will accumulate to a greater degree than under 'normal-ventilation' conditions, thereby accelerating acidification.

10.5 Technology space

Human use of space in the deep sea is not limited to its capacity to receive waste and buffer our planet from the effects of fossil fuel burning. The relatively stable, unobstructed, unoccupied, and freely available space of the deep sea is suitable for the deployment of a global network of telecommunications cables that link continents and islands. These same features and the ocean's sound-transmission properties are also exploited for the deployment of deep-ocean military surveillance infrastructure, and fixed scientific observatories on the seafloor and in the water column.

10.5.1 Submarine telecommunication cables—connecting the continents

Transoceanic communications cables represent the first use of space in the deep sea for technological deployments. The invention and development of the telegraph in the 1830s and 1840s quickly led to the laying of underwater telegraph cables, first connecting Britain to Europe and Ireland, and then across the Atlantic in the 1850s and 1860s (Chapter 1). The first trans-Pacific telegraph cables were deployed in the early twentieth century. Transoceanic telephone cable technology became available in the 1950s, and extensive deployments began in the 1960s, adding to the global network of submarine telecom cables. The development of fibre-optic cable technologies in the 1980s paved the way for transoceanic telecommunications cables to support the growth of the Internet. Currently, 99 per cent of data transmission between continents is carried by approximately 400 submarine cables with a total length of approximately 1 million km. This network has been estimated to support $10 trillion/day of financial transactions (Maza 2018).

The main point of interaction of subsea cables with marine organisms is on the deep seabed, where the majority of cable lies with little or no entrenchment. Recent studies have indicated that cables pose minimal impacts on life in these environments (Ramirez et al. 2011; Taormina et al. 2018). Comparisons of sediment cores taken adjacent to and away from cables showed few significant differences in infaunal diversity or abundance. Exposed cables do provide attachment substrata for sessile organisms that are typically unable to colonise the surrounding soft sediments (first recognised by Wyville-Thomson 1873; see Chapter 1). Kogan et al. (2006) suggested there was minimal impact of submarine cables, although anemones increased in abundance, and some fish were more common, which was interpreted as a response to greater complexity in the environment. However, Scott et al. (2018) have shown that the electromagnetic field

generated by underwater cables may have an impact on animals that settle on cables or use them as a refuge. Data from 1877 to 1955 showed a total of sixteen cable faults caused by the entanglement of various cetaceans. Such deadly entanglements have entirely ceased with improved techniques for placement of modern coaxial and fibre-optic cables, which have less tendency to self-coil when lying on the seabed.

10.5.2 Deep-ocean military and scientific infrastructure

By their very nature, military infrastructures in the deep sea remain secret, although both sides in the Cold War would have used the deep sea to establish 'listening stations', especially for submarines, as these became the dominant mechanism for the delivery of nuclear weapons. There are no definitive reports of military activity in the deep sea, although Thurber et al. (2014) suggest that passive military activity, such as underwater listening stations, is widespread along continental margins but also present in canyons, on seamounts, and at the midocean ridges. The vastness of the deep ocean together with limited civilian access ensures that undetected military use of the deep seabed is likely to continue well into the future.

'Stand-alone' deep-sea scientific instruments have developed rapidly in the digital age. These are deployed by free-falling or lowering from a surface ship, and are recovered by acoustic command; others are deposited on the seabed, by submersible or remotely operated vehicles (ROVs), for a set period of time such as 1 year, and subsequently recovered. The ecological impacts of instrument deployments are likely to be the same as any small-scale disturbance at the seabed, with potential substratum disturbance where the equipment is installed and the introduction of a new hard substratum on which animals could settle. Some of the first data on growth rates in deep-sea invertebrates came from a recording of a single barnacle in the view of a time-lapse camera (Lampitt 1990).

For the study and monitoring of midwater ecosystems, there will undoubtedly be an increased number of instruments such as gliders, bouys, Autonomous Underwater Vehicles (AUVs), and other types of marine drones in the near future. These instruments, along with benthic moorings, are designed to have minimal impact on local environmental conditions, as scientists want to acquire 'normal' data. However, consideration of the use of increasing numbers of autonomous vehicles in the deep ocean and inherent legal implications of this use of space is also an emerging topic (Danovaro et al. in press).

Recently, subsea telecommunications technology has been used to support cabled observing stations in the deep sea that are connected to shore stations that provide power and real-time communications (Tunnicliffe et al. 2008; Juniper et al. 2013; Juniper et al. 2019). These cabled observatories support the long-term, high-resolution time-series study of a broad range of ocean and Earth processes, from geophysics to biology (Juniper et al. 2019). Cabled observatories are also being deployed in the deep sea by high-energy particle physicists to study neutrinos originating from outer space that are detected by moored arrays of optical and acoustic detectors. The AMADEUS (ANTARES Modules for the Acoustic Detection Under the Sea) system in the deep Mediterranean was the first such system (Aguillar et al. 2018). Future arrays or 'neutrino telescopes' are planned for other deep-sea locations (Adrián-Martínez et al. 2016; Boehmer et al. 2019).

As cabled observatories increase in number in the deep sea (Juniper et al. 2019), it will be important to consider their environmental impact. Most of these installations are relatively small in scale and unlikely to introduce significant substratum surfaces to the deep sea. One exception will be future neutrino telescopes that could deploy hundreds of sensor moorings occupying many cubic kilometres of deep-sea volume (Adrián-Martínez et al. 2016; Boehmer et al. 2019). Cabled observation maintenance using ROVs results in localised disturbance of the seabed around cabled-observatory platforms that needs to be considered, particularly in ecologically sensitive areas that are often priority sites for observatory installation (Juniper et al. 2019). While most cabled-observatory sensors are passive, active acoustic instruments (sonars and acoustic doppler current profilers) and camera lights respectively introduce measurable levels of artificial sound and light directly into the deep sea. Some observatories are voluntarily establishing limited-duty cycles for lights and active acoustic instruments, but best practices remain to be established (Juniper et al. 2013; Juniper et al. 2019).

10.6 Conclusion

We have attempted in this chapter to make a case for considering space as an important component of the natural capital of the deep sea by reviewing historic and current use of space in the deep-water column and seabed. These usages are many, not always obvious, and in some cases, are quantitatively significant and have environmental effects that are measurable despite our limited knowledge of deep-sea ecosystems. There is even a developing, increasingly coherent international governance system regulating use of space in the deep sea in the form of broad conventions such as UNCLOS and individual agreements addressing specific problems such as radioactive waste disposal.

Are we making high-value use of space in the deep sea? Certainly, deploying seabed telecommunication cables across ocean basins represents a usage that provides substantial, near-term economic and societal returns on investment, and has relatively minimal environmental impact. Similar arguments, based on the value of information returns, can be made for the deployment of scientific and possibly military sensor platforms in the deep sea. But high-value as they may be, cables and sensor platforms represent a negligible use of space in the deep sea. What about the deep sea as a permanent repository for solid and liquid waste generated on land? For example, consideration is being given to the sequestration of CO_2 in deep-sea sediments, and feasibility studies are underway (Teng et al. 2018). Consciously and otherwise, we continue to use 1 billion km^3 of deep-ocean water and 240 million km^2 of seabed as 'empty space' to absorb waste, heat, CO_2, and noise generated by growing industrial activity and populations. One might argue that this represents low-value use of the natural capital of the deep sea and that it has enabled industrialisation to continue further along an unsustainable trajectory than would have occurred had there been no such waste buffer available. Yet, until such a time as we are able to quantitatively weigh deep-sea ecosystem services and genetic resources lost to waste disposal and climate change, against the economic and environmental costs of improving materials recycling and energy production on land, we cannot objectively address this question. Presently, global society still has little appreciation of the value of the living resources of the deep sea, and even science would have difficulty quantifying their importance to the planetary life-support system upon which we all depend.

Acknowledgements

SKJ acknowledges research support from the Natural Sciences and Engineering Research Council (Canada) and the British Columbia Leadership Chair in Ocean Ecosystems and Global Change, and additional support from Ocean Networks Canada. Contributions from KT and YR were supported by Ocean Networks Canada through funding from the Canada Foundation for Innovation and the Department of Fisheries and Oceans (Canada). Lynn Wong assisted with the literature review.

References

Agardy, T., Aguilar, N., Cañadas, A., et al. (2007). *A Global Scientific Workshop on Spatio-Temporal Management of Noise*. https://www.researchgate.net/profile/Gianni_Pavan/publication/230818878_A_Global_Scientific_Workshop_on_Spatio-Temporal_Management_of_Noise/links/0912f504f1fa08622f000000/A-Global-Scientific-Workshop-on-Spatio-Temporal-Management-of-Noise.pdf.

Aguilar, J. A., Al Samarai, I., Albert, A., et al. (2018). AMADEUS—The Acoustic Neutrino Detection Test System of the ANTARES Deep-sea Neutrino Telescope. Nuclear Instruments and Methods in Physics Research Section A: Accelerators, Spectrometers, Detectors and Associated Equipment Volumes 626–27, 2011, 128–143. doi: HYPERLINK "https://arxiv.org/ct?url=https%3A%2F%2Fdx.doi.org%2F10.1016%2Fj.nima.2010.09.053&v=32718a9d" 10.1016/j.nima.2010.09.053.

Ahrens, M. J. and Morrisey, D. J. (2005). Biological Effects of Unburnt Coal in the Marine Environment. *Oceanography and Marine Biology: An Annual Review*, 43, 69–122. doi: 10.1201/9781420037449.ch3.

Al-Sid-Cheikh, M., Rowland, S.J., Stevenson, K., et al. (2018). Uptake, Whole-Body Distribution, and Depuration of Nanoplastics by the Scallop *Pecten maximus* at Environmentally Realistic Concentrations. *Environmental Science and Technology*, 52 (24), 14480–6. doi: HYPERLINK "https://doi.org/10.1021/acs.est.8b05266"doi.org/10.1021/acs.est.8b05266.

Anderson, J.C., Park, B.J., and Palace, V.P. (2016). Microplastics in Aquatic Environments: Implications for Canadian Ecosystems. *Environmental Pollution*, 218, 269–80. doi: d HYPERLINK "https://doi.org/10.1016/j.envpol.2016.06.074"oi.org/10.1016/j.envpol.2016.06.074.

Andrady, A. L. (2011). Microplastics in the Marine Environment. *Marine Pollution Bulletin*, 62, 1596–605.

Angiolillo, M. (2019). Debris in Deep Water: An Environmental Evaluation, in C. Sheppard (ed.) *World Seas: An Environmental Evaluation. Volume III: Ecological Issues and Environmental Impacts*. Amsterdam, Netherlands: Elsevier, pp. 251–68.

Barber, B. M., Odean, T., and Zhu, N. (2009). Systematic Noise. *Journal of Financial Markets*, 12, 547–69.

Barnes D. K. A., Gallani, F., Thompson, R. C., and Barlaz, M. (2009). Accumulation and Fragmentation of Plastic Debris in Global Environments. *Philosophical Transactions of the Royal Society B*, 364 (1526). https://doi.org/10.1098/rstb.2008.0205.

Barnes, D. K. A. and Milner, P. (2005). Drifting Plastic and Its Consequences for Sessile Organism Dispersal in the Atlantic Ocean. *Marine Biology*, 146, 815–25.

Baulch, S. and Perry, C. (2014). Evaluating Impacts of Marine Debris on Cetaceans. *Marine Pollution Bulletin*, 80, 210–21.

Bergmann, M. and Klages, M. (2012). Increase of Litter at the Arctic Deep-sea Observatory HAUSGARTEN. *Marine Pollution Bulletin*, 64, 2734–41.

Besseling, E., Foekeme, E. M., Van Franeker, J. A., et al. (2015). Microplastic in a Macrofilter Feeder: Humpback Whale *Megaptera novaeangliae*. *Marine Pollution Bulletin*, 95, 248–52.

Bollmann, M., Bosch, T., Colijin, F., et al. (2010). *World Ocean Review: Living with the Oceans*. Hamburg, Germany: Maribus.

Borghi, V. and Porte, C. (2002). Organo-tin Pollution in Deep-Sea Fish from the Northwestern Mediterranean. *Environmental Science and Technology*, 36, 4224–8.

Borsani, J. F., Clark, C. W., Nani, B., and Scarpiniti, M. (2007). Fin Whale Avoid Loud Rhythmic Low-frequency Sounds in the Liguruan Sea. Abstracts of the International Conference on the Effects of Noise on Aquatic Life, Nyborg, Denmark (CD-ROM). Cited in Panigada, S., Pavan, G., Borg, J.A., Galil, B.S., Vallini, C. *Biodiversity Impacts of Ship Movement, Noise, Grounding and Anchoring*, pp. 10–41 in A. Abdullaand O. Linden, O. (eds.) (2008). *Maritime Traffic Effects on Biodiversity in the Mediterranean Sea, Volume 1—Review of Impacts,*

Priority Areas and Mitigation Measures. Gland, Switzerland and Malaga, Spain: IUCN.

Briggs, K. B., Ricardson, M. D., and Young, D. K. (1996). The Calcification and Structure of Megafaunal Assemblages in the Venezuela Basin, Caribbean Sea. *Journal of Marine Research*, 54, 705–30.

Browne, M. A., Galloway, T., and Thompson, R. C. (2007). Microplastic—An Emerging Contaminant of Potential Concern? *Environmental Toxicology and Chemistry*, 3, 559–61.

Bu-oolayan, A. H. and Al-Yakoob, S. (1998). Lead, Nickel and Vanadium in Seafood: An Exposure Assessment for Kuwaiti Consumers. *Science of the Total Environment*, 223, 81–6.

Canals, M., Puig, M., Durrieu de Madron, X., et al. (2006). Flushing Submarine Canyons. *Nature*, 444, 354–7.

Carpenter, E. J. and Smith, K. L. Jr (1972). Plastics on the Sargasso Sea Surface. *Science of the Total Environment*, 155, 1240–1.

Carton, G. and Jagusiewicz, A. (2009). Historic Disposal of Munitions in U.S. and European Coastal Waters, How Historic Information Can Be Used in Characterizing and Managing Risk. *Marine Technology Society Journal*, 43(4), 16–32. doi:10.4031/MTSJ.43.4.1

Chapman, N.R. and Price, A. (2011). Low Frequency Deep Ocean Ambient Noise Trend in the Northeast Pacific Ocean. *Journal of the Acoustical Society of America*, 129(5), EL161–65. doi: 10.1121/1.3567084.

Charmasson, S., Sarradin, P. M., Le Faouder, A., et al. (2009). High Levels of Natural Radioactivity in Biota from Deep-sea Hydrothermal Cents: A Preliminary Communication. *Journal of Environmental Radioactivity*, 100, 522–6.

Chen, C. A., Lui, H., Hsieh, C., et al. (2017). Deep Oceans May Acidify Faster than Anticipated due to Global Warming. *Nature Climate Change* 7, 890–4 doi:10.1038/s41558-017-0003-y.

Cherry, R. D. and Heyraud, M. (1982). Evidence of High Natural Radiation Doses in Certain Mid-water Oceanic Organisms. *Science*, 218, 54–6.

Cole, M., Lindique, P., Halsband, C., and Galloway, T. S. (2011). Microplastics as Contaminants in the Marine Environment: A Review. *Marine Pollution Bulletin*, 62, 2588–97.

Courtene-Jones, W., Quinn, B., Ewins, C., Gary, S.F., and Narayanaswamy, B. E. (2019). Con-sistent Microplastic Ingestion by Deep-sea Invertebrates Over the Last Four Decades (1976–2015), A Study from the North East Atlantic. *Environmental Pollution*, 244, 503–12. doi: doi.org/10.1016/j.envpol.2018.10.090.

Danovaro, R., Fanelli, E., Aguzzi, J., et al. (in press). Ecological Variables for Developing a Global Deep-ocean Monitoring and conservation strategy. *Nature Ecology and Evolution*.

Davidson, P., Jones, H., and Large, J. H. (2003). The Nuclear Hazards of the Recovery of the Nuclear Powered Submarine Kursk. Contract report from Large & Associates, Consulting Engineers. http://www.large-associates.com/kurskpaper.pdf. Accessed 18 March 2020.

Depledge, M. H., Galgani, F., Panti, C., et al. (2013). Plastic Litter in the Sea. *Marine Environmental Research*, 92, 279–81.

Derraik, J. G. B. (2002). The Pollution of the Marine Environment by Plastic Debris: A Review. *Marine Pollution Bulletin*, 44, 842–52.

Desforges, J-P., Galbraith, M., Dangerfield. N., and Poss, P. S. (2014). Widespread Distribution of Microplastics in Subsurface Seawater in the NE Pacific Ocean. *Marine Pollution Bulletin*, 29, 94–9.

Devault, D. A., Winterton, P., and Bellvert, B. (2016). Ship Breaking or Scuttling? A Review of Environmental, Economic and Forensic Issues for Decision Support. Environmental Sci-ence and Pollution Research, 24 (33). doi: 10.1007/s11356-016-6925-5.

Dimitrakakis, E., Hahladakis, J., and Gidarakos, E. (2014). The 'Sea Diamond' Shipwreck: En-vironmental Impact Assessment in the Water Column and Sediments of the Wreck Area. *In-ternational Journal of Environmental Science and Technology*, 11, 1421–32.

Durrieu de Madron, X., Houpert, L., Puig, P., Sanchez-Vidal, A., and Testor, P. (2013). Interaction of Dense Shelf Water Cascading and Open-sea Convection in the Northwestern Mediterranean During Winter 2012. *Geophysical Research Letters*, 40, 1379–85.

EPA (2019). Ocean Disposal Map. https://www.epa.gov/ocean-dumping/ocean-disposal-sites. Accessed 30 November.

Feldt, W., Kanisch, G., and Vobach, M. (1989). Deep-sea Biota of the NE Atlantic and Their Radioactivity, in F. Nyffeler and W. Simmons (eds.) *Interim Oceanographic Description of the North-East Atlantic Site for the Disposal of Low-level Radioactive Waste, Volume 3*. Paris: OECD Nuclear Energy Agency, pp. 178–204.

Frias, J. P., Otero, V., and Sobral, P. (2014). Evidence for Microplastics in Samples of Zooplankton from Portuguese Coastal Waters. *Marine Environmental Research*, 95, 89–95.

Gage, J. D. and Tyler, P. A. (1991). *Deep-sea Biology: A Natural History of Organisms at the Deep-sea Floor*. Cambridge: Cambridge University Press.

Galgani, F., Souplet, A., and Cadiou, Y. (1996). Accumulation of Debris on the Deep Sea Floor off the French Mediterranean Coast. *Marine Ecology Progress Series*, 142, 225–34.

Galgani, F., Hanke, G., and Maes, T. (2015). Global Distribution, Composition and Abundance of Marine Litter, in M. Bergmann, L. Gutow, and M. Klanges (eds.) *Marine Anthropogenic Litter*. Dordrecht, Netherlands: Springer, pp. 29–56.

Galgani, F., Burgeot, T., Bocquene, G., et al. (1995a). Distribution and Abundance of Debris on the Continental Shelf of the Bay of Biscay and in Seine Bay. *Marine Pollution Bulletin*, 30, 58–62.

Galgani, F., Leaute, J. P., Moguedet, P., et al. (2000). Litter on the Sea Floor Along European Coasts. *Marine Pollution Bulletin*, 40, 516–27.

Galgani, F., Fleet, D., Van Franeker, J., et al. (2010). *Marine Strategy Framework Directive: Task Group 10 Report, Marine Litter*. European Commission Joint Research Centre Scientific and Technical Report Luxembourg: Office for Official Publications of the European Communities. doi: 10.2788/86941.

Garcia-Rivera, S., Lizaso, J. L. S., and Millan, J. M. B. (2017). Composition, Spatial Distribution and Sources of Macro-marine Litter of the Gulf of Alicante Seafloor (Spanish Mediterranean). *Marine Pollution Bulletin*, 121, 249–59. doi: 10.1016/j.marpolbul.2017.06.022.

Grassle, J. F. (1991). Effects of Sewage Sludge on Deep-sea Communities (abstr.). *EOS*, 72, 84.

Greig, R. A., Wenzloff, D. R., and Pearce, J. B. (1976). Distribution and Abundance of Heavy Metals in Finfish, Invertebrates and Sediments Collected at a Deepwater Disposal Site. *Marine Pollution Bulletin*, 7(10), 185–7.

Gregory, M. R. and Ryan, P. (1997). Pelagic Plastics and Other Seaborne Persistent Synthetic Debris, in J. M. Coe and D. B. Rogers (eds.) *Marine Debris: Sources, Impact and Solutions*. New York: Springer Verlag, pp. 46–66.

Grimes, D. J., Singleton, F. L., Stemmler, J., et al. (1984). Microbiological Effects of Wastewater Effluent Discharge into Coastal Waters of Puerto Rico. *Water Research*, 18, 613–19.

Hidalgo-Ruz, V., Gutow, L., Tompson, R. C., and Thiel, M. (2012). Microplastics in the Marine Environment: A Review of the Methods Used for Identification and Quantification. *Environmental Science and Technology*, 46, 3060–75.

Hoibraten, S., Thoresen, P. E., and Haugen, A. (1997). The Sunken Nuclear Submarine *Komsomolets* and Its Effects on the Environment. *Science of the Total Environment*, 202, 67–78.

ICES (1995). Underwater Noise of Research Vessels: Review and Recommendations, in ICES Cooperative Research Report 1995. Copenhagen, Denmark: ICES. Cooperative Research Report No. 209. pp. 61.

Jewett, S. C. (1976). Pollutants of the Northeast Gulf of Alaska. *Marine Pollution Bulletin*, 7, 169.

Jones, R. J. (2007). Chemical Contamination of a Coral Reef by the Grounding of a Cruise Ship in Bermuda. *Marine Pollution Bulletin*, 54, 905–11.

Juniper, S. K., Metabos, M., Mihaly, S. F., et al. (2013). A Year in Barkley Canyon: A Time-series Observatory Study of Mid-slope Benthos and Habitat Dynamics Using the NEPTUNE Canada Network. *Deep-Sea Research II*, 92, 114–23.

Juniper, S. K., Thornborough, K., Douglas, K., and Hillier, J. (2019). Remote Monitoring of a Deep-sea Marine Protected Area: The Endeavour Hydrothermal Vents. *Aquatic Conservation: Marine and Freshwater Ecosystems*, 29 (S2), 84–102. doi: 10.1002/aqc.3020.

Katsanevakis, S., Verriopoulos, G., Nicolaidou, A., and Thessalou-Legaki, M. (2007). Effect of Marine Litter on the Benthic Megafauna of Coastal Soft Bottoms: A Manipulative Field Experiment. *Marine Pollution Bulletin*, 54, 771–8.

Kidd, R. B. and Huggett, Q. J. (1981). Rock Debris on Abyssal Plains in the Northeast Atlantic: a Comparison of Epibenthic Sledge Hauls and Photographic Surveys. *Oceanologica Acta*, 4, 99–104.

Koelmans, A. A., Besseling, E., and Foekema, E. M. (2014) Leaching of Plastic Additives to Marine Organisms. *Environmental Pollution*, 187(2014), 49–54.

Kogan, I., Paull, C. K., Kuhnz, L. A., et al. (2006). ATOC/Pioneer Seamount Cable After 8 Years on the Seafloor: Observations, Environmental Impact. *Continental Shelf Research*, 26, 771–87.

Kho, T. T. B. (1982). Remarks by Tommy T. B. Kho (Singapore), President of the Third Unit-ed Nations Conference on the Law of the Sea. 6–11 December. New York.

Koutsodendris, A., Papatheodorou, G., Kougiourouki, O., and Georgiadis, M. (2008). Benthic Marine Litter in Four Gulfs in Greece, Eastern Mediterranean; Abundance, Composition and Source Identification. *Estuarine and Coastal Shelf Science*, 77, 501–12.

Kunc, H. P., Sigwart, J. D., Lyons, G. N.m and McLaughlin, K. E. (2014). Anthropogenic Noise Affects Behavior Across Sensor Modalities. *The American Naturalist*, 184. doi: 10.1086/677545.

Lampitt, R.S. (1990). Directly Measured Rapid Growth of a Deep-sea Barnacle. *Nature, London*, 345, 805–7.

Lebreton, L. C. M., van der Zwet, J., Damsteeg, J-W., et al. (2017). River Plastic Emissions to the World's Oceans. *Nature Communications*, 8, 15611.

Lee, D-I., Cha, H-S., and Jeong, S-B. (2006). Distribution Characteristics of Marine Litter on the Sea Bed of the East China Sea and the South Sea of Korea. *Estuarine and Coastal Shelf Science*, 70, 187–94.

Le Quéré, C., et al. (2018). Global Carbon Budget 2018. Earth System Science Data, 10, 2141-2194. https://doi.org/10.5194/essd-10-2141-2018.

Lin, C-H. and Hu, J-H. (2007). SAMHO BROTHER Benzene Ship Accident. Marine Pollution Bulletin, 54(8), 1285–6. doi: doi.org/10.1016/j.marpolbul.2007.03.013.

Lindsey, R. and Dahlman, L. (2020). Climate Change: Global Temperature. NOAA Climate. https://www.climate.gov/news-features/understanding-climate/climate-change-global-temperature.

Lusher, A. L, Hernandez-Milan, G., O'Brien, J., et al. (2015). Microplastic and Macroplastic Ingestion by a Deep Diving, Oceanic Cetacean: The True's Beaked Whale Mesoplodon mirus. Environmental Pollution, 199, 185–91.

Madurell, T., Orejas, C., Requena, S., et al. (2012). The Benthic Communities of the Cap de Creus Canyon, in M. Wurtz (ed.) Mediterranean Submarine Canyons: Ecology and Governance. Gland, Switzerland: IUCN, pp. 123–32.

Mattsson, K., Johnson, E. V., Malmendal, A., et al. (2017). Brain Damage and Behavioural Disorders in Fish Induced by Plastic Nanoparticles Delivered Through the Food Chain. Sci-entific Reports, 7, 11452. https://doi.org/10.1038/s41598-017-10813-0.

Maza, C. (2018). Will Russia Cut Underwater Internet Cables? Newsweek, 2 March 2018. https://www.newsweek.com/will-russia-cut-underwater-internet-cables-military-leaders-warn-suspicious-868695?utm_source=yahoo&utm_medium=yahoo_news&utm_campaign=rss&utm_content=/rss/yahoous/news.

McDonald M. A., Mesnick, S. L., and Hildebrand J. A. (2006). Biogeographic Characterization of Blue Whale Song Worldwide: Using Song to Identify Populations. Journal of Cetacean Research and Management, 8, 55–65.

Mevenkamp, L., Stratmann T., Guilini, K., et al. (2017). Impaired Short-Term Functioning of a Benthic Community from a Deep Norwegian Fjord Following Deposition of Mine Tailings and Sediments. Frontiers in Marine Science, 30 May. doi.org/10.3389/fmars.2017.00169.

Mirlean, N., Baraj, B., Niencheski, L.F., Baisch, P., and Robinson, D. (2001). The Effect of Accidental Sulphuric Acid Leaking on Metal Distributions in Estuarine Sediment of Patos Lagoon. Marine Pollution Bulletin, 42, 1114–17.

Miyake, H., Shibata, H., and Furushima, Y. (2011). Deep-sea Litter Study Using Deep-sea Observation Tools. In: Omori, K., Guo, X., Yoshie, N., Fujii, N., Handoh, I.C., Isobe, A., Tanabe, S. (eds.) Interdisciplinary Studies on Environmental Chemistry—Marine Environmental Modeling and Analysis. Terrapub, Ehime University, Japan, 261–9.

Moore, B. C. J., Alcantara, J. I., and Glasberg, B. P. (2002). Behavioural Measurement of Level-dependent Shifts in the Vibration Pattern on the Basilar Membrane. Hearing Research, 163, 101–10.

Moore, C. J., Moore, S. L., Leecaster, M. K., and Weisberg, S. B. (2001). A Comparison of Plastic and Plankton in the North Pacific Central Gyre. Marine Pollution Bulletin, 42, 1297–300.

Mordecai, G., Tyler, P. A., Masson, D. G., and Huvenne, V. A. I. (2011). Litter in Submarine Canyons off the West Coast of Portugal. Deep-Sea Research II, 58, 2489–96.

Naval Technology (2014). Peril in the Depths—The World's Worst Submarine Disasters. Naval-technology.com, 6 March. https://www.naval-technology.com/features/featureperil-in-the-depths-the-worlds-worst-submarine-disasters-4191027/.

Nies, H. (1989). Plutonium and ^{137}Cs in the Water Column of the NE Atlantic, in W. Niffler W. (ed.) Interim Oceanographic Description of the NE Atlantic for the Disposal of Low Level Radioactive Waste. Paris: Nuclear Energy Agency, pp. 77–81.

Nieukirk, S. L., Stafford, K., Mellinger, D. K., and Fox, C. G. (2004). Low-frequency Whale and Seismic Airgun Sounds Recorded in the Mid-Atlantic Ocean. Journal of the Acoustical Society of America, 115, 1832. doi: doi.org/10.1121/1.1675816.

Nihoul, C. C. (1991). Dumping at Sea. Ocean and Shoreline Management, 16, 313–26. Norwegian Institute for Water Research (2014). Microplastics in Marine Environments: Occur-rence, Distribution, and Effects. Oslo, Norway: NIVA.

O'Brine, T. and Thompson, R. C. (2010). Degradation of Plastic Carrier Bags in The Marine Environment. Marine Pollution Bulletin, 60, 2279–83.

Orejas, C., Gori, A., Lo Iacono, C., et al. (2009). Cold-water Corals in the Cap de Creus Canyon, Northwestern Mediterranean: Spatial Distribution, Density and Anthropogenic Impact. Marine Ecology Progress Series, 397, 37–51.

OSPAR (2017). Dumping and Placement of Dredged Material. OSPAR Intermediate Assess-ment - Pressures from Human Activities. OSPAR Assessments Portal. https://oap.ospar.org/en/ospar-assessments/inter-mediate-assessment-2017/pressures-human-activities/dumping-and-placement-dredged-material.

Osterberg, C., Carey, A. G., and Curl, H. (1963). Acceleration of Sinking Rates of Radionucleides in the Ocean. Nature, 200, 1276–7.

Peele, E. R., Singleton, F. L., Deming, J. W., Cavani, B., and Colwell, R. R. (1981). Effects of Pharmaceutical Wastes on Microbial Populations in Surface Waters at the Puerto Rico Dump Site in the Atlantic Ocean. Applied and Environmental Microbiology, 41, 873–9.

Peng, G., Bellerby, R., Zhang, F., Sn, X., and Li, D. (2020). The Ocean's Ultimate Trashcan: Hadal Trenches as

Major Depositories for Plastic Pollution. *Water Research*, 168, https://doi.org/10.1016/j.watres.2019.115121.

Pham, C. K., Ramirez-Llodra, E., Alt, C. H. S., et al. (2014). Marine Litter Distribution and Density in European Seas from the Shelves to Deep Basins. *PLoS ONE*, 9, 12.

Prego, R. and Cobelo-Garcia, A. (2004). Cadmium, Copper and Lead Contamination of the Seawater Column on the Prestige Shipwreck (NE Atlantic Ocean). *Analytica Chimica Acta*, 524, 23–6.

Puig, P., Palanques, A., Orange, D. L., Lastras, G., and Canals, M. (2008). Dense Shelf Water Cascades and Sedimentary Furrow Formation in the Cap de Creus Canyon, Northwestern Mediterranean Sea. *Continental Shelf Research*, 28, 2017–30.

Puig, P., Canals, M., Company, J. B., et al. (2012). Ploughing the Deep Sea Floor. *Nature*, 489, 286–90.

Purkey, S. G. and Johnson, G. C. (2010). Warming of Global Abyssal and Deep Southern Ocean Waters Between the 1990s and 2000s: Contributions to Global Heat and Sea Level Rise Budgets. *Journal of Climate*, 23, 6336–51.

Raaymakers, E. S. and Gregory, C. (eds.) (2002). 1st East Asia Regional Workshop on Ballast Water Control and Management, Beijing, China, 31 Oct.–2 Nov. 2002: Workshop Report. GloBallast Monograph Series.

Ramirez-Llodra, E., De Mol, B., Company, J. B., Coll, M., and Sardà, F. (2013). Effects of Natural and Anthropogenic Processes in the Distribution of Marine Litter in the Deep Mediterranean Sea. *Progress in Oceanography*, 118, 273–87.

Ramirez-Llodra, E., Tyler, P. A., Baker, M. C., et al. (2011). Man and the Last Great Wilderness: Human Impact on the Deep Sea. *PLoS ONE*, 6(8), e22588.

Ramirez-Llodra, E., Trannum, H. C., Evenset, A., et al. (2015). Submarine and Deep-sea Mine Tailing Placements: A Review of Current Practices, Environmental Issues, Natural Analogs and Knowledge Gaps in Norway and Internationally. *Marine Pollution Bulletin*, 97, 13–35.

Richardson, W. J., Greene, C. R. Jr, Malme, C. I., and Thomson, D. H. (1995). *Marine Mammals and Noise*. San Diego, CA: Academic Press.

Rhein, M., Rintoul, S. R., Aoki, S. et al. (2013). Observations: Ocean, in T. F. Stocker, D. Qin, G-K. Plattner, et al. (eds.) *Climate Change 2013: The Physical Science Basis. Contribution of Working Group I to the Fifth Assessment Report of the Intergovernmental Panel on Climate Change. Cambridge*: Cambridge University Press, pp. 255–316.

Riley, J. P., and Chester, R. (1971). *Introduction to Marine Chemistry*. London: Academic Press.

Roemmich, D. and Wunsch, C. (1984). Apparent Seasonal Changes in the Climatic State of the Deep North Atlantic. *Nature*, 307, 447–50.

Rogowska, J. and Namieśnik, J. (2009). The Assessment of the Marine Environment Risk Due to the Presence of Substances of Shipwrecks Origin—Analytical Problems. *Analityka*, 3, 52–4.

Romeo, T., Pietro, B., Peda, C., et al. (2015). First Evidence of Presence of Plastic Debris in Stomach of Large Pelagic Fish in the Mediterranean Sea. *Marine Pollution Bulletin*, 95, 358–61.

Rotllant, G., Abad, E., Sarda, F., et al. (2006). Dioxin Compounds in the Deep-sea Rose Shrimp *Aristeus antennatus* (Risso, 1816) throughout the Mediterranean Sea. *Deep-Sea Research I*, 53, 1895–906.

Ryan, P. G. (2015). A Brief History of Marine Litter, in M. Bergmann, L. Gutow, and M. Klages (eds) *Marine Anthropogenic Litter*. Cham, Switzerland: Springer, pp. 1–25. doi.org/10.1007/978-3-319-16510-3_1.

Salvado, J. A., Grimalt, J. O., Lopez, J. F., et al. (2012a). Role of Dense Shelf Water Cascading in the Transfer of Organochlorine Compounds to Open Marine Waters. *Environmental Science and Technology*, 46, 2624–32.

Salvado, J. A., Grimalt, J. O., Lopez, J. F., et al. (2012b). Transformation of PBDE Mixtures During Sediment Transport and Resuspension in Marine Environments (Gulf of Lion, NW Mediterranean Sea). *Environmental Pollution*, 168, 87–95.

Santos-Echandia, J., Prego, R., and Cobelo-Garcia, A. (2005). Copper, Nickel, and Vanadium in the Western Galician Shelf in Early Spring After the Prestige Catastrophe: Is There Seawater Contamination? *Analyical and Bioanalytical Chemistry*, 382, 360–5.

Schlining, K., Von Thun, S., Kuhnz, L., et al. (2013). Debris in the Deep: Using a 22-year Video Annotation Database to Survey Marine Litter in Monterey Canyon, Central California, USA. *Deep-Sea Research I*, 79, 96–105.

Scott, K., Harsany, P., and Lyndon, A. R. (2018). Understanding the Effects of Electromagnetic Field Emissions from Marine Renewable Energy Devices (MREDS) on the Commercially Important Crab, Cancer pagurus (L.). *Marine Pollution Bulletin*, 131, 560–8.

Southall, B. L. (2005). Final Report of the National Oceanic and Atmospheric Administration (NOAA) International Symposium "Shipping Noise and Marine Mammals: A Forum for Science, Management, and Technology". Arlington, VA: NOAA.

Southall. B. L. and Scholik-Schlomer, A. (2008). Potential Application of Vessel-quieting Technology on Large Commercial 244 Vessels, in Final Report of the National Oceanic and Atmospheric Administration (NOAA) International 245 Conference. 1–2 May 2007. Silver Spring, MD: NOAA Fisheries.

Strafella, P., Fabi, G., Spagnolo, A., et al. (2015). Spatial Pattern and Weight of Seabed Marine Litter in the Northern and Central Adriatic Sea. *Marine Policy*, 91, 120–7.

Taomina, B., Bald, J., Want, A., et al. (2018). A Review of Potential Impacts of Submarine Power Cables on the Marine Environment: Knowledge Gaps, Recommendations and Future Directions. *Renewable and Sustainable Energy Reviews*, 96, 380–91.

Taylor, M. L., Gwinnett, C., Robinson, L. F., and Woodall, L. C. (2016). Plastic Microfibre Ingestion by Deep-sea Organisms. *Scientific Reports*, 6, 33997.

Takahashi S., Tanabe S., and Kubodera T. (1997). Butyltin Residues in Deep-sea Organisms Collected from Suruga Bay, Japan. *Environmental Science and Technology*, 31, 3103–9.

Teng, Y. and Zhang, D. (2018). Long-term Viability of Carbon Sequestration in Deep-sea Sediments. Science Advances, 4(7), eaao6588. doi: 10.1126/sciadv.aao6588.

The Maritime Executive (2015). Military Ordnance Dumped in Gulf of Mexico. https://www.maritime-executive.com/article/military-ordinance-dumped-in-gulf-of-mexico. Accessed 18 March 2018.

Thiel, H. (2003). Anthropogenic Impacts on the Deep Sea, in P. A. Tyler (ed.) *Ecosystems of the World*. Amsterdam, Netherlands: Elsevier, pp. 427–71.

Thompson, R. C., Swan, S. H., Moore, C. J., and vom Saal, F. S. (2009). Our Plastic Age. *Philosophical Transactions of the Royal Society B*, 364, 1073–976.

Thompson, R. C., Olsen, Y., Mitchell, R. P., et al. (2004). Lost at Sea: Where Is All the Plastic? *Science*, 304, 838.

Thurber, A. R., Sweetman, A. K., Narayanaswamy, B. E., et al. (2014). Ecosystem Function and Services Provided by the Deep Sea. *Biogeosciences*, 11, 3941–63.

Tubau, X., Canals, M., Lastras, G., et al. (2015). Marine Litter on the Floor of the Submarine Canyons of the Northwestern Mediterranean Sea: The Role of Hydrodynamic Processes. *Progress in Oceanography*, 134, 379–403.

Tyack, P., Gordon, J., and Thompson, D. (2003). Controlled Exposure Experiments to Determine the Effects of Noise on Marine Mammals. *Marine Technology Society Journal*, 37, 41–53.

UNEP (2009). *Marine Litter: A Global Challenge*. Nairobi, Kenya: UNEP.

Van Cauwenberghe, L., Vanreusel, A., Mees, J., and Janssen, R. (2013). Microplastic Pollution in Deep-sea Sediments. *Environmental Pollution*, 182, 495–9.

VanderZwaag, D. L. (2015). The International Control of Ocean Dumping: Navigating From Permissive to Precautionary Shores, in R. Rayfuse (ed.) *Research Handbook on International Environmental Law*. Cheltenham: Edward Elgar, pp. 132–47. Available at: https://www.e-elgar.com/shop/research-handbook-on-international-marine-environmental-law.

Van Dover, C. L., Grassle, J. F., Fry, B., Garit, R. H., and Starczak, V. R. (1992). Stable Isotope Evidence for Entry of Sewage Derived Organic Material into a Deep-sea Food Web. *Nature, London*, 360, 153–6.

Vartanov, R. and Hollister, C. D. (1997). Nuclear Legacy of the Cold War. Russian Policy and Ocean Disposal. *Marine Policy*, 21, 1–15.

Vert, M., Doi, Y., Hellwich, K-H., et al. (2012). Terminology for Biorelated Polymers and Applications (IUPAC Recommendations 2012). *Pure and Applied Chemistry*, 84, 377–410.

Vieira, R., Raposo, I., Sobral, P., et al. (2014). Lost Fishing Gear and Litter at Gorringe Bank (NE Atlantic). *Journal of Sea Research*, 100, 91–8. http://dx.doi.org/10.1016/j.seares.2014.10.005.

Wale, M. A., Simpson, S. D., and Radford, A. N. (2013). Noise Negativity Affects Foraging and Antipredator Behaviour in Shore Crabs. *Animal Behaviour*, 86, 111–18.

Watters, D. L., YokLavich, M. M., Love, M. S., and Schroeder, D. M. (2010). Assessing Marine Debris in Deep Seafloor Habitats off California. *Marine Pollution Bulletin*, 60, 131–8.

Watts, G. (2007). Measurement of Airborne Sound Insulation of Timber Noise Barriers: Comparison of In Situ Method CEN/TS 1793–5 with Laboratory Method EN 1793–2. *Applied Acoustics*, 68, 421–36.

Woodall, L. C., Sanchez-Vidal, A., Canals, M., et al. (2014). The Deep Sea as a Major Sink of Microplastic Debris. *Royal Society Open Science*, 1, 140317.

Wyville Thomson, C. (1873). *The Depths of the Sea*. London: MacMillan and Co.

CHAPTER 11

A holistic vision for our future deep ocean

Eva Ramirez-Llodra, Maria Baker, and Paul Tyler

11.1 Challenges and possibilities for a healthy ocean

Healthy oceans are essential to maintain a healthy planet. Amongst other processes, the ocean regulates climate by absorbing excess anthropogenic heat and CO_2 (Chapter 9); it regenerates and recycles nutrients that support productivity (Chapter 8), provides food and other biological and nonbiological resources to humans (Chapters 2, 4, 5, 6, 7, and 10), and offers a variety of societal and cultural benefits (Chapters 7, 8, and 10). But the ocean is facing many challenges that need urgent attention. Robust scientific data and innovative technological, policy, and industrial solutions are essential to support a sound managemezznt of the deep-ocean natural capital, both within and beyond national jurisdiction (Chapter 3), to ensure future healthy and productive oceans (Figure 1).

Deep-ocean biodiversity, ecosystem functioning, and the services they provide (provisioning, regulating, supporting, and cultural; Thurber et al. 2014) are becoming increasingly important as humans turn to these remote systems for resource replenishment, new products, and new uses. New technologies and improvements in capacity development and collaboration across all nations will vastly improve our knowledge of the deep ocean, the riches therein, and the ecosystem vulnerabilities. World-wide human connectivity, facilitated by digital infrastructure, increasingly enables opportunities to collaborate across disciplines, sectors, and geographical boundaries to ensure that we understand as much as is feasible in order to manage properly human activities in this highly complex system. An interlinked web of existing and new international agreements (e.g. treaty on biodiversity beyond national jurisdiction; Rabone et al. 2019) is contributing to improved management and protection of the deep sea realm.

11.2 Cumulative and synergistic interactions

As with many systems on Earth, there is a delicate ecological balance in the deep ocean that must be maintained. Understanding the interactions of the different components of natural capital in the deep sea (i.e. any use of the deep ocean and its seabed, and the stock of renewable and nonrenewable biotic and abiotic resources that may be of benefit to humankind, Chapter 1) is complex, as many of the variables are interlinked and have cumulative and synergistic effects on the ecosystem (Figure 1). The impacts on deep-sea ecosystems from single stressors arising from the use of the ocean's natural capital have been described in detail in the chapters of this book. Some interactions amongst stressors have also been discussed. Here, we provide a short summary of the overall synergies among the different activities and processes where one or more impacts acting synergistically can result in a magnified effect on the ecosystem (Ramirez-Llodra et al. 2011). These synergies are particularly intense in the upper continental margins (200 to 400 m depth), where a higher number of activities have been co-occurring for decades (e.g. fishing, hydrocarbon exploitation, and pollution), and which are accentuated by the overarching effects of climate change and ocean acidification (Stramma et al. 2010; Ramirez-Llodra et al. 2011).

Eva Ramirez-Llodra, Maria Baker, and Paul Tyler, *A holistic vision for our future deep ocean* In: *Natural Capital and Exploitation of the Deep Ocean.* Edited by: Maria Baker, Eva Ramirez-Llodra and Paul Tyler, Oxford University Press (2020). © Oxford University Press. DOI: 10.1093/oso/9780198841654.003.0011

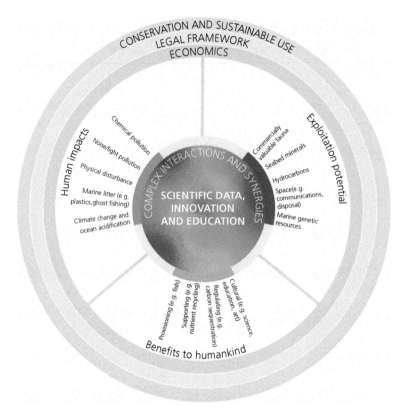

Figure 1 Components of deep-sea natural capital, indicating interactions and the overarching roles of economics, the legal framework and conservation and sustainability. Science and innovative solutions are shown to be central requirements for future management. Image by E. Mackay, National Institute of Water and Atmospheric Research (NIWA), New Zealand and the editors.

The capacity of the ocean for climate regulation, through the absorption of excess atmospheric temperature and CO_2, is resulting in a continuous increase in sea temperature and ocean acidification. Because the environmental stressors related to this regulating process (e.g. sea warming, decreased pH, deoxygenation, increased stratification, and modified currents) affect the oceans globally, it is with these variables and associated ecosystem responses (e.g. species distribution changes, changes in primary productivity, food-web modifications, and shifting communities) that most synergistic or additive processes are observed. Warmer temperatures, increased acidification, and hypoxia interact amongst themselves, causing reduced species tolerance to these same stressors (Gattuso et al. 2015; Sweetman et al. 2017). The synergistic effect of the different stressors related to climate change and ocean acidification has the potential to

reduce biodiversity and species richness significantly, facilitating the spread of opportunistic species. This will result in shifts towards more simple ecosystems (Hall-Spencer and Harvey 2019) and the potential loss of functional diversity (Teixido et al. 2018). This may severely impact the services these ecosystems support, such as nutrient recycling, detoxification, or biomass production amongst others (Gattuso et al. 2015; Sweetman et al. 2017; Hall-Spencer and Harvey 2019).

The resulting physiological stress on species subjected to synergistic or additive effects of different stressors further decreases the resilience of species and communities that are exposed to additional physical or chemical disturbances from other activities. Examples of such activities include: damage from trawling, chemical and plastic pollution, drill cuttings from hydrocarbon or terrestrial mineral exploitation, and in the future, effects of deep-sea mining and the consequent large sediment plumes (Levin and Le Bris 2015). For example, commercially fished populations for which individual fitness is reduced because of modified environmental conditions (e.g. warmer temperatures, lower pH, and changes in food-chain dynamics) may be pushed to levels where the populations cannot be maintained. There is evidence that the recovery, after benthic trawling of some cold-water coral reefs may be significantly hindered by slower skeletal growth and weaker skeletons of the corals caused by ocean acidification (Guinotte et al. 2006).

Modified large-scale circulation resulting from warmer oceans will contribute to increased hypoxia and affect nutrient upwelling, with subsequent effects on species distribution that can impact fisheries (Gattuso et al. 2015). Modified circulation and increased stratification, as well as the increased frequency of dense shelf-water cascading events or severe coastal storms, can also affect the transport and accumulation of litter and other pollutants, with potential effects on the deep-water biological communities, including on commercially valuable species or species that provide major ecosystem functions supporting important ecosystem services (Pham et al. 2014; Avio et al. 2016).

Stressors arising from different industries competing for space can also have cumulative impacts. This is the case for fishing and mining, where deep-water fishing resources and phosphorites co-occur on some upper continental margins (e.g. New Zealand, Namibia, South Africa, and Mexico), while commercially valuable fish and manganese crusts co-occur over some seamounts. The simultaneous exploitation of these resources could result in severe effects for the habitat and associated faunal communities (including fish nursery grounds), particularly if environmental and resource management efforts are not conducted in an integrated manner (Ramirez-Llodra et al. 2011). The effects of trawling or seabed mining on benthic communities may result, also, in the loss of marine genetic resources (MGR) with important bioengineering or biomedical potential before we have even discovered them.

The synergistic or additive effects of different stressors, including the global effects of climate change and ocean acidification, need to be taken into account in management and conservation measures, such as area-based management tools (including MPAs) and environmental impact assessments, that acknowledge the connectivity and changing conditions of the ocean—both within and beyond national jurisdiction (Tiller et al. 2019).

11.3 Advancing science in policy

The complexity of the deep-ocean environment makes it challenging to tease out the most important priority factors for consideration in effective management design. Inclusion of current peer-reviewed science is crucial during the formation of rules for the future use of our deep ocean. To achieve effective incorporation of scientific evidence, scientific knowledge must be translated for nonscientific audiences including policymakers, negotiators, lawyers, conservation groups, industry, and civil society.

Communication channels to enable science to inform legal and regulatory processes are constantly improving, but ongoing efforts to avoid 'dumbing-down' science findings whilst ensuring translation of science to jargon-free, understandable language are required. Policy briefs and interventions that offer tangible and practical options for use of scientific information in policy are useful. Media exposure is increasing, aided by a wide range of groups (including nongovernmental organizations) and platforms, from television, radio, and newspapers to the latest social media channels. In turn, this strong driver is pushing deep-ocean management further up the political agenda.

Humans can gain long-term benefit from the natural capital of the deep sea, as long as care and consideration for the environment, both in spatial and temporal terms, is given to exploration and exploitation. Vigilant management can result in increased biological productivity of the ocean and hence longevity of the exploitation potential (Costello et al. 2016; Costello et al. 2019). In contrast, the abiotic resources are finite, and their regeneration is on geological timescales. These reserves must be managed in a way that will ensure they will remain available for many generations to come.

Translation of scientific findings is essentially interpretation, and so care must be taken to convey accurate understanding of results and conclusions. What may be an aesthetic choice in delivery, such as using a particular graph or image to illustrate a point, may ultimately have serious consequences in decision-making if taken out of context. Additionally, it is essential that communication is bi-directional, with scientists understanding policymakers to ensure that the input provided is in a form that can be used. Building long-term connections and gaining trust and respect on all sides is crucial for effective information exchange. Science communication requires listening and learning from all angles and should not be a one-way street. Historically, too few scientists have become involved in policy discussions and informal conversations that lead to greater awareness of issues by policy makers and their constituents.

In 2013, the Deep-Ocean Stewardship Initiative (DOSI) was formed in order to aid the delivery of deep-sea science to inform policy. This has grown into a substantial network of hundreds of academics and practitioners, many of whom are scientists (including postgraduate students), but many network members are also experts from the fields of law, policy, economy, conservation groups, industry, and other academic disciplines. This community shares information and learns from one another. There are numerous focus groups within DOSI, the topics of which have been covered in this book. For example, the DOSI Minerals Working Group has official observer status at the International Seabed Authority and regularly inject scientific information to help guide the development of mineral exploitation regulations for areas beyond national jurisdiction (ABNJ). The BBNJ Working Group is intimately involved in the discussions and negotiations for an international legally binding instrument for the conservation and sustainable use of marine biological diversity of ABNJ under UNCLOS. Table 1 highlights some of the key high-level initiatives relating to the deep ocean that scientists can inform and be informed by.

Interest in science to policy work is undoubtedly growing. The more scientists that gain an understanding of the policy agenda and formulate some of their scientific research accordingly, the better. Ultimately there is an urgent need for the development and effective implementation of legal and policy measures to address all current and future

Table 1 Key high-level initiatives addressing deep-ocean issues informed by deep-sea science.

Initiative	Discipline
International Seabed Authority (ISA)	Deep-seabed mining
Biodiversity Beyond National Jurisdiction New Instrument (BBNJ process)	Biodiversity in areas beyond national jurisdiction (ABNJ)
UN Framework Convention on Climate Change (UNFCCC) Conference of the Parties (COP), including the Intergovernmental Panel on Climate Change (IPCC) for Oceans and Cryosphere	Climate change
Convention on Biological Diversity (CBD) COPs	Biodiversity
UN General Assembly Fisheries: Food and Agriculture Organisation (FAO), Regional Fisheries management Organizations (RFMO)	Fishing
London Convention & London Protocol (International Maritime Organisation, IMO)	Dumping waste
National Oil & Gas Regulatory Agencies	Oil and gas
UN Decade of Ocean Science for Sustainable Development (2021–2030)	Multidisciplinary
UN 2030 Agenda for Sustainable Development (and 17 SDGs)	Multidisciplinary

exploitation of natural capital. Consideration should also be given to the cumulative effects of climate change and interacting uses, as outlined above.

It is important that scientists work with multiple stakeholders to identify issues and research needs that contribute to enhancing knowledge and the science needed for decision-making. For the deep ocean, although our knowledge rate is increasing, uncertainty persists as often it is only possible to gain a snap-shot view. A globally integrated network of systems that can observe the deep ocean effectively and over long time scales would improve understanding of the state of the deep ocean, characterising existing conditions and quantifying its response to climate variability and human disturbance. This science would support development and planning for sustainable oceans. The Deep Ocean Observing Strategy (or DOOS) is working towards this vision. Other strategies are under discussion, including by the recently convened DOSI Decade of Deep Ocean

Science Working Group. With the Intergovernmental Oceanographic Commission at the helm, the UN Decade of Ocean Science for Sustainable Development is almost upon us (2021–2030). It hopes to provide a common framework to ensure that ocean science can fully support countries' actions to manage the oceans sustainably and, more particularly, to achieve the UN 2030 Agenda for Sustainable Development through its seventeen Sustainable Development Goals (SDGs) and associated 169 targets. To address the lack of fundamental knowledge, the deep ocean has been highlighted as a priority. Under the UN Decade banner, DOSI aims to promote research effort on a global scale to understand the role of the deep-sea ecosystems in ocean health and resilience and enable conservation and sustainable use.

Hopes and aspirations for the future include the continuation of improved communications at all levels and among different bodies. This will enhance capacity development and technology transfer in all nations, especially those with deep-sea resources in their national waters, thereby increasing societal awareness and addressing as many of the deep-sea unknowns as is possible. Such a holistic vision will help towards securing a healthy future for our deep-ocean ecosystems and their long-term natural capital.

Acknowledgements

We would like to thank all of the authors of this book for providing the detailed information that is at the base of this chapter. ERLL would like to acknowledge the support from NIVA and REV Ocean. MB would like to acknowledge support from Arcadia—a charitable fund of Lisbet Rausing and Peter Baldwin.

References

Avio, C. G., Gorbi, S., and Regoli, F. (2016). Plastics and Microplastics in the Oceans: From Emerging Pollutants to Emerged Threat. *Marine Environmental Research*, 128, 2–11. https://doi.org/10.1016/j.marenvres.2016.05.012.

Costello, C., Ovando, D., Clavelle, T., et al. (2016). Global Fishery Prospects Under Contrasting Management Regimes. *Proceedings of the National Academy of Sciences USA*, 113, 5125–9. https://doi.org/10.1073/pnas.1520420113.

Costello, C., Cao, L., Gelcich, S., et al. (2019). *The Future of Food from the Sea*. Washington, DC: World Resources Institute. https://oceanpanel.org/sites/default/files/2019-11/19_HLP_BP1_ESA4_web.pdf.

Gattuso, J. P., Magnan, A., Bille, R., et al. (2015). Contrasting Futures for Ocean and Society from Different Anthropogenic CO_2 Emissions Scenarios. *Science*, 349, 1–10.

Guinotte, J. M., Orr, J., Cairns, S., et al. (2006). Will Human-induced Changes in Seawater Chemistry Alter the Distribution of Deep-sea Scleractinian Corals? *Frontiers in Ecology and the Environment*, 4, 141–6. doi: https://doi.org/10.1890/1540-9295(2006)004[0141:WHCISC]2.0.CO;2.

Hall-Spencer, J. M. and Harvey, B. P. (2019). Ocean Acidification Impacts on coastal Ecosystem Services Due to Habitat Degradation. *Emerging Topics in Life Sciences*, 3 (2), 197–206. doi: https://doi.org/10.1042/ETLS20180117.

Levin, L. A. and Le Bris, N. (2015). The Deep Ocean Under Climate Change. *Science*, 350, 766–8.

Pham, C., Ramirez-Llodra, E. Alt, C., et al. (2014). Marine Litter Distribution and Density in European Seas, from the Shelves to Deep Basins. *PLoS ONE*, 9(4), e95839.

Rabone, M., Harden-Davies, H., Collins, J. E., et al. (2019). Access to Marine Genetic Resources (MGR): Raising Awareness of Best-Practice Through a New Agreement for Biodiversity Beyond National Jurisdiction (BBNJ). *Frontiers in Marine Science*, 6, 520. doi: 10.3389/fmars.2019.00520.

Ramirez-Llodra, E., Tyler, P. A., Baker, M. C., et al. (2011). Man and the Last Great Wilderness: Human Impact on the Deep Sea. *PLoS ONE*, 6(8), e22588.

Stramma, L., Schmidt, S., Levin, L. A., and Johnson, G. C. (2010). Ocean Oxygen Minima Expansions and Their Biological Impacts. *Deep-Sea Research I*, 210, 587–95. doi: https://doi.org/10.1016/j.dsr.2010.01.005.

Sweetman, A. K., Thurber, A. R., Smith, C. R., et al. (2017). Major Impacts of Climate Change on Deep-sea Benthic Ecosystems. *Elementa Science of the Anthropocene*, 5, 4. doi: https://doi.org/10.1525/elementa.203.

Teixidó, N., Gambi, M. C., Parravacini, V., et al. (2018). Functional biodiversity loss along natural CO_2 gradients. *Nature Communications*, 9, 5149. https://doi.org/10.1038/s41467-018-07592-1.

Thurber, A. R., Sweetman, A. K., Narayanaswamy, B. E., et al. (2014). Ecosystem Function and Services Provided by the Deep Sea. *Biogeosciences*, 11, 3941–63.

Tiller, R., De Santo, E., Mendenhall, E., and Nyman, E. (2019). The Once and Future Treaty: Towards a New Regime for Biodiversity in Areas Beyond National Jurisdiction. *Marine Policy*, 99, 239–42. https://doi.org/10.1016/j.marpol.2018.10.046.

Name index

Note: Tables and figures are indicated by an italic *t* and *f* following the page number.

Subject index

Note: Tables and figures are indicated by an italic *t* and *f* following the page number.